Understanding the Universe

Springer
New York
Berlin
Heidelberg
Barcelona
Budapest
Hong Kong
London
Milan
Paris
Santa Clara
Singapore
Tokyo

James B. Seaborn

Understanding the Universe

An Introduction to Physics and Astrophysics

Springer

James B. Seaborn
Department of Physics
University of Richmond
Richmond, VA 23173

Library of Congress Cataloging-in-Publication Data
Seaborn, James. B.
Understanding the universe : an introduction to physics and astrophysics / James B. Seaborn.
p. cm.
Includes bibliographical references and index.
ISBN 0-387-98295-7 (alk. paper)
1. Astrophysics. I. Title.
QB461.S35 1997
523.01—dc21 97-23858

Printed on acid-free paper.

Production managed by Steven Pisano; manufacturing supervised by Johanna Tschebull.
Photocomposed by The Bartlett Press, Inc., Marietta, GA.
Printed and bound by Maple-Vail Book Manufacturing Group, York, PA.
Printed in the United States of America.

9 8 7 6 5 4 3 2 1

ISBN 0-387-98295-7 Springer-Verlag New York Berlin Heidelberg SPIN 10633986

To Jill, Carol, Richard, Thomas, and Katrina

and

To Gwen

Preface

For the last eighteen years, I have been teaching an introductory course in astrophysics. The course is intended for nonscience majors satisfying a general education requirement in natural science. It is a physics course with applications in astronomy. The only prerequisite is the high school mathematics required for admission to the university. For a number of years, I used an astronomy text, which I supplemented with lecture notes on physics. There are many good astronomy texts available, but this was not a satisfactory state of affairs, since the course is a physics course. The students needed a physics text that focused on astronomical applications. Over the last few years, I have developed a text which my students have been using in manuscript form in this course. This book is an outgrowth of that effort.

The purpose of the book is to develop the physics that describes the behavior of matter here on the earth and use it to try to understand the things that are seen in the heavens. Following a brief discussion of the history of astronomy from the Greeks through the Copernican Revolution, we begin to develop the physics needed to understand three important problems at a level accessible to undergraduate nonscience majors: (1) the solar system, (2) the structure and evolution of stars, and (3) the early universe. All of these are related to the fundamental problem of how matter and energy behave in space and time.

The first part of the book is directed toward the development of aspects of Newtonian mechanics needed to understand the properties and motions of the planets and other objects that make up the solar system. Applications are made to phenomena on the earth as well as in the sky.

A chapter on electricity and magnetism discusses the importance of electric charge as a fundamental property of matter and lays the foundation for a later discussion of the modern theory of the structure of matter. Electric charge as the source of electromagnetic radiation, including light, is established in this chapter.

To appreciate how stars form and evolve, some understanding of the microscopic structure of matter is required. The failure of classical physics to account for certain phenomena involving the interaction of radiation and matter is established. We begin with Planck's introduction of discreteness in nature to match the black body radiation measurements that were available around the end of the nineteenth century. The photoelectric effect is explained through Einstein's quantization of the electromagnetic field. Bohr's application of these quantum ideas to a model of the hydrogen atom accounted for the observed spectrum of hydrogen. Bohr's picture is extended to the many-electron atom to provide a qualitative understanding of the uniqueness and discreteness of atomic spectra and the principal features of the Periodic Table of the Chemical Elements. With this background, one is able to understand for example, how radiation from stars and stellar spectra are produced, how spectra are related to characteristics of stars, and why the spectra differ from one star to the next.

A typical main-sequence star is a brilliant ball of hot gas powered by nuclear fusion. An appreciation of how such a system functions calls for an understanding of the behavior of gases and some knowledge of nuclear physics. A chapter on the kinetic theory of gases and the development of the ideal gas law is followed by a discussion of nuclear structure and nuclear reactions including nuclear fusion. These ideas are applied to a description of the birth and evolution of a star like the sun. The physical characteristics and processes that appear at different epochs in the life of a star are described and discussed. The evolutionary track of a typical star is connected with stars in various stages of evolution as they are represented on a Hertzsprung–Russell diagram.

The last part of the book deals with cosmology. The measurements of cosmic distances and the redshifts of the distant galaxies led Hubble to the famous law that bears his name and to the conclusion that the universe is expanding. The standard model of the Big Bang, particularly as it applies to the early universe, is discussed and observational data supporting the Big Bang are presented.

James B. Seaborn

Contents

Preface vii

1. Watchers of the Heavens 1

1.1. Celestial Motions 1
1.2. Early Astronomy 2
1.3. The Copernican Revolution 5
Exercises 13

2. The Stuff Moves Around 16

2.1. Fundamental Properties 16
2.2. Derived Quantities 17
2.3. Scalars and Vectors 21
Exercises 24

3. Eyes on the Skies 33

3.1. Wave Phenomena 33
3.2. Reflection and Refraction of Light 35
3.3. The Doppler Effect 47
Exercises 48

4. Newton Puts It All Together 54

4.1. Newton's Laws of Motion 54
4.3. Weight 60
Exercises 62

5. Running the Machine 69

5.1. Mechanical Work 69
5.2. Energy 70
5.3. Collisions 74

5.4. Power 75
Exercises 75

6. Off the Straight and Narrow 82

6.1. Uniform Circular Motion 82
6.2. Centripetal Force 84
Exercises 85

7. The Gravity of It All 89

7.1. Early Speculations 89
7.2. Newton's Law of Universal Gravitation 92
7.3. Measurement of the Force Constant 94
7.4. Escape Velocity 96
7.5. Planetary Atmospheres 97
Exercises 97

8. Round and Round She Goes 104

8.1. Orbital Motion 104
8.2. Artificial Satellites in Earth Orbit 107
Exercises 108

9. As the World Turns 113

9.1. Mechanical Equilibrium 113
9.2. The Origin of the Solar System 120
Exercises 122

10. Let There Be Light! 127

10.1. Electricity 127
10.2. Magnetism 132
10.3. Induced Electric Currents 137
10.4. Electromagnetic Radiation and Light 138
10.5. The Earth's Magnetic Field 139
10.6. Electric Potential Energy 141
Exercises 142

11. What's the Matter? 148

11.1. Thomson's Model of the Atom 149
11.2. The Discovery of Radioactivity 150
11.3. The Geiger–Marsden Experiment 151
11.4. Rutherford's Model of the Atom 153
Exercises 154

12. Hot Stuff **156**

12.1. Radiation from Hot Matter 156
12.2. The Failure of Classical Physics 158
12.3. Planck's Quantum Hypothesis 159
Exercises 160

13. Einstein's Bundles **161**

13.1. The Photoelectric Effect 161
13.2. Momentum of Light 165
13.3. Equivalence of Mass and Energy 166
Exercises 168

14. The Great Dane **172**

14.1. The Structure of the Atom 172
14.2. The Man from Copenhagen 174
14.3. Comets 179
Exercises 182

15. Sugar and Spice and Everything Nice . . . **187**

15.1. The Many-Electron Atom 189
15.2. Atomic Spectra 193
15.3. The Periodic Table of the Elements 194
Exercises 195

16. The Starry Messenger **196**

16.1. Intrinsic Brightness 196
16.2. Stellar Spectra 203
16.3. The Hertzsprung–Russell Diagram 207
16.4. Binary Stars 209
Exercises 213

17. The Sun Is a Gas **218**

17.1. Boyle's Law 219
17.2. A Molecular Model of the Gas 220
17.3. The Ideal Gas Law 222
17.4. Absolute Temperature 224
Exercises 227

18. The Sun Is a Nuclear Furnace 233

18.1. The Structure of the Atomic Nucleus 233
18.2. Nuclear Reactions 242
Exercises 245

19. No More to Wonder What You Are . . . 249

19.1. The Evolution of a Star Like the Sun 249
19.2. The Evolution of a Heavier Star 256
19.3. The Stuff Between the Stars 259
19.4. Solar Neutrinos 261
Exercises 263

20. The Flight of the Galaxies 266

20.1. The Nebulae 266
20.2. Variable Stars and Cosmic Distances 267
20.3. Hubble's Discovery 269
20.4. The Structure of a Spiral Galaxy 273
Exercises 274

21. The Big Picture 277

21.1. The Cosmological Principle 277
21.2. The Quasars 278
21.3. The Cosmic Background Radiation 279
21.4. A Final Word 281
Exercises 282

Appendix A. Linear Graphs 285

Exercises 287

Appendix B. Physical and Astronomical Data 289

Appendix C. Useful Formulas 291

Appendix D. The Chemical Elements 292

Appendix E. The Brightest Stars in the Sky 295

Bibliography 296

Index 297

1

Watchers of the Heavens

1.1 Celestial Motions

When we look with our naked eye at the night sky, what do we see? There were no telescopes until the beginning of the seventeenth century, so we see essentially the same things the ancients saw. We see the stars rise on the eastern horizon, move together across the sky in the course of the night, and set in the west. The so-called "fixed stars" appear to keep the same relative positions during their nightly motion. Ancient starwatchers imagined they could recognize in certain groups of these stars patterns that they could associate with mythological events and characters. These patterns are called "constellations." The ancient ones still carry their classical names (Orion, Virgo, Pegasus, etc.). They are used by modern astronomers to give approximate locations of celestial objects in the sky. For example, the statement, "The moon is in Virgo" is comparable to saying "Richmond is in Virginia." In both cases, a large surface (sky, earth) is divided into many small pieces and the rough position of a still smaller object (moon, Richmond) is given by specifying the piece (Virgo, Virginia) in which it is located. The precise position of a point on a spherical surface is given by specifying two coordinate angles. For points on the surface of the earth, these are the *latitude* and *longitude*. The analogous angular coordinates on the celestial sphere are called *declination* and *right ascension*.

By observing over longer time intervals of days or weeks, we can detect other motions of celestial objects. The moon is the most prominent object that appears in the night sky. At certain times, it is seen as a very thin crescent in the western sky just after sundown. Over the next few days, the bright part of the moon grows in size and brightness ("waxing moon") reaching a maximum ("full moon") about fifteen days later. It then diminishes in apparent size and brightness ("waning" moon) reaching a minimum at "new moon" after another fifteen days. Then the series of phases begins all over again. During this cycle, the moon is also seen to move relative to the fixed stars, tracing out a path through the set of twelve constellations known as the *Zodiac*.

The sun, which dominates the daytime sky, also exhibits a double motion. Each day we see it rise on the eastern horizon in the morning, move westward, reaching its highest point in the sky (the *meridian*) around noon, and then set in the west in the evening. If we look carefully at the stars on the eastern horizon just before sunrise, we find that on a given day the stars there rise about four minutes earlier than they did the day before. In addition to its daily westward motion across the sky, the sun is moving slowly eastward on the celestial sphere, making a complete circuit through the constellations of the Zodiac in about 365 days. The path that the sun follows on this slow, eastward drift among the stars is called the *ecliptic* for a reason that we shall understand shortly.

There are five other objects seen in the night sky that to the naked eye look like stars (albeit unusually bright ones), but differ from stars in that they move around relative to the fixed stars. Because of this behavior, they are called "wandering stars" or *planets* (from a Greek word meaning *wanderer*). The motions of these objects are rather complicated and an explanation of their peculiar behavior required a major revolution in physical thought, as we shall see.

1.2 Early Astronomy

As far as we know, people in all societies have had an interest in astronomy. No doubt the principal motives have varied from group to group. In agricultural societies, tracking the celestial motions is important for making an accurate calendar that keeps the days of the year in step with the seasons. Seafaring nations or desert tribes require a familiarity with the heavens in order to find their way over large, trackless stretches of the earth's surface. In exercising their influence over the laity, members of priestly orders would find it useful to be able to put their predictions of astronomical alignments on surer ground.

1.2.1 The Greeks

The Greeks are particularly important, because they appear to be the first to have developed a mechanical model to describe the structure and operation of the universe. They were the first real cosmologists as opposed to others who offered mystical or mythological accounts concerning the nature of the cosmos.

We begin in the fourth century B.C. with Plato who ascribed to the celestial objects two attributes: they are spheres attached to transparent spherical shells, and their motions are circular. Spheres and circles—these two geometrical constructs underlie the theoretical astronomy and cosmology of the Greeks. Aristotle, Plato's pupil, formulated from these notions a mechanical description of the cosmos. In Aristotle's view, the universe comprised a set of nested, transparent spheres centered on the earth. Rotating with different speeds and on different axes, these spheres carried the celestial objects—sun, moon, stars, and planets—around the earth producing the phenomena seen in the sky. To give a reasonable approximation to the celestial motions, Aristotle's model required a total of 55 crystal spheres. In succeeding centuries, a parade of innovators refined and improved Aristotle's system culminating in the synthesis of Claudius Ptolemy, a Greek who lived in

Alexandria in the second century A.D. By Ptolemy's time, all but one of the crystalline spheres had disappeared from the model. Only the outermost one carrying the fixed stars remained. The other celestial objects—still spheres themselves—orbited around the earth on complicated paths that consisted of combinations of circles. It was a fairly complex system, but given the relatively poor data of the time, it provided a reasonable approximation to the celestial motions—particularly if one didn't look too closely or for too long a time. The earth was fixed at the center of the universe, and all celestial bodies circled the earth. This is the essence of the cosmology of the Greeks—the Ptolemaic system.

1.2.2 The Earth Is Round

The robustness of the Greeks' astronomy was rooted in their efforts to construct a model of the universe to match what they saw in the sky and to put this into mathematics. Mathematics was for them a tool of science. They required their theories to confront experimental observation—the ultimate test of a theory of nature. An outstanding example is the discovery late in the third century B.C. that the earth is round. Even more remarkable, a fairly accurate measurement of the earth's *size* was obtained.

Eratosthenes flourished in the waning years of the third century and the beginning of the second century B.C. He had read that at noon on the day of the summer solstice (21 June on our calendar) an image of the sun could be seen reflected from the bottom of a vertical well at the ancient town of Syene in Egypt. The sun was directly overhead at that place at that time.

At the same time — noon on the day of the solstice — some distance to the north in Alexandria where Eratosthenes lived, a vertical pole cast a shadow. That is, the sun was *not* directly overhead in Alexandria. With this observation, Eratosthenes was able to establish that Alexandria and Syene lie on a *curved* earth and not a flat one. This is illustrated in Figs. 1.1 and 1.2. The flat earth model in Fig. 1.1 predicts no shadow at the solstice for a vertical pole either at Alexandria or at Syene. The curved earth model in Fig. 1.2 predicts no shadow for a vertical pole at Syene

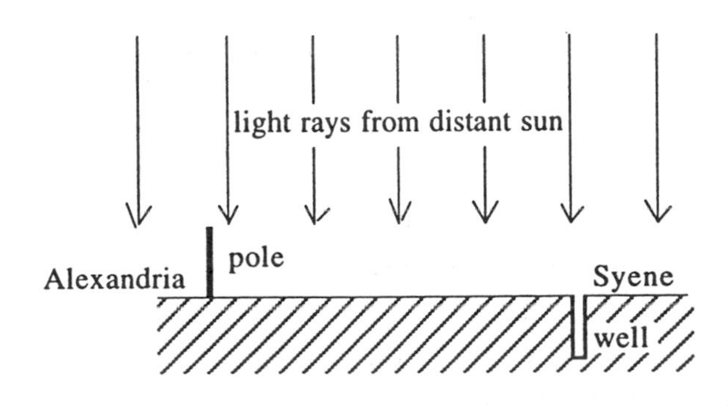

FIGURE 1.1. The flat earth model.

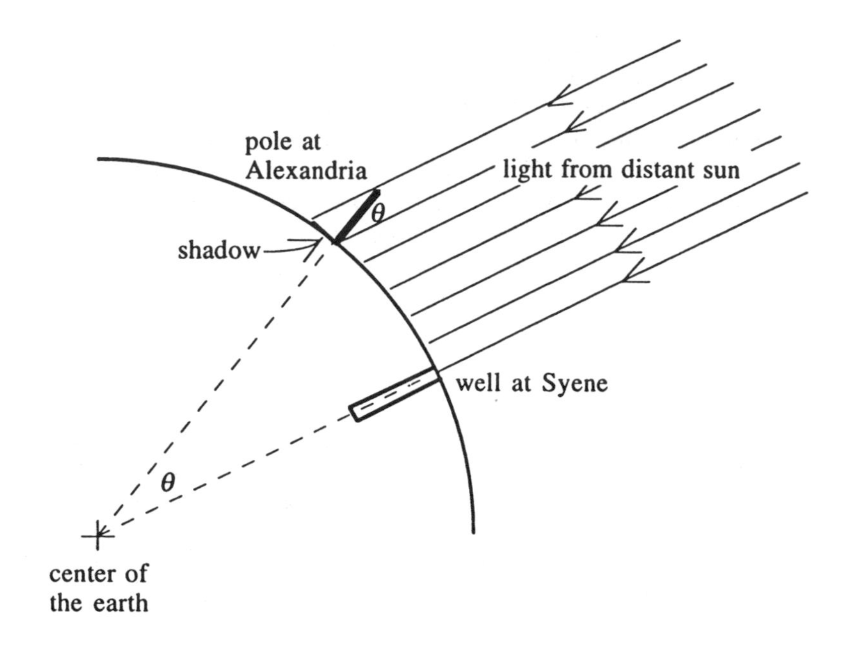

FIGURE 1.2. The curved earth model.

at the solstice, while a vertical pole at Alexandria at the same time does cast a shadow.

Eratosthenes was also able to obtain a measurement of the size of the earth. His observations led him to conclude that the earth is round and he assumed that it is a sphere. From measurements of the height of the pole at Alexandria and the length of the shadow it cast at the time of the solstice, he could deduce the angle of the sun (denoted by θ in Fig. 1.2) with respect to the vertical. He found a value corresponding to about 7°. A simple geometrical construction shows that the angle of the sun is equal to the angle at the center of the earth subtending an arc from Alexandria to Syene as illustrated in Fig. 1.2. By hiring someone to step off the distance from Alexandria to Syene, he obtained the last piece of information he needed to determine the size of the earth. The measured distance was about 5,000 stadia. The *stadium* was a Greek unit of length with a value between 100 and 200 meters. Thus, the circumference of the earth is given by

$$\frac{\text{circumference of the earth}}{\text{distance from Alexandria to Syene}} = \frac{2\pi r}{5{,}000 \text{ stadia}} = \frac{360^\circ}{7^\circ},$$

from which we find the radius of the earth $r = 41{,}000$ stadia. The exact value of the stadium used by Eratosthenes is unknown, but even with this uncertainty, his measurement of the size of the earth was remarkably accurate.

1.2.3 Ptolemy and the Arabs

Ptolemy's view of the heavens prevailed for about 1,400 years. The Romans, political successors to the Greeks, made little contribution themselves to theoretical astronomy. Their interest in science was more inclined toward practical engineering enterprises with military applications—bridges, roads, viaducts, forts, and so on. With the fall of the Roman Empire, a dark age fell on Europe in which little scientific activity is evident.

The Arabs were good observational astronomers. The clear desert sky was especially suited for observing the celestial spectacle. Many of the brighter stars visible in the night sky derive their names from Arabic. The definite article *al*, for example, appears in many star names (Altair, Aldebaran, Algol, Almach, etc.). While the darkness lay over Europe, the Arabs preserved and translated into Arabic the works of Aristotle, Ptolemy, and the Greek philosophers. The learning of the Greeks returned home to Europe with the Saracen invasion in the eighth and ninth centuries. By the thirteenth century the monks had rediscovered the complete works of Aristotle and translated them into Latin. Largely through the influence of Thomas Aquinas, Ptolemy's geocentric system along with Aristotle's science became intertwined with and incorporated into the doctrine of the Roman Catholic Church. Ptolemy's system with the earth at the center of the universe fit well with the Church's view of man as the focus of creation.

1.3 The Copernican Revolution

In 1543, *De revolutionibus orbium coelestium* ["On the Revolutions of the Celestial Spheres"] was published. This is the famous work of the Polish cleric Nicolas Copernicus in which he proposed a new arrangement for the universe. In his system, the earth is not the center of the cosmos. The sun is. The earth itself is a planet and is at the center only of the moon's orbit. The universe according to Copernicus is illustrated in Fig. 1.3, adapted from Copernicus's own drawing.

1.3.1 Tycho and Kepler

In reality, Copernicus's model was much more complicated than his drawing implies. Although conceptually simpler, in practice it turned out to be as complicated as Ptolemy's. His system called for circular orbits for the planets as they travel around the sun. To match the observed behavior of the planets, it was necessary to resort to *combinations* of circles—a device used by Ptolemy—thereby spoiling the simplicity.

This was the principal flaw in Copernicus's model—the requirement that the paths be circular, a relic from the fourth century B.C. laid on astronomy by Plato himself. Overcoming two millennia of prejudice in favor of circles was not easy. It was accomplished largely through the complementary work of two men—almost

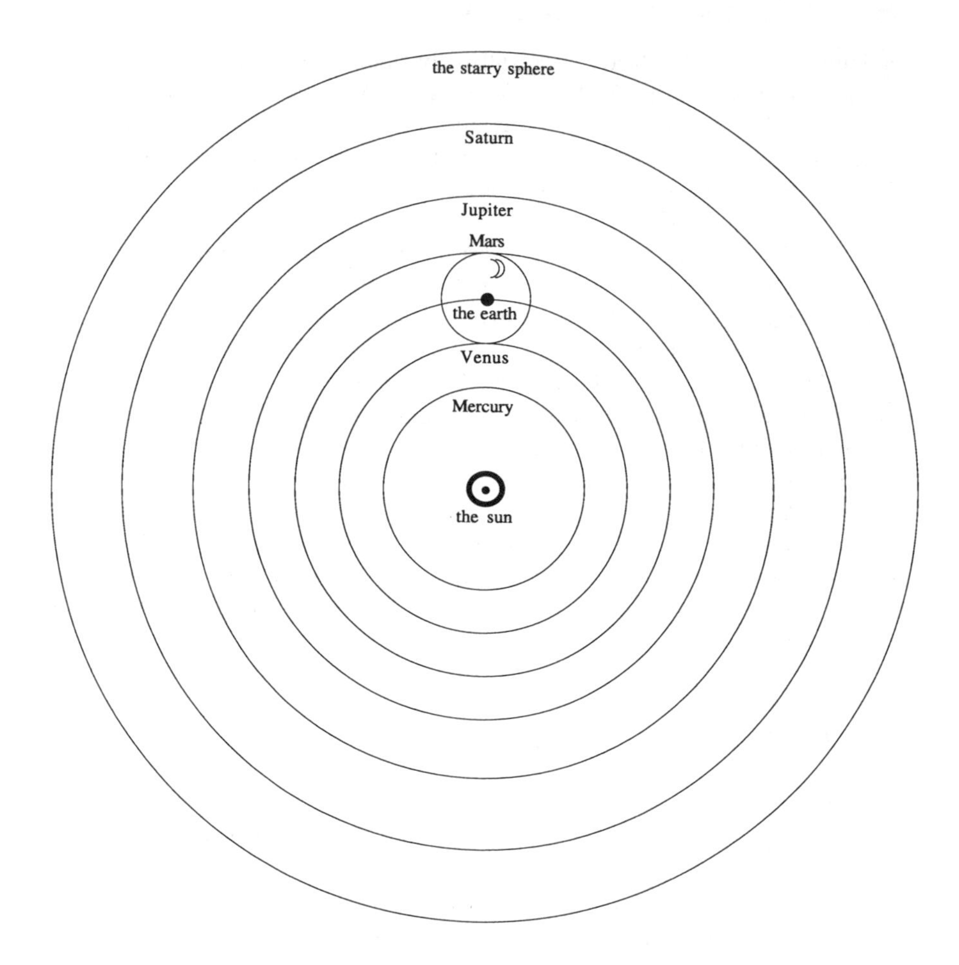

FIGURE 1.3. The Copernican universe.

against their own wills. One, Tycho Brahe, was a Danish astronomer and the other, Johannes Kepler, a German mathematician.

Tycho Brahe was a keen observer of the night sky. In the late sixteenth century, there were no telescopes, but Brahe accumulated a superb collection of instruments for measuring positions on the celestial sphere. Such instruments were very costly to construct, but Brahe belonged to the nobility, was wealthy, and could afford such extravagances. In the last quarter of the century, first in Denmark and later in Prague, Brahe and his assistants collected and recorded extensive and precise observational data on the positions of the objects in the heavens. Toward the end of his life, he was joined in Prague by Kepler, an able mathematician who occupied

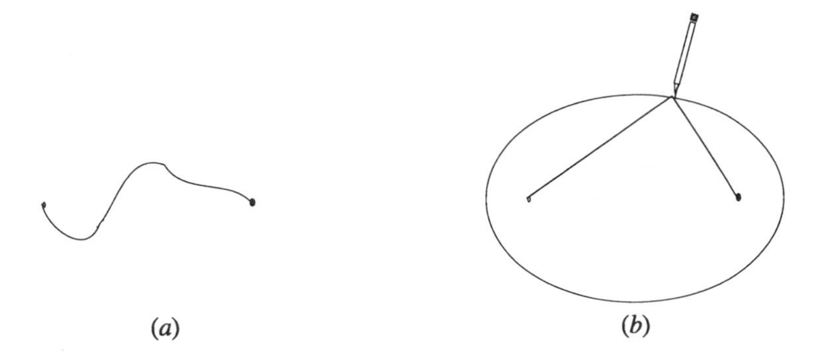

FIGURE 1.4. Construction of an ellipse.

a position on the socio-economic spectrum at the opposite end from Tycho Brahe. Kepler deduced from the data that the orbits of the planets are not circles at all, but *ellipses*.

An ellipse is a closed geometrical figure that can be constructed by the following procedure. The ends of a string are fixed at two points in a plane as illustrated in Fig. 1.4*a*. By keeping the point of a sharp pencil against the string so that the string remains taut and moving the pencil over the plane, the point of the pencil will trace out an ellipse as in Fig. 1.4*b*. The points at which the ends of the string are fixed are called the *foci* (singular, *focus*) of the ellipse. The axis drawn from one edge of the ellipse to the other and passing through both focal points is called the *major axis*. Half of this distance—from the geometrical center to the edge—is the *semimajor axis* and is denoted by b. See Fig. 1.5.

The relationship between Brahe and Kepler was a stormy one, but in the end Kepler was able to extract from the extensive and precise observations of Brahe three generalizations known as Kepler's laws of planetary motion.

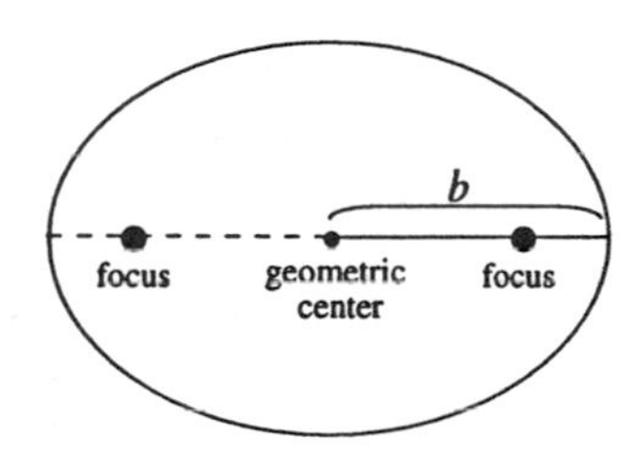

FIGURE 1.5. Features of an ellipse.

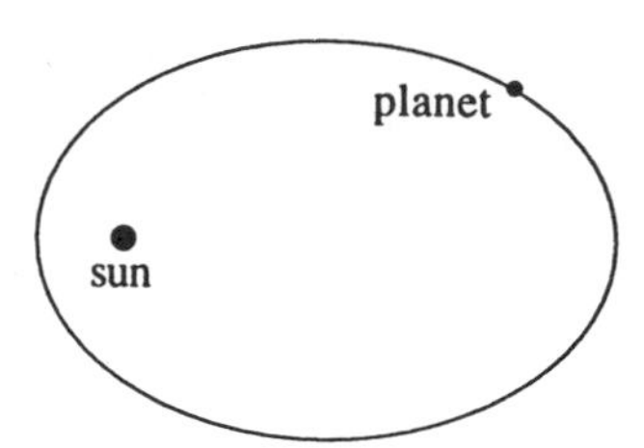

Kepler's First Law

The path of each planet is an ellipse with the sun at one focus.

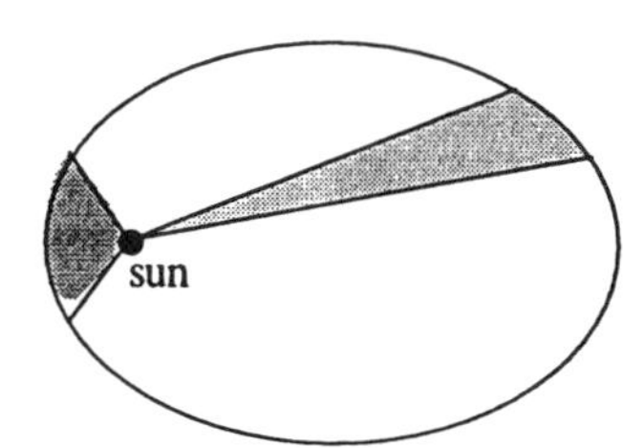

Kepler's Second Law

The line from the sun to the planet sweeps out equal areas in equal times.

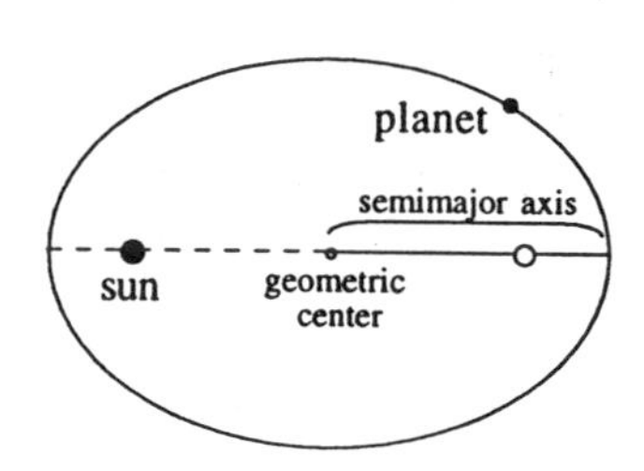

Kepler's Third Law

The square of the orbital period of a planet is proportional to the cube of the semimajor axis of the orbit.

Kepler's third law can be written symbolically as

$$T^2 = K_{\text{sun}} b^3, \tag{1.1}$$

where T is the period of the orbit, that is, the time required for the planet to complete one orbit around the sun. The constant K_{sun} represents some property of the sun and therefore has the same value for all of the planets.

A circle is a special case of an ellipse with the two foci located at the same point. This is nearly the case for the planetary orbits, but not quite. It was Kepler's original plan to use Brahe's observations to find the right combinations of circles for Copernicus's model of the heavens. For many months, he worked at this task. Brahe's data were simply too good to allow him to get away with circles. In desperation, he resorted to ellipses and found the key to the problem. These two qualities of Brahe's data—they were *extensive* and they were *precise*—drove Kepler inexhorably to elliptical orbits for the planets.

The length of the semimajor axis of the earth's orbit is defined to be one *astronomical unit* (AU). This is a convenient unit for expressing distances within the solar system. Orbital data for the planets are given in Table 1.1.

EXAMPLE. The semimajor axis of the orbit of Uranus is 19.2 AU. Calculate the time it takes for Uranus to make a complete orbit around the sun.

Table 1.1. Orbital Data for the Planets[a]

planet	semimajor axis (AU)	period (years)
Mercury	0.39	0.24
Venus	0.72	0.62
Earth	1.00	1.00
Mars	1.52	1.88
Jupiter	5.20	11.86
Saturn	9.54	29.46
Uranus	19.19	84.01
Neptune	30.06	164.79
Pluto	39.53	247.7

[a] Kenneth R. Lang, *Astrophysical Data: Planets and Stars*, Springer-Verlag, New York, 1991, p. 41.

SOLUTION. First, apply Kepler's third law, Eq. (1.1) to the earth to obtain $K_{\text{sun}} = 1\ \text{yr}^2/\text{AU}^3$. Then, for Uranus Eq. (1.1) becomes

$$T^2 = K_{\text{sun}} b^3 = (1\text{yr}^2/\text{AU}^3)(19.2\ \text{AU})^3 = 7,078\ \text{yr}^2.$$

Thus, for the period of Uranus $T = 84.1$ yr.

1.3.2 Lunar Phases, Seasons, and Eclipses

Now let's apply the Copernican system to account for certain phenomena connected with the celestial motions, for example the phases of the moon, seasons, and eclipses. The earth describes a nearly circular orbit around the sun. The earth also spins on its axis, giving rise to night and day. These two motions define natural time intervals. The *day* is associated with the rotational motion of the earth. We define a *solar day* to be the time interval between successive appearances of the sun at the same position in the sky (for example, the time between successive sunrises). Since this varies a little bit over the course of the year, we take the *average* over the year and call that the *mean solar day*. That is the twenty-four-hour day we are familiar with. The time required for the earth to complete its orbit is one *year*—about $365\frac{1}{4}$ days (but not quite).

There is one other natural time interval that has played a more important role in some other cultures than it has in ours. That is the month—the time required for the moon to pass through a complete set of phases—about $29\frac{1}{2}$ days. With respect to the stars, the moon takes about $27\frac{1}{3}$ days to orbit the earth. During this time, the earth advances along its orbit around the sun. The moon shines on the earth light it is reflecting from the sun.

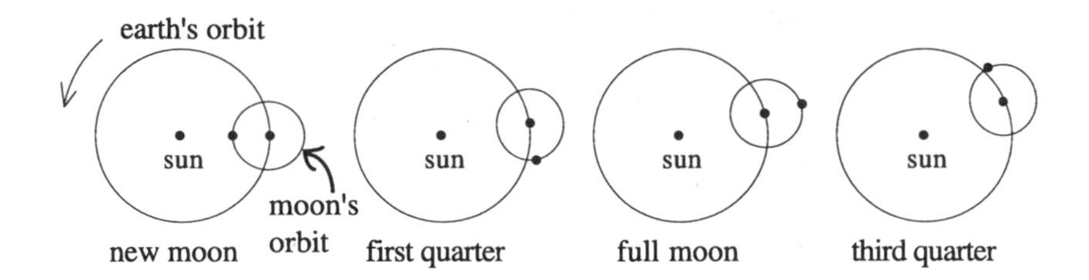

FIGURE 1.6. Lunar phases.

At the beginning of the month (new moon), the illuminated side of the moon is turned away from the earth as seen in Fig. 1.6. About eight days later, the moon has traveled to a point relative to the sun and the earth such that about half of the illuminated part of the moon is turned toward the earth. This is first quarter. Another seven or eight days later, the moon is directly opposite the sun as seen from the earth and the full face of the illuminated part of the moon is seen from the earth. This is the full moon.

The difference between the *synodic* period (time for a complete set of lunar phases) and the *sidereal* period (time for the moon to make a complete orbit with respect to the stars) is illustrated in Fig. 1.7.

We experience seasons on earth because the earth's rotation axis is not perpendicular to the plane of the earth's orbit around the sun. The spin axis of the earth is tilted at an angle of $23\frac{1}{2}^\circ$ with respect to a line perpendicular to the orbital plane. As a result, during some part of the earth's orbit the sun shines more directly on the northern hemisphere of the earth than on the southern (summer in the northern). In addition, during summer there are more hours of daylight in which the sun shines on a given part of the earth's surface. The geometry is illustrated in Fig. 1.8.

It is energy (light) from the sun that warms the earth. The more intense the radiation (energy per unit area per unit time) on a given area and the longer the exposure time, the warmer the area. An energy source (the sun) emits energy in all directions. The energy emitted into a given cone remains in the cone, but spreads out over an ever increasing area, thereby diminishing the intensity with distance from the source.

If at a given distance from the source we intercept this cone with an absorbing plane, the energy will be concentrated in a smaller area (more intense) when the plane is perpendicular to the direction from the source. Tilting the plane relative to this orientation results in the light being spread out over a larger area, reducing the intensity. With the earth's spin axis tilted toward the sun, the northern hemisphere receives more intense radiation while in the southern hemisphere the radiation is spread out over a larger area due to the more oblique incidence of the light and is therefore less intense. This effect is shown qualitatively in Fig. 1.9.

Eclipses are produced when light from the sun is blocked by the earth or the moon, casting a shadow on the other body. For example, a *solar* eclipse is seen on earth at the time of the new moon when the moon is directly between the earth and

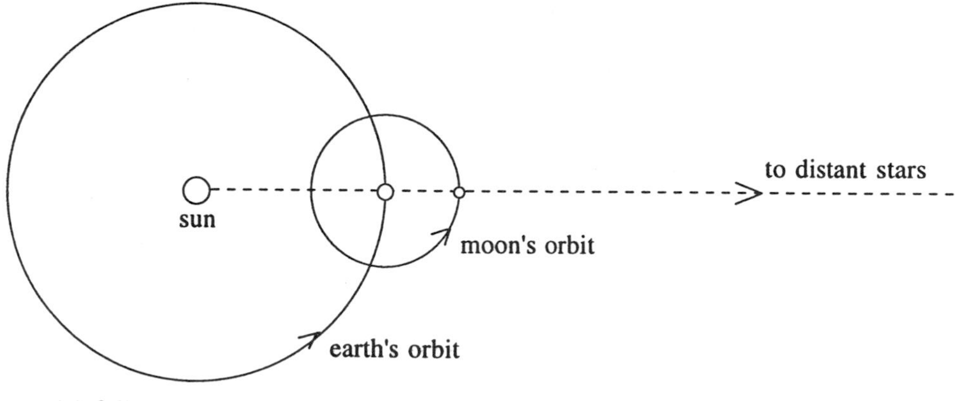

(*a*) full moon

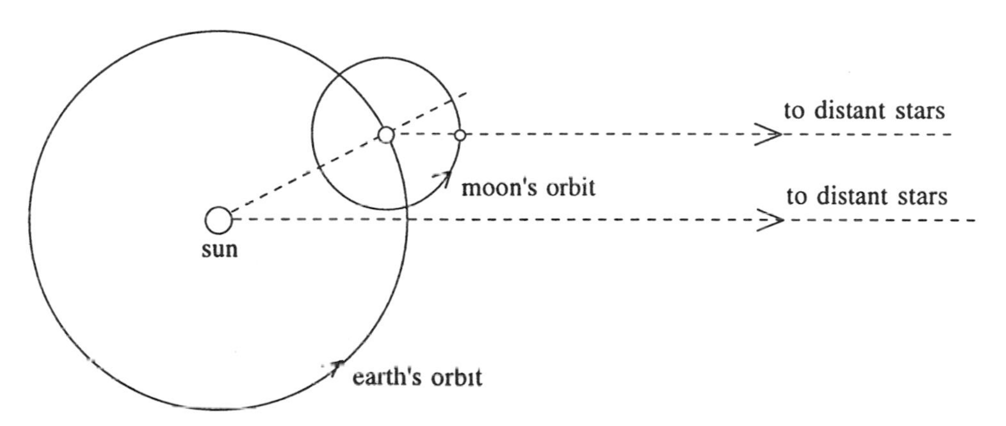

(*b*) 27.3 days later, moon has completed one orbit, but moon is not full

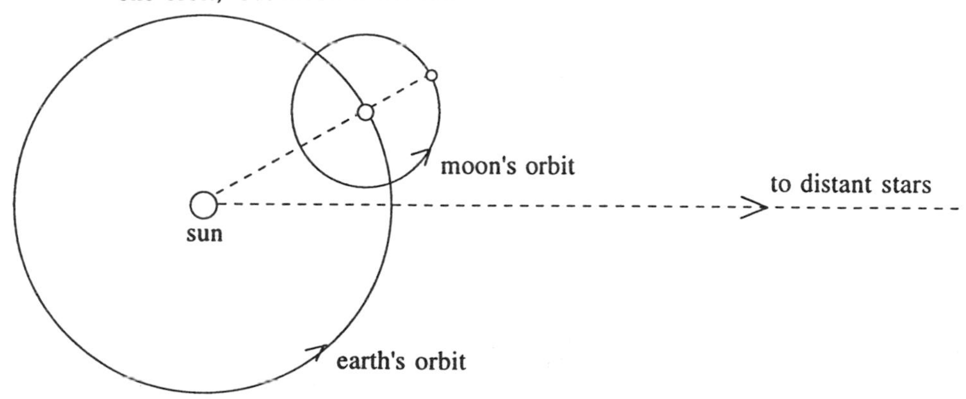

(*c*) at 29.5 days, moon is full again

FIGURE 1.7. Lunar periods.

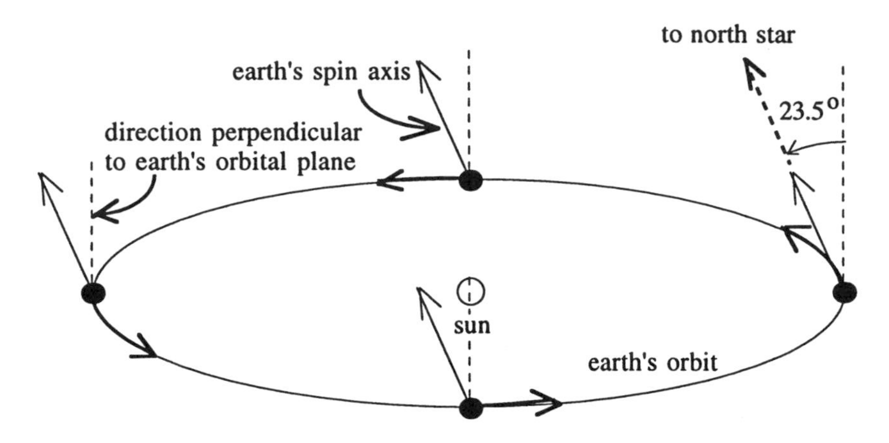

FIGURE 1.8. The reasons for the seasons.

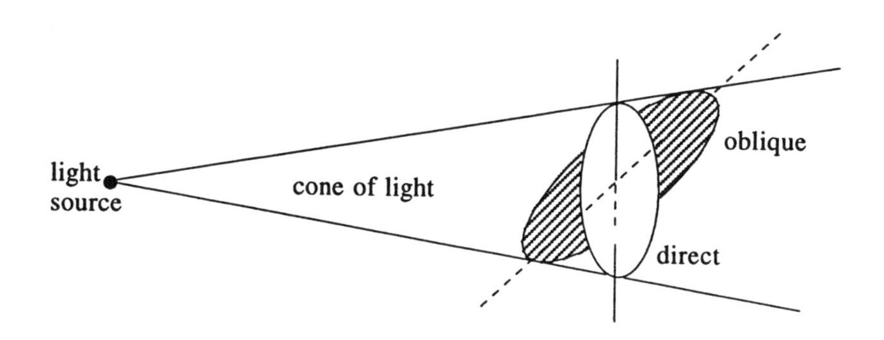

FIGURE 1.9. Radiation intensity.

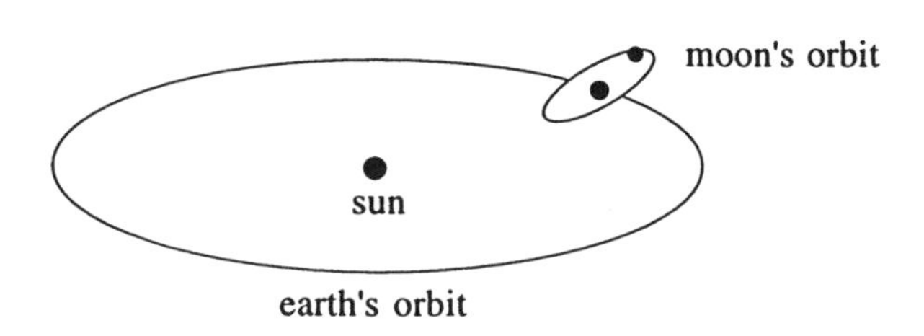

FIGURE 1.10. Eclipses do not occur every month.

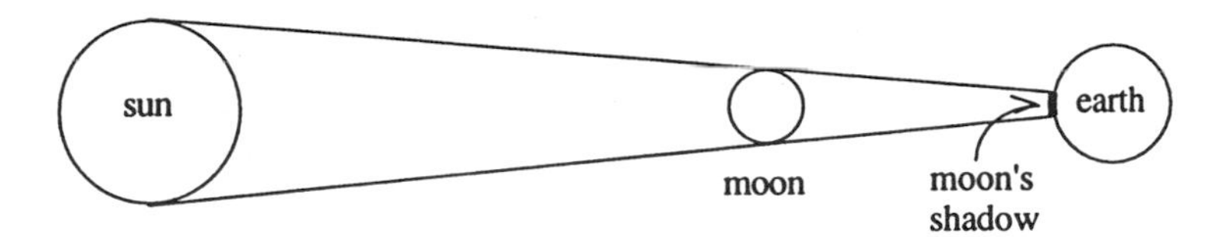

FIGURE 1.11. Solar eclipse geometry.

the sun and the moon casts a shadow on the earth. To a viewer in this shadow, the light from the sun is blocked out. The moon comes between the earth and the sun once a month at the time of the new moon.

Why do we not see a solar eclipse every month? The explanation lies in the fact that the orbit of the moon around the earth and the orbit of the earth around the sun do not lie in the same plane. The moon's orbital plane is tilted about 5° with respect to the orbital plane of the earth as shown qualitatively in Fig. 1.10.

The moon crosses the plane of the earth's orbit twice each month. If one of these crossings occurs precisely at the time of the new moon, a solar eclipse will occur and will be visible on earth to those observers in the moon's shadow. The geometry for a solar eclipse is shown in Fig. 1.11. At other times, the new moon is below or above the plane of the earth's orbit and casts no shadow on the earth.

A similar situation occurs when the full moon is in the earth's orbital plane and the earth casts a shadow on the moon. In this case, observers on the nighttime side of the earth see a *lunar* eclipse. Because the earth is so much larger than the moon, the moon is completely engulfed in the earth's shadow. Also because of the difference in sizes of the moon and the earth, the alignment need not be so precise for a lunar eclipse to occur as for a solar eclipse. Hence, lunar eclipses occur more frequently than solar eclipses.

The plane of the earth's orbit is called the *ecliptic* plane, because *eclipses* occur when the moon is on this plane at new moon or full moon. From the point of view of an earth-based observer, the path of the sun through the constellations is also called the ecliptic for the same reason. When the moon crosses this path in the sky at new moon or full moon, eclipses are seen on earth. The two uses of ecliptic are equivalent.

Exercises

1.1. Assume that Eratosthenes's measurements of the distance from Alexandria to Syene and the angle of the sun at Alexandria at noon on the day of the solstice were accurate. Use these data to determine the length (in meters) of the stadium he used to express the size of the earth. For the actual size of the earth, see Appendix B.

1.2. What was the main defect in Copernicus's model of the heavens? Explain clearly why this was a problem.

1.3. Mars, as seen from the earth, varies greatly in brightness over time intervals of the order of months. Use drawings to explain clearly why this is so.

1.4. As seen from the earth, Venus is brightest at its crescent phase. The moon is *not* brightest at its crescent phase. Use words and drawings to explain clearly the difference.

1.5. From the earth, Mercury and Venus are only seen just after sunset or just before sunrise. Use words and drawings to explain clearly why this is so.

1.6. The planet Neptune requires about 165 years to make a complete orbit around the sun. Use Kepler's third law to estimate Neptune's average distance from the sun.

Answer: (4.5×10^9 km)

1.7. Suppose an object is discovered to move in a circular orbit around the sun and that it makes one complete orbit in 583 years. How far is this object from the sun?

Answer: (69.8 AU)

1.8. On New Year's day in 1801, a Sicilian astronomer discovered a faint celestial object that appeared to move relative to the fixed stars. Later, he gave this object the name Ceres, after the Roman agricultural goddess. Before the end of that year, the famous German mathematician Karl Friedrich Gauss established that the object was in a closed orbit around the sun at a mean distance of 2.77 AU. How long does it take Ceres to make a complete orbit around the sun?

1.9. Suppose a new planet is discovered to be moving around the sun in an elliptical orbit with a semimajor axis 1.16×10^{10} km. Estimate the orbital period of this planet.

Answer: (680 years)

1.10. Because of a certain peculiarity observed in Mercury's orbit, astronomers in the last century speculated that an undiscovered planet orbited the sun inside the orbit of Mercury. The anomaly in Mercury's orbit was attributed to this planet, which would be close to the sun and difficult to see from the earth. Because of its proximity to the fiery sun, this putative planet was called Vulcan. Suppose Vulcan moves in a circular orbit of radius 0.22 AU. Estimate the time that it would take Vulcan to make a complete orbit around the sun.

Answer: (38 days)

1.11. A minor planet (asteroid) travels in a circular orbit around the sun, making a complete orbit in 5.59 years. Use Kepler's third law to estimate how close this object comes to the earth. What is its maximum distance from the earth? A drawing will be helpful.

Answer: (3.2×10^8 km; 6.2×10^8 km)

1.12. Some scientists have suggested that mass extinctions of life on earth—such as the one that killed off the dinosaurs 65 million years ago—occur periodically at intervals of 26 million years. One interesting, albeit controversial, theory that has been proposed to account for these periodic catastrophes calls for a very small, dim companion star orbiting around the sun. This star will periodically disturb a distant cloud of matter that surrounds the sun and send a hail of comets hurtling through the inner solar system. At such times, the comets would be so numerous that a collision with the earth would be highly probable. Estimate the length of the major axis of the orbit of this object.

Answer: (175,000 AU)

1.13. In 1977, an object, called Chiron, with a diameter of about 200 km was discovered in an elliptical orbit around the sun. At its closest approach to the sun (perihelion), Chiron is 8.49 AU from the sun. At its greatest distance from the sun (aphelion), it is 18.91 AU away. From these data, calculate the orbital period of Chiron. Explain clearly your reasoning. A drawing showing Chiron's orbit around the sun will help here.

Answer: (50.7 years)

1.14. It is not possible to design an accurate calendar? Why not?

1.15. Explain clearly and completely why the moon shows phases

1.16. It takes a little over 27 days for the moon to complete an orbit around the earth, but over 29 days are required for the moon to pass through a complete set of phases. By means of words and drawings, use the heliocentric (sun-centered) view of the universe to give a clear and complete explanation of this difference.

1.17. Use words and drawings to explain clearly and completely why we experience seasons on the earth.

1.18. What is the ecliptic? Why is it so named?

1.19. Explain clearly and completely how a solar eclipse is produced. Make a drawing showing all of the relevant objects. Under what circumstances do solar eclipses occur? Why do solar eclipses (compared to lunar eclipses) occur relatively rarely at a given point on the earth's surface?

1.20. Make a qualitatively correct drawing similar to Fig. 1.11 to illustrate how a total lunar eclipse is produced. Based on your drawing, explain clearly why a lunar eclipse may be seen by an observer anywhere on the nighttime side of the earth. Explain clearly why lunar eclipses occur more frequently than solar eclipses.

2

The Stuff Moves Around

Cosmology, as the term itself suggests, is a study of the cosmos—the universe—especially those aspects that have to do with its origin, its large scale structure, and its evolution. We shall consider these lofty notions more carefully toward the end of our study here. In the meantime, we shall look at smaller-scale features of the universe, particularly the structure and evolution of stars and, closer to home, the star system of which we are a part—our solar system.

2.1 Fundamental Properties

We begin by recognizing that the universe has three fundamental properties or qualities. Two of these are *space* and *time*. It is important to recognize that space is not "external" to the universe. The universe does not occupy a volume of space. Rather, space is an attribute of the universe itself, not something the universe is "in." Similar statements can be made about time. Time and space are not defined apart from the universe. The third fundamental quality of the universe is *matter* (or, equivalently, *energy**). Our universe is not an empty one. It contains matter and our concern here will be with the way matter behaves in space and time.

Our first step is to make these fundamental notions quantitative. This is necessary in order to explore the universe in a systematic way and to acquire some understanding of its contents and behavior. Therefore, for each of these qualities we introduce a quantity that allows us to attach numbers to that quality—a quantity that allows us to "measure" it. In Table 2.1, we list in the first column the qualities and in the second column the corresponding quantities.

In general, matter has several properties through which it can manifest itself. One of these is called *inertia*, which is the intrinsic ability that matter has to resist changes in its motion. It is this property that we focus on here. We shall come to a fuller discussion of it later, but for now we simply identify *mass* as the quantity we use to measure inertia.

---

*As expressed by Einstein's famous relation $E = mc^2$.

Table 2.1. Fundamental Qualities

quality	quantity
space	length
time	time interval
matter	mass

To be able to compare measurements of a given quantity made at different times or on different parts of the universe and to be able to communicate the results of our observations to other people, we need a system of standard units for these fundamental quantities. There are two such systems in general use today, and the basic units in each are given in Table 2.2.

Notice that in the two systems two different quantities, mass and weight, are given as measures of inertia. In the metric system, the mass unit (kilogram) is the familiar one, whereas in the English system, it is the weight unit (pound) that is more common. The physical distinction between mass and weight will become clear presently. (See Chapter 4.)

2.2 Derived Quantities

Very often we are interested in a feature of the universe that involves more than one of these fundamental notions. This leads us to define quantities that are combinations of the fundamental ones. For example, the matter in the universe is moving around in space and time. Here we have an interesting quality of the universe (*motion*), which we wish to measure.

We may be able to tell simply by looking that one object moves faster than another, but we want to be able to say how much faster it is moving. We wish to compare the motions of two objects (or the same object at different times) in a quantitative way. The definition of the quantity is at our disposal, but we should make it in a way that is consistent with our intuition. In particular, we want the quantity which measures motion to be larger when the motion is faster and smaller when the motion is slower. For simplicity, consider an object moving in a straight

Table 2.2. Systems of Units of Measurement

	space	time	inertia	
system	length	time interval	mass	weight
metric (mks)	meter	second	kilogram	
English	foot	second		pound

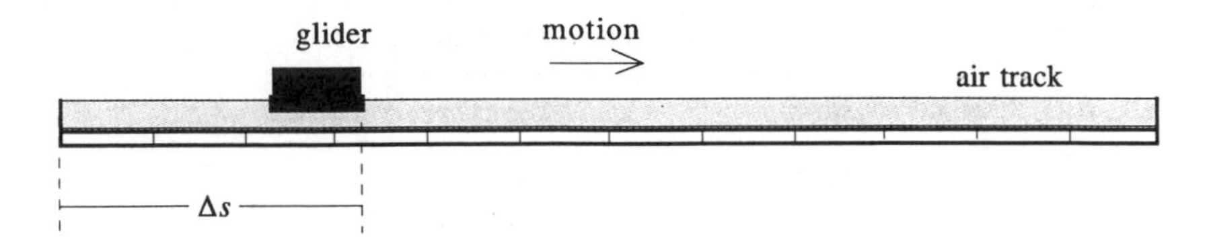

FIGURE 2.1. Uniform motion on a frictionless track.

line, say a block of wood sliding on a smooth, horizontal surface. Clearly, motion involves a change of position (Δs) of the block during some time interval (Δt).*

We call the quantity that measures motion *velocity* and define the average velocity of the block during the time interval Δt as†

$$\bar{v} = \frac{\Delta s}{\Delta t}. \tag{2.1}$$

We shall need a more general definition of velocity later, but for the present discussion this one will suffice.

As an example, let us consider the motion of a metal glider sliding along a straight, horizontal air track as shown in Fig. 2.1. The cushion of air on which the glider floats reduces the friction enough that for our purpose we can neglect it.

The times (t) at which the leading edge of the glider reaches certain points on the track are observed. The data are given in the table in Fig. 2.2 where Δs is the

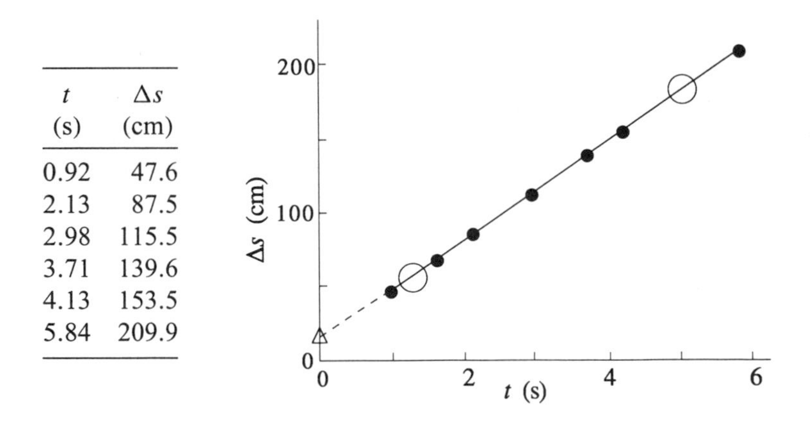

t (s)	Δs (cm)
0.92	47.6
2.13	87.5
2.98	115.5
3.71	139.6
4.13	153.5
5.84	209.9

FIGURE 2.2. Uniform motion.

*The Greek letter Δ will be used to indicate the difference between two values of a given quantity and is read "change in." For example, Δt is read "change in t."

†The bar on top of a quantity denotes the average value of the quantity.

distance of the leading edge of the glider from the end of the track as indicated in Fig. 2.1.

In Fig. 2.2, we plot the data and visually fit the result to a straight line. We choose two arbitrary points on the line (circled in Fig. 2.2) and obtain for the slope the value 33 cm/s.* Since the data are consistent with a straight line, it is clear that the motion is uniform and the slope is just the (constant) velocity with which the glider moves along the track. Notice that, in general, the slope carries units (in this case, cm/s) and that it can be interpreted physically (here it is the average velocity of the glider).

We have extended the line (dashed part) to the point where it intercepts the vertical (Δs) axis. The value of the vertical intercept can be read directly off the graph at this point (marked by the open triangle in Fig. 2.2). We find it to be about 17 cm. Since the vertical intercept represents at $t = 0$ the distance of the leading edge of the glider from the end of the track, it is clear that the vertical intercept tells us where the glider was located on the track (17 cm from the left-hand end) when the timing device (stopwatch, etc.) was started. So we find that the parameters that determine a linear graph may have units and may also have a direct physical interpretation.

Let us raise one end of the air track in Fig. 2.1 and allow the glider to slide down the track as in Fig. 2.3. We see that the glider picks up speed as it moves down the track. The motion changes with time—a new quality. To make it quantitative, we define a quantity called *acceleration*, denoted by a, which is the rate at which the motion (velocity) changes with time. Over a given time interval Δt, the average acceleration is

$$\bar{a} = \frac{\Delta v}{\Delta t}. \tag{2.2}$$

Now let us make more extensive measurements for the glider on the air track. We release the glider from rest at the higher end of the track and measure the time it takes for it to slide down to the lower end. We repeat this procedure several times, releasing the glider from rest at various distances from the lower end of the track. The data are given in the table in Fig. 2.4. If we plot Δs vs t, we do *not* get

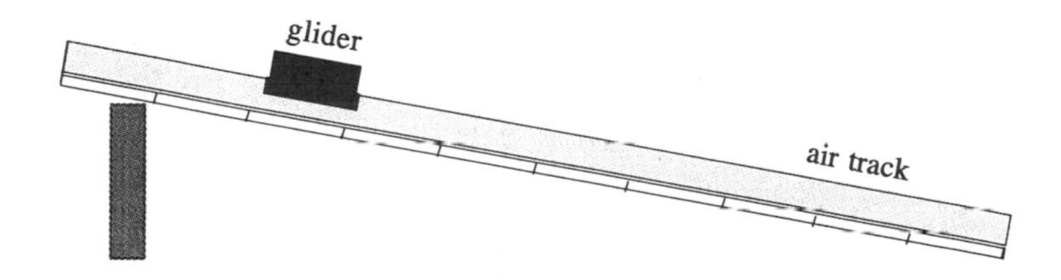

FIGURE 2.3. Changing motion.

*See Appendix A for the treatment of linear graphs.

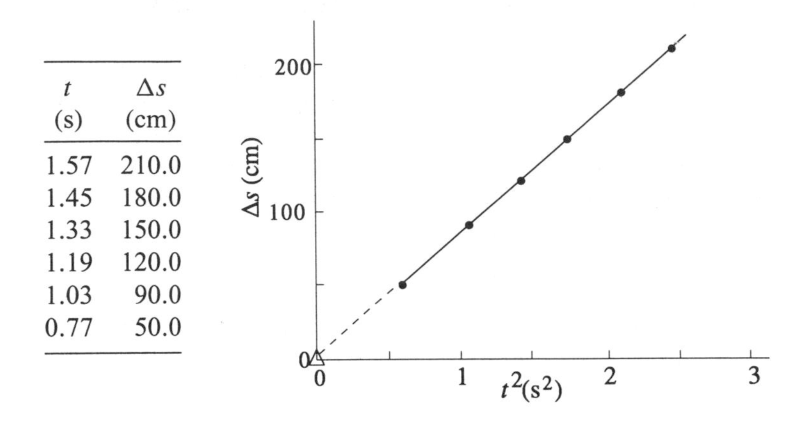

t (s)	Δs (cm)
1.57	210.0
1.45	180.0
1.33	150.0
1.19	120.0
1.03	90.0
0.77	50.0

FIGURE 2.4. Acceleration.

a straight line for the graph. This is to be expected, since we already know that the velocity is not constant. However, if we plot Δs vs t^2, as in Fig. 2.4, we find that the data *are* consistent with a linear graph. The graph extends to the point (0,0), which simply tells us that in a time interval of zero, the glider will have traveled zero distance. From the data, we obtain the experimental result,

$$\Delta s = mt^2, \tag{2.3}$$

where m is the (constant) slope of the line.

What is the connection between this result and the definition of acceleration in Eq. (2.2)? Let's make a simple assumption. It may turn out to be a false assumption, but to make some progress, let's assume that the acceleration of the glider is uniform (i.e., the velocity changes at a constant rate). In this case, $\bar{a} = a =$ constant. For each run, the glider starts with initial velocity v_0 at time $t = 0$. Therefore, $\Delta v = v - v_0$ and $\Delta t = t$. Thus,

$$v = v_0 + at, \tag{2.4}$$

where v is the velocity at time t. To compare with Eq. (2.3), we need a relation between Δs and t. According to Eq. (2.1), the *average* velocity over a given run is

$$\bar{v} = \frac{\Delta s}{t} = \frac{s - s_0}{t}. \tag{2.5}$$

The glider starts with velocity v_0 at $t = 0$ and accelerates uniformly to final velocity v at the bottom of the track. Therefore, $\bar{v}$ must have a value somewhere between the two extremes of initial velocity v_0 and final velocity v. Since the acceleration is uniform, $\bar{v}$ is exactly halfway between the two,

$$\bar{v} = \frac{v_0 + v}{2}. \tag{2.6}$$

On substituting Eq. (2.6) into Eq. (2.5) and rearranging, we get

$$s = s_0 + \tfrac{1}{2}(v + v_0)t \tag{2.7}$$

Finally, we replace v in Eq. (2.7) by $v_0 + at$ from Eq. (2.4) to obtain

$$s = s_0 + v_0 t + \tfrac{1}{2}at^2. \tag{2.8}$$

Since the glider in the example starts from rest, $v_0 = 0$ and Eq. (2.8) reduces to

$$s - s_0 = \Delta s = \tfrac{1}{2}at^2.$$

With our assumption that a is a constant, we see that this expression is consistent with the empirical result in Eq. (2.3). Comparing this last equation with Eq. (2.3) shows that $m = \frac{1}{2}a$, which gives us a physical interpretation of the slope of the graph in Fig. 2.4, namely, the slope of the graph is equal to one-half the acceleration of the glider.

EXAMPLE. The driver of a car on a straight highway takes her foot off the accelerator pedal and slows uniformly from 120 km/h to 90 km/h in twenty seconds. How far does she travel during the time the car is slowing down?

SOLUTION. First, calculate the acceleration from Eq. (2.4),

$$\begin{aligned} v &= v_0 + at \\ 90\,\frac{\text{km}}{\text{h}}\left(\frac{1\ \text{h}}{3{,}600\ \text{s}}\right)\left(\frac{1{,}000\ \text{m}}{\text{km}}\right) & \\ &= 120\,\frac{\text{km}}{\text{h}}\left(\frac{1\ \text{h}}{3{,}600\ \text{s}}\right)\left(\frac{1{,}000\ \text{m}}{\text{km}}\right) + a(20\ \text{s}) \\ a &= -0.417\ \text{m/s}^2, \end{aligned}$$

where the minus sign means that the acceleration is in the direction opposite to the initial velocity (which we took to be positive). With this acceleration, we obtain from Eq. (2.8),

$$s = s_0 + v_0 t + \tfrac{1}{2}at^2$$

$$\begin{aligned} s &= 0 + 120\,\frac{\text{km}}{\text{h}}\left(\frac{1\ \text{h}}{3{,}600\ \text{s}}\right)\left(\frac{1{,}000\ \text{m}}{\text{km}}\right)(20\ \text{s}) + \frac{1}{2}(-0.417\ \text{m/s}^2)(20\ s)^2 \\ &= 583\ \text{m}. \end{aligned}$$

Thus, the car travels about 580 m while slowing from 120 km/h to 90 km/h.

2.3 Scalars and Vectors

In our work, we shall encounter two general kinds of quantities—*scalars* and *vectors*.

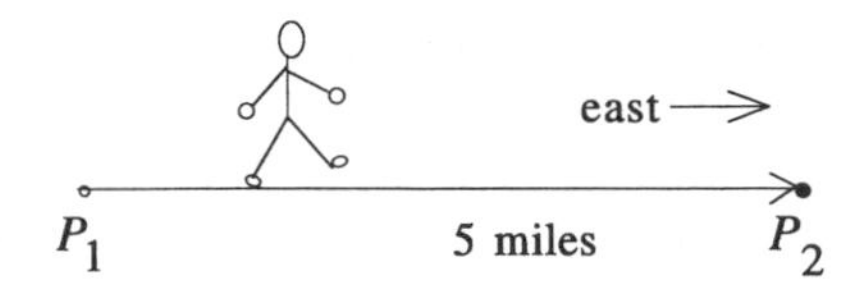

FIGURE 2.5. Hiker's displacement.

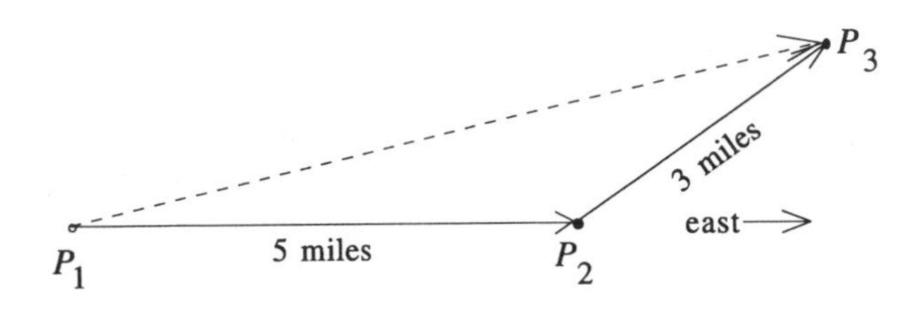

FIGURE 2.6. Resultant displacement.

- Scalars have magnitude only. Examples are mass, temperature, energy, density.
- Vectors have direction as well as magnitude. Examples include velocity, acceleration, force.

To illustrate the difference, let's imagine a hiker who starts at point P_1 and walks five miles due east to a point P_2 as in Fig. 2.5. We say that his *displacement* (a vector) is five miles due east, while the *distance* he traveled (a scalar) is five miles.

Now suppose he makes a second displacement to a point P_3 which is three miles from P_2 in a direction 35° north of east as in Fig. 2.6. What is the total distance the hiker traveled—the single distance equivalent to the two parts of the trip? It is eight miles ($5 + 3 = 8$). The rule for adding scalars is clear. We add them algebraically. What is the resultant *displacement*—the single displacement equivalent to the two displacements the trip comprises? This displacement from P_1 to P_3 is indicated by the dashed line in Fig. 2.6. Clearly, it is less than eight miles (about 7.6 miles) and in a direction about 13° north of east. This example suggests a general way to add vectors.

In this book, we shall use italic type for representing scalars (e.g., m for mass) and boldface roman type to denote a vector (e.g., **F** for force). Pictorially, we represent a vector by a directed line segment with the length of the line corresponding to the magnitude of the vector and the arrowhead on the end indicating its direction.

This is illustrated for two vectors **A** and **B** in Fig. 2.7a. The vector **A** has twice the magnitude of **B**, therefore the line representing **A** is two times as long as that for **B**. To add **B** to **A** to get a third vector **C**,

$$\mathbf{A} + \mathbf{B} = \mathbf{C},$$

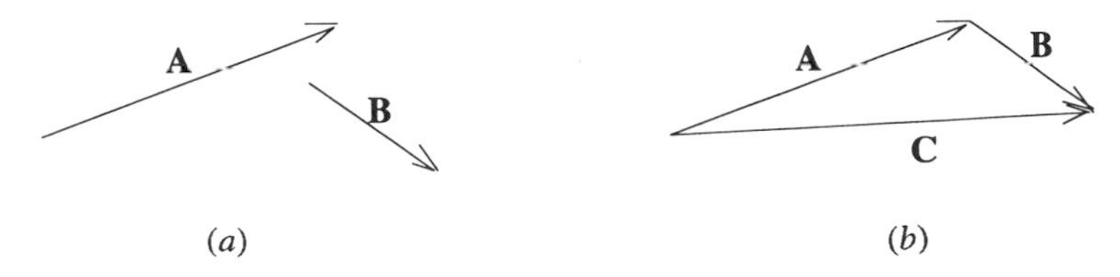

FIGURE 2.7. Vector addition.

we keep **B** pointing in the same direction, but move it in the plane until the tail of **B** touches the head of **A** as in Fig. 2.7*b*. The vector drawn from the tail of **A** to the head of **B** is the resultant **C**. For our work, this discussion of vectors will be sufficient.

We are now ready to generalize our definitions of velocity and acceleration. The average velocity of an object is given by

$$\bar{\mathbf{v}} = \frac{\Delta \mathbf{s}}{\Delta t}, \tag{2.9}$$

where $\Delta\mathbf{s}$ is the displacement of the object in the time interval Δt. The *instantaneous* velocity $\mathbf{v}$ (velocity at a given instant) is obtained from this definition in the limit as Δt approaches zero.

Similarly, the average acceleration is

$$\bar{\mathbf{a}} = \frac{\Delta \mathbf{v}}{\Delta t}. \tag{2.10}$$

In the limit $\Delta t \longrightarrow 0$, $\bar{\mathbf{a}}$ becomes the instantaneous acceleration $\mathbf{a}$. An important thing to note here is that we can accelerate an object not only by changing its speed, but also by changing its *direction* of motion, even if the speed remains constant. In general, both speed and direction change. A car traveling around a circular track at a constant speed of 30 mph is accelerating because its direction of motion is continuously changing.

EXAMPLE. A woman sets out in the direction 29° N of E and walks 1,325 meters over a flat, open field. She then turns and heads in the direction 43° N of W and marches off 1,550 meters in this direction. Finally, she makes another turn and walks 1,175 meters in the direction 57° S of W. Find the total distance she walked and her resultant displacement.

SOLUTION. The total distance she traveled is simply the sum of the magnitudes of the individual displacements,

$$\text{distance} = 1{,}325 \text{ m} + 1{,}550 \text{ m} + 1{,}175 \text{ m} = 4{,}050 \text{ m}.$$

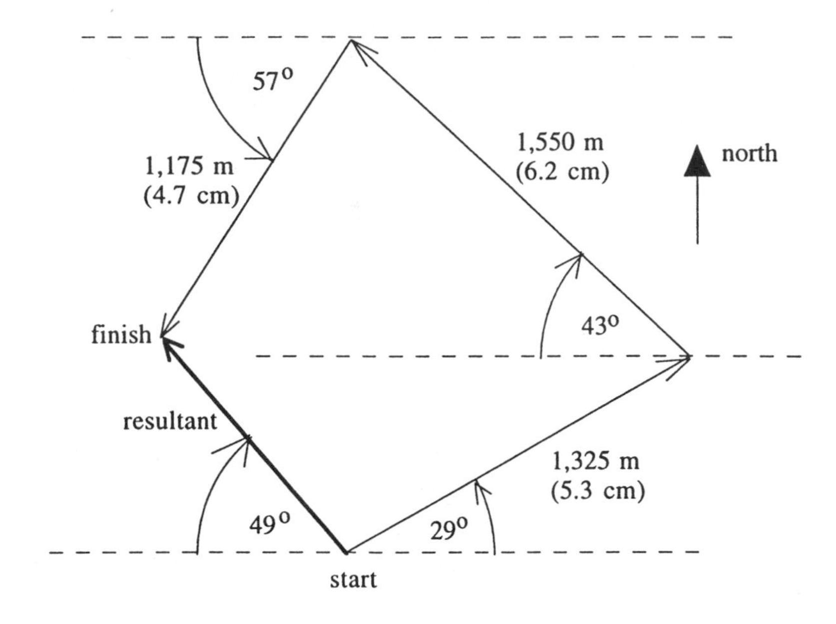

FIGURE 2.8. Resultant of three displacements.

To find the displacement, we use a ruler and protractor to arrange the three displacement vectors according to the addition rule in Fig. 2.7. Remember that the lengths of the vectors represent the magnitudes. We choose a scale that is convenient for fitting the drawing on the paper. For example, with 4 cm on the drawing representing 1,000 m on the actual hike the magnitudes of the separate displacements are related to the lengths of the corresponding vectors on the drawing by (1) 1,325 m = 5.3 cm, (2) 1,550 = 6.2 cm, and (3) 1,175 m = 4.7 cm. The scale drawing is shown in Fig. 2.8.

We construct the resultant by drawing the vector from the tail of the first vector (start) to the head of the third vector (finish). Measuring the length of the vector and the angle this vector makes with the east-west line, we find from the scale drawing that the resultant displacement is 940 m in the direction 49° N of W.

Exercises

2.1. A small sailboat is observed to move directly away from a dock. Its distance from the dock at various times is measured by observing the time it passes each of several buoys along its route. From the graph below, determine the velocity of the boat. How long after the boat left the dock was the timer started?

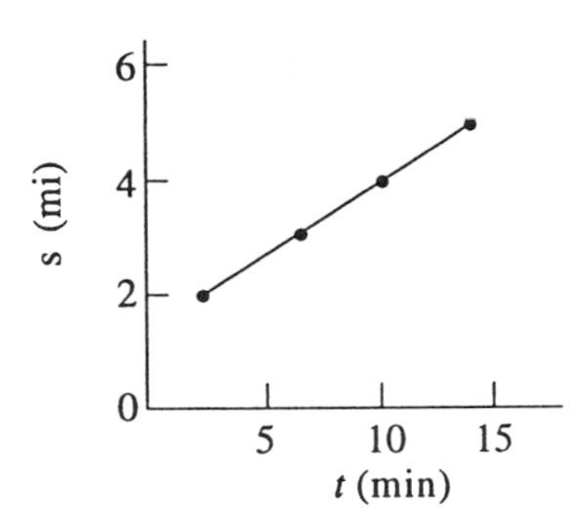

2.2. Two women walk out of the clubhouse, climb into a golf cart, and set off at constant speed on a straight path over smooth, level ground. Along the way, the passenger starts a stopwatch and begins to record the time they pass certain markers along their route. Each marker gives the distance between the marker and the clubhouse. The data are given in the graph. *From the graph*, determine the velocity of the cart. Write an equation giving the position s of the cart relative to the clubhouse in terms of the time t shown on the stopwatch. How long after they left the clubhouse did they start the stopwatch? How far are they from the clubhouse when the stopwatch reads exactly 7 minutes?

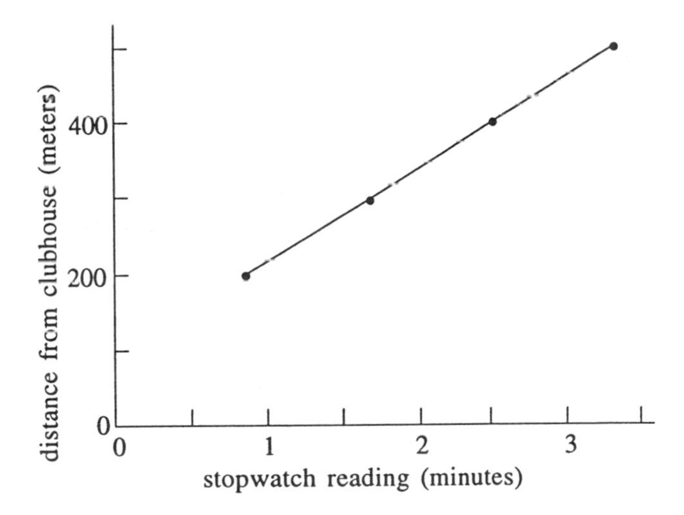

2.3. A bicycle rider leaves the courthouse in her village and pedals at a constant speed along a straight, level road to the courthouse in the next town, which is thirty miles away. In the afternoon, she begins to notice the time according to her wristwatch when she passes certain mileposts along the way. These posts give the distance from her courthouse. The data are shown graphically below. From the graph, determine the speed (in mi/min) at which the rider travels. To the nearest minute according to her watch, at what time did she leave the courthouse?

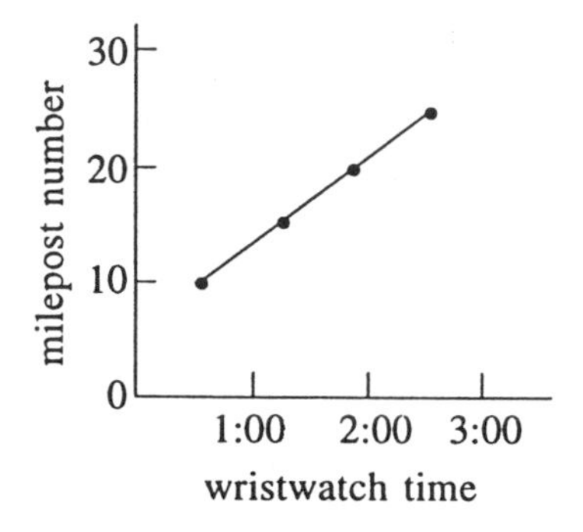

2.4. A car is traveling at a constant speed of 25 mi/h. Can you say that the car is not accelerating? Explain clearly.

2.5. Assume that the earth moves at constant speed in a circular orbit around the sun. Estimate the orbital speed of the earth.

Answer: (108,000 km/h)

2.6. A horse-drawn sleigh is driven east over a straight, level road which passes through two villages 21 miles apart. (Easton is due east of Weston.) The driver notes the time (on a stopwatch) that she passes certain mileposts, which give the distance from the Post Office in Weston. The data are shown in the graph. From the graph, determine where the sleigh was, relative to the Weston Post Office, when the driver started her stopwatch. Write your answer in a complete sentence. From the graph, obtain an equation that gives the position s of the sleigh at time t. Use your equation to find the time (according to her stopwatch) that she reaches Easton. Explain clearly and completely your reasoning in arriving at each answer.

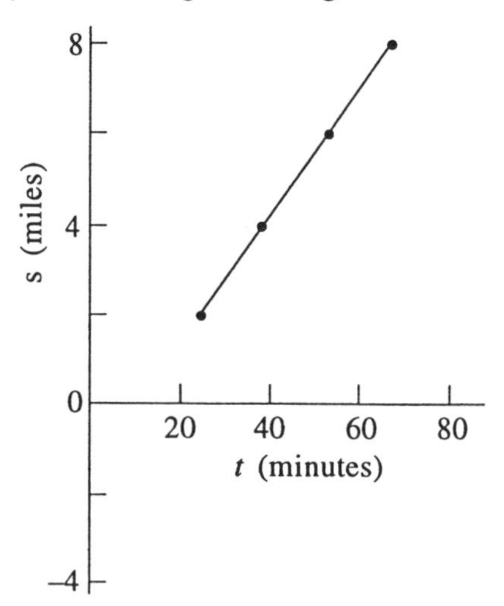

2.7. To a good approximation, Venus travels at constant speed in a circular orbit around the sun. Calculate the distance Venus moves in one day along its orbital path.

Answer: (3 million km)

2.8. A body is experiencing a constant acceleration of 3 m/s^2. How much speed does the body gain each second? If it starts from rest, what is its speed after 2 seconds? What is its speed after 5 seconds?

2.9. A bicycle rider increases her speed from 5 to 10 mi/h in the same time that a car goes from 50 to 55 mi/h. Which, if either, has the greater acceleration? Explain clearly your reasoning.

2.10. An air track is inclined slightly with respect to the horizontal so that a glider starting from rest will slide with constant acceleration over the full length of the track (1.30 m) in 3.21 seconds. Calculate the speed of the glider as it reaches the lower end of the track.

Answer: (0.81 m/s)

2.11. Starting from rest, a drag racer reaches a speed of 240 km/h over a distance of 420 m on a straight, level track. Calculate the acceleration of the racer.

Answer: (5.29 m/s^2)

2.12. The takeoff speed of a certain airplane is 155 km/h. Starting from rest, the aircraft requires a run of 750 m to become airborne. How much time is required for this takeoff roll?

Answer: (34.8 s)

2.13. A go-cart starts from rest on a straight, level strip and accelerates uniformly until it reaches its top speed of 58 km/h fourteen seconds later. How far from its starting point has the go-cart traveled one minute after it started moving?

Answer: (854 m)

2.14. A rocket-propelled sled traveling with a speed of 327 km/h is brought to rest uniformly over a distance of 140 meters. Calculate the stopping time and the acceleration.

Answer: (3.08 s; 29.5 m/s^2)

2.15. A boy on a sled slides down a long, straight slope. His position relative to some point on the slope is observed at various times. The data are shown graphically below. *From the graph*, determine the acceleration of the sled. Write an equation for the position s of the sled at time t. Assuming that the motion continues with the same acceleration, use your equation to find the position of the sled when the timer reads 80 seconds.

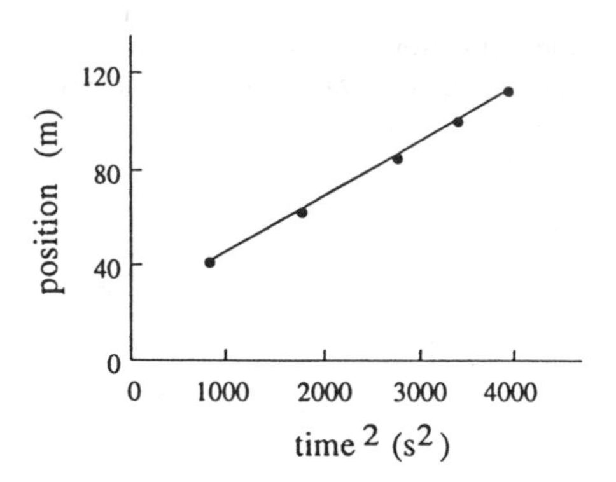

2.16. A car traveling 60 km/h can be stopped in a distance of 29.3 meters. How much distance is required to stop the car if it is traveling 135 km/h (typical speed on a German autobahn)?

Answer: (148 m)

2.17. A rocket is launched vertically upward and reaches a maximum speed of 4,800 km/h at an altitude of 240 km at which point its engine is cut off. Find the time it takes the rocket to reach this altitude and the average acceleration (in m/s^2) of the rocket.

Answer: (6 min; 3.70 m/s^2)

2.18. A passenger train sits on a straight, level track with the locomotive sitting some distance from the terminal. The train begins to move with a constant acceleration along the track. At the instant that it begins to move, the engineer starts a stopwatch. Position markers indicating the distance from the terminal are located at intervals along the track. After some time, the engineer begins to note the time (according to his stopwatch) at which he passes certain of these markers. His data are shown in the graph below. How fast (in km/h) is the train traveling when the locomotive is 1,200 m from the terminal? How far from the terminal was the locomotive when the train started moving?

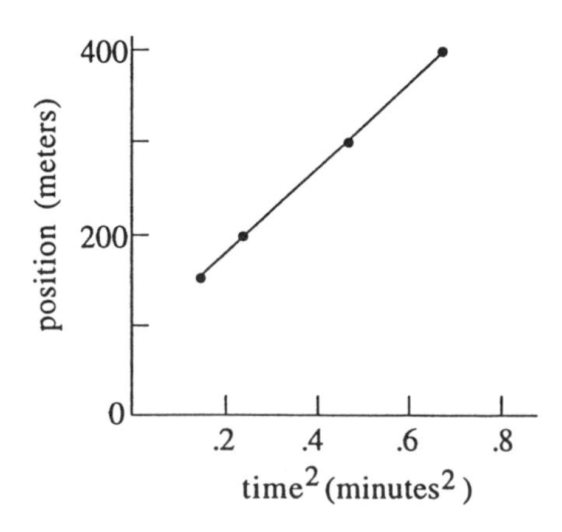

2.19. A car is traveling on a straight, level road at a speed of 72 km/h. The driver suddenly hits the brake and stops in 14.3 seconds. How far does the car travel in this time?

Answer: (143 m)

2.20. An aircraft carrier is sailing with a speed of 27 km/h with no wind. Assisted by a horizontal catapult, an aircraft is launched over the bow of the ship. The uniform acceleration of the aircraft is 34.3 m/s^2. Its airspeed as it leaves the catapult is 168 km/h. Calculate the length of the catapult.

Answer: (30.9 m)

2.21. A small block moving with a constant acceleration is observed at one instant to be moving to the right with a speed of 4.87 m/s. It is seen 4.17 s later 2.46 m to the right of the position where it was first observed and now moving to the left. How fast is the block moving when this second observation is made?

Answer: (3.69 m/s)

2.22. A bullet is fired with a velocity of 400 m/s into a wooden door that is 2.68 cm thick. The bullet passes clear through the door and emerges from the other side $8.70 \times 10^{\ 5}$ second after it first struck the door. How fast is the bullet moving as it leaves the door? Assume that the bullet changes speed uniformly as it passes through the door.

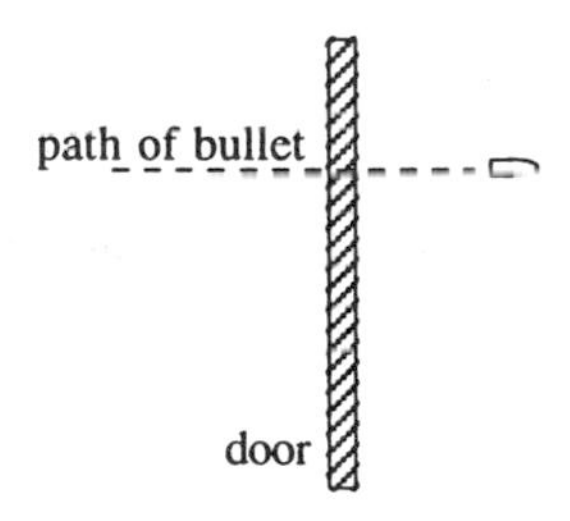

Answer: (216 m/s)

2.23. A car on a straight stretch of road is traveling at 135 km/h. The driver suspects a radar trap ahead and slows his car uniformly to 88 km/h in seven seconds. How far does the car travel in this time?

Answer: (217 m)

2.24. A windsurfer is sailing along at 12.8 km/h when his sail catches a sudden gust of wind and increases his speed uniformly to 17.3 km/h in 11.6 seconds. How far does the gust carry him in this time?

Answer: (48.5 m)

2.25. A particle moving with a constant acceleration is observed at one instant to be moving to the left with a speed of 13.8 m/s. Eight seconds later it is observed to be moving to the right with a speed of 22.7 m/s. How far apart are the two points where the particle was observed? Is the second point to the right or to the left of the first point?

Answer: (35.6 m; to the right)

2.26. The pilot of an aircraft in level flight on a straight course reduces the aircraft speed uniformly from 183 km/h to 152 km/h in seventeen seconds. How far does the aircraft travel during this time interval?

Answer: (791 m)

2.27. A small block slides along a straight path on a rough, horizontal surface. It is observed at one instant to be moving to the right with a speed of 0.932 m/s. Five seconds later it is still moving to the right, but it has slowed down to 0.560 m/s. Assuming the acceleration is constant, how far does the block travel in this time interval?

Answer: (3.73 m)

2.28. On an interplanetary journey, a spacecraft is moving with a constant velocity of 17,100 km/h. The engine is fired for 46.9 s, accelerating the craft uniformly to 17,850 km/h without changing the direction of motion. How far does the spacecraft travel during the time the engine is running?

Answer: (227.7 km)

2.29. Starting from rest, a small cart is pulled along a straight track. At various times t observed on a stopwatch, the position s of the cart is measured relative to one end of the track. The data are given in the graph. From the graph, obtain an equation for the position s in terms of the time t. Use this equation to find where the cart is located relative to the end of the track and how fast it is moving when the stopwatch reads 1.68 s.

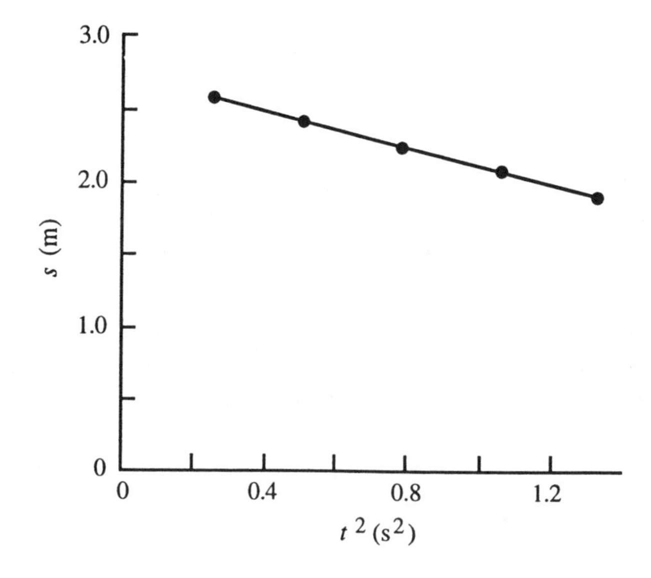

2.30. A girl on a sled is given a push that starts her down a long, straight slope with an initial speed of 0.833 m/s. She reaches the bottom of the slope 23.8 s later with a speed of 1.944 m/s. Assuming that the acceleration of the sled is constant, what is the length of the slope?

Answer: (33.0 m)

2.31. A car starts from rest with a constant acceleration of 1.87 m/s^2. When it reaches a speed of 70 km/h, it stops accelerating and continues traveling at this constant speed. How long does it take the car to travel one kilometer from its starting point?

Answer: (56.6 s)

2.32. A deputy sheriff is traveling east along a straight stretch of narrow, two-lane highway at a constant speed of 85 km/h. In her rearview mirror, she sees a small car coming up behind her at high speed. As the car whizzes past her, it is traveling at its top speed of 145 km/h. The county line intersects this highway 5.5 km east of the point where the speeder draws even with the deputy. The deputy has no authority to issue a speeding ticket unless she apprehends the culprit within her own county. If he makes it over the county line before she can stop him, she will not be able to give him a speeding ticket. As the speeder draws even with her, the deputy begins to accelerate uniformly at 0.445 m/s^2 until she overtakes him. She shoots past the speeder, whips in front of him, and drives down the middle of the road so that he cannot pass. They both immediately begin to reduce speed at uniform rates until they stop. The distance from the point where she overtakes him to the point where they stop is 1,350 meters. Does the speeder get a ticket? How fast was the deputy traveling when she passed the speeder?

2.33. A sailboat leaves its dock and sails in succession to several points on the shore of a large lake. To reach the first point, the boat sails 1.43 mi in a direction 37° N of E. From there, it heads 26° S of W to a point 2.37 mi away. The last leg of the journey is over a distance of 1.64 mi at 70° N of E. By means of a scale drawing, find the displacement of the boat relative to its dock.

Answer: (1.43 mi, 72° N of W)

2.34. A military tank starts at its home base and travels 14.4 mi in a direction 62° N of E. The tank then heads 17° S of W and travels 11.2 mi in that direction. Finally, it turns and travels 21.9 mi in the direction 36° S of E. From a scale drawing, determine the resultant displacement of the tank from its home base.

Answer: (14.2 mi, 14° S of E)

2.35. A hiker sets out from his base camp early in the morning and travels 8.5 mi in the direction 17° N of E and stops for lunch. After lunch, he heads 75° N of W and travels 7.4 mi in that direction before he stops to rest. After resting for a while, he picks up his gear and walks 6.5 mi in the direction 14° N of

E. Find the hiker's resultant displacement (magnitude and direction) from his base camp.

Answer: (16.8 mi, 41.8° N of E)

2.36. A lone horseman leaves the village of Polena and travels to Ascoti, which is 4.96 mi away in the direction 21.3° N of E. At Ascoti, he turns and heads in the direction 40.6° N of E toward Tonio, 3.10 mi from Ascoti. He leaves Tonio traveling in the direction 28.7° S of W and rides for 7.95 miles. At this point, darkness overtakes him and he stops for the night. All of the roads between the villages are straight paths over level ground. How far does he travel on this journey? At the end of the journey, where is the rider relative to Polena? Write your answer in a complete sentence.

2.37. Sorgenloch is 2.3 km due south of Nieder Olm. Stadecken is 3.2 km due west of Nieder Olm. Essenheim is 1.9 km in a direction 22° E of N from Stadecken. Find the location of Essenheim relative to Sorgenloch.

Answer: (4.8 km, 58° N of W)

3

Eyes on the Skies

Much of what we know about the universe at large has come to us in the form of light—from stars, planets, galaxies, and other collections of luminous and illuminated matter. Light is just a small portion of a continuous spectrum of electromagnetic radiation—a portion to which our eyes happen to be sensitive and which our brains can distinguish as separate colors. Electromagnetic radiation is emitted by matter. Later we shall see in some detail how that comes about.

3.1 Wave Phenomena

Certain features of electromagnetic radiation, and light in particular, can be understood in the framework of a wave model. In this picture, we regard light as a wave with two mutually perpendicular parts—an electric field component and a magnetic field component. All waves, even complicated ones like light, have certain features in common. For simplicity, let us look at some of these characteristics for a wave on the surface of a large, shallow pool of water. Such a wave can be produced by inserting a tiny probe at a fixed point in the water and vibrating the probe up and down in the water at a regular rate. As a result of the vibration, concentric circular ripples emanate from the point at which the probe strikes the water and propagate across the surface of the pool. At a given instant, as seen from above the water surface, the loci of points marking the wave crests are concentric rings as shown in Fig. 3.1.

Because of the regular rate of vibration of the probe, the crests are uniformly spaced. The distance between successive crests is a characteristic of the wave called the *wavelength* which we shall denote by the Greek letter λ. The crests are also uniformly spaced in time. That is, if we focus on a particular point on the surface, a wave crest will appear at that point at regular intervals. The time interval between successive crests at a given point is defined to be the *period* of the wave and is denoted by T. Related to the period is the *frequency* with which crests appear at a given point. The frequency is represented by the Greek letter ν and is clearly the reciprocal of the period (i.e., $\nu = 1/T$).

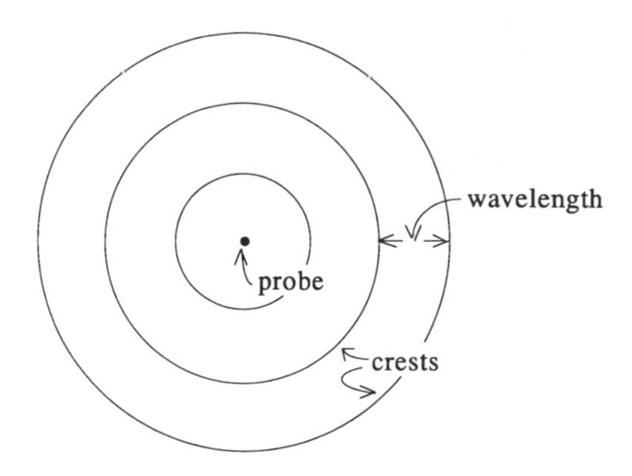

FIGURE 3.1. Surface waves on a shallow pool of water.

We have said that the waves *propagate* over the surface of the water. Let's examine this notion a little more carefully. As time increases, the boundary between the disturbed part of the water surface and the undisturbed part moves farther away from the probe. If we look at a cross section of the water in the region of this boundary we see a series of uniformly spaced crests up to a point and after that a smooth, undisturbed water surface. In Fig. 3.2, the upper drawing illustrates this at a time t_1 where the boundary is at position S_1. At a later time t_2, the boundary has moved farther out to position S_2 as shown in the lower drawing. In the time interval $\Delta t = t_2 - t_1$, the edge of the wave has moved a distance $\Delta S = S_2 - S_1$. Thus, we can associate with the wave a *propagation velocity* v defined by

$$v = \frac{\Delta S}{\Delta t}.$$

Since this holds for any distance and corresponding time interval, we can choose the distance to be the wavelength (i.e, $\Delta S = \lambda$) with the corresponding time

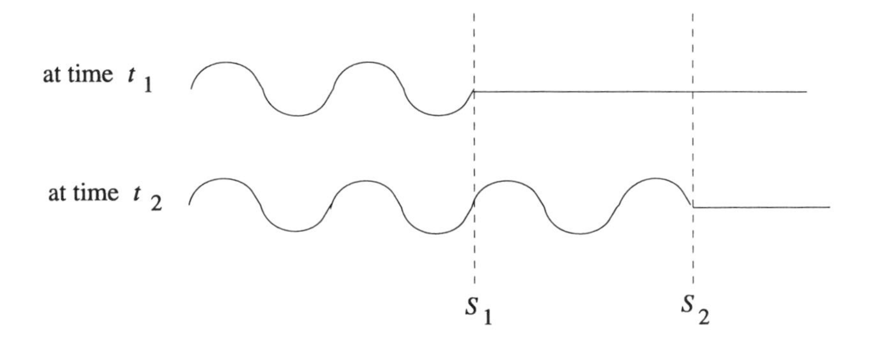

FIGURE 3.2. Propagation of a water wave.

interval equal to the period ($\Delta t = T = 1/\nu$), in which case the propagation velocity is

$$v = \lambda \nu. \tag{3.1}$$

This is a general relationship connecting the propagation velocity of a wave with its wavelength and frequency. It holds for any kind of wave including electromagnetic waves.

3.2 Reflection and Refraction of Light

In regions far from the source, the wavefronts (loci of points on the same crest, trough, etc.) are approximately represented by planes as illustrated in Fig. 3.3*a*.

A directed line segment drawn perpendicular to the wavefronts is called a *ray* and indicates the direction of propagation of the wave. In Fig. 3.3*b*, we show light incident in air on the plane surface of a block of glass. Some of the light is reflected from the surface back into the air and some is transmitted through the boundary and propagates in the glass with a change in the direction of propagation (*refraction*). By measuring the angle between the incident ray and a line perpendicular to the glass surface at the point of incidence (this is the *angle of incidence* and is denoted by i) and the angle of reflection i' (i.e., the angle between the reflected ray and the line perpendicular to the boundary), we always find that $i' = i$. This relationship is called *the law of reflection*. The surface need not be a plane. The law of reflection holds for light reflected from *any* smooth surface regardless of its shape.

3.2.1 Mirrors

An astronomical telescope is a light-gathering instrument. The quality of the telescope is measured by the sharpness of the image of the source producing the light.

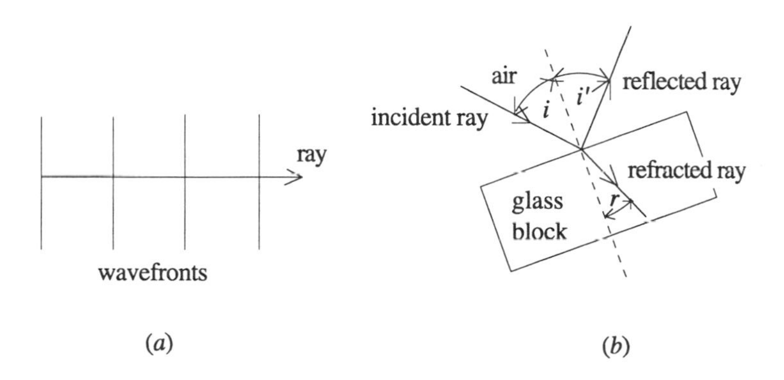

FIGURE 3.3. Reflection and refraction at the surface of a glass block.

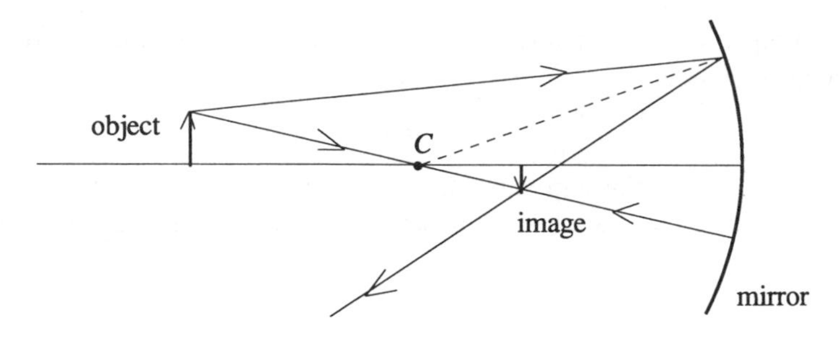

FIGURE 3.4. Reflection in a spherical mirror.

The telescope has two main optical components: an objective (lens or mirror) to collect the light and an eyepiece (usually a lens) to focus the light and form an image. To understand how optical images are produced, let's consider a curved mirror in the shape of the inside of a spherical shell as shown in Fig. 3.4.

Light diverges in all directions from a small object (represented by a vertical arrow in Fig. 3.4) standing in front of the mirror. The drawing shows two typical light rays diverging from the tip of the object and striking the mirror. The corresponding reflected rays are drawn consistent with the law of reflection. These reflected rays are seen to converge in front of the mirror and contribute (along with all other rays diverging from the tip of the object and incident on the mirror) to the image of the tip of the arrow as shown. Similarly, light rays that come from other points on the object and strike the mirror contribute to corresponding points on the image.

Notice that the image is formed *in front* (on the reflecting side) of the mirror with the reflected light actually passing through the image position. Such an image can be captured on a screen placed at the image position. An image formed in this way is a *real* image. This is to be contrasted with the image formed *behind* the mirror. For example, Fig. 3.5 shows how the image is formed by a plane mirror (the kind you see in most bathrooms). Again, we follow the light paths for two rays diverging from the tip of the object and falling on the mirror. The corresponding reflected rays *appear* to come from a point *behind* the mirror where the image is formed. It is clear that the rays that form the image are reflected off the front of

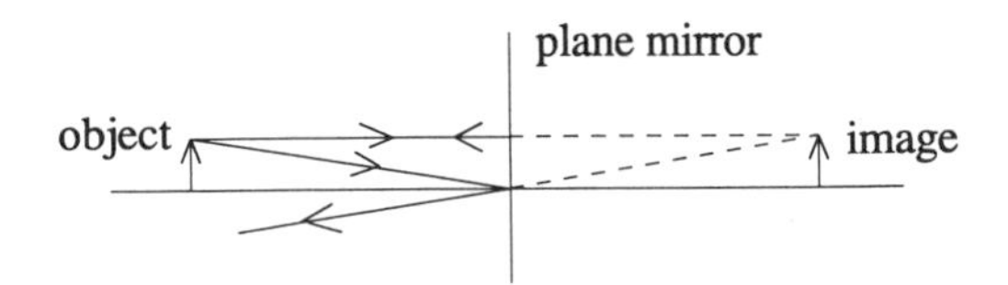

FIGURE 3.5. Ray diagram for a plane mirror.

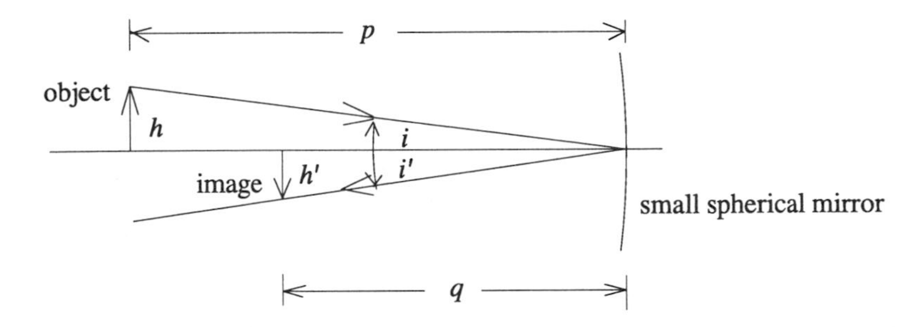

FIGURE 3.6. Definitions of parameters for image formation in mirrors.

the mirror and never actually reach the image position. An image formed in this way is called a *virtual* image.

We can also see that the image in Fig. 3.4 is *inverted*, that is, its orientation is opposite to that of the object. The image in Fig. 3.5 is *erect*, having the same orientation as the object. Suppose we place a small luminous object at various distances in front of a small spherical mirror similar to the arrangement shown in Fig. 3.6. The image position can be determined by placing a small white card in front of the mirror and moving it toward or away from the mirror until a sharp image of the object appears on the card. For each object position, we measure the mirror-to-object distance (denoted by p in Fig. 3.6) and the mirror-to-image distance q. The data are given in the table in Fig. 3.7. A plot of q versus p does not yield a linear relationship. However, if we plot the *reciprocals* of these quantities, we do get a linear graph as shown in Fig. 3.7. From any two arbitrary points on the line, we find that the slope of the graph is -1. Thus, the equation for this line

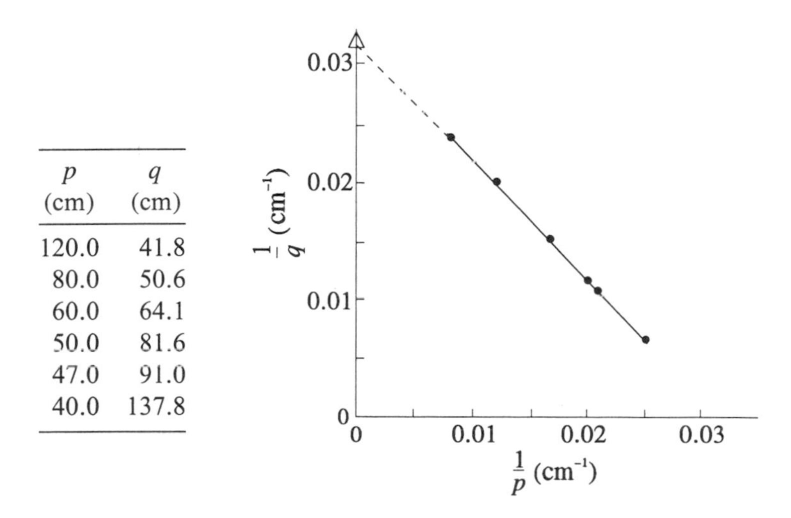

p (cm)	q (cm)
120.0	41.8
80.0	50.6
60.0	64.1
50.0	81.6
47.0	91.0
40.0	137.8

FIGURE 3.7. Object and image distances for a spherical mirror.

(see Appendix A) is,

$$\frac{1}{q} = -\frac{1}{p} + \text{vertical intercept}. \tag{3.2}$$

Notice that the vertical intercept has the dimension of a reciprocal length (cm^{-1}). Since this is a property of the mirror, we define this length to be the *focal length* of the mirror and denote it by f. With this definition, Eq. (3.2) can be written,

$$\frac{1}{q} = -\frac{1}{p} + \frac{1}{f}. \tag{3.3}$$

For any small, spherical mirror with focal length f, this equation gives the position of the image for any small object. The vertical intercept of the graph in Fig. 3.7 has the value 0.032 cm^{-1} corresponding to a focal length $f = 31$ cm for the mirror in the example.

The mirror formula, Eq. (3.3), can be applied to any small mirror if we use the following sign convention:

- $p > 0$ for an object *in front* of the mirror;
- $q > 0$ for an image formed *in front* of the mirror;
- $f > 0$ for a *concave* mirror.

Invoking the law of reflection, $i' = i$, we see from Fig. 3.6 that the object height h and the image height h' are corresponding sides of similar right triangles. Two other corresponding sides of these triangles have lengths p and q, respectively. Therefore,

$$-\frac{h'}{h} = \frac{q}{p}. \tag{3.4}$$

In Eq. (3.4), we have introduced a minus sign, because the image is inverted relative to the object. That is, if we take "up" to be positive, then, because both q and p are positive in Fig. 3.6, we see that $h > 0$ and $h' < 0$ according to Eq. (3.4), meaning that the image is pointing "down."

EXAMPLE. A small object 1.9 mm tall stands 23.6 cm to the left of a small, concave spherical mirror with a focal length of 35 cm. Find the position, height, and character of the image.

SOLUTION. With these data and the sign convention above, Eq. (3.3) is

$$\frac{1}{q} = -\frac{1}{23.6\text{ cm}} + \frac{1}{35.0\text{ cm}},$$

from which we obtain $q = -72.4$ cm. Thus, the image is formed 72.4 cm *to the right* of the mirror, which means that it is a *virtual* image. From this result, we get for Eq. (3.4),

$$-\frac{h'}{1.9\text{ mm}} = \frac{-72.4\text{ cm}}{23.6\text{ cm}},$$

which yields $h' = 5.8$ mm. The image is upright and 5.8 mm tall.

EXAMPLE. A small object 0.23 cm tall stands 41 cm to the left of a small, convex spherical mirror with a focal length of 28 cm. Find the position, height, and character of the image.

SOLUTION. Using the data according to the sign convention, Eq. (3.3) becomes

$$\frac{1}{q} = -\frac{1}{41.0 \text{ cm}} + \frac{1}{-28.0 \text{ cm}},$$

from which we obtain $q = -16.6$ cm. This is a virtual image 16.6 cm to the right of the mirror. For Eq. (3.4), we have

$$-\frac{h'}{0.23 \text{ cm}} = \frac{-16.6 \text{ cm}}{41.0 \text{ cm}}.$$

This implies an image that is 0.093 cm tall with the same orientation as the object.

We can also find the image position, height, and character for a small object in front of a small, spherical mirror by a geometrical method. Consider a small, concave mirror as illustrated in Fig. 3.8*a*. Because the mirror is small, we can approximate it by a straight, vertical line. We draw a long horizontal line perpendicular to the mirror and passing through its center. This line is the *principal axis*. A small object represented by the vertical arrow is placed to the left of the mirror as shown.

Light diverges in all directions from the tip of the arrow. Consider a ray from the tip of the arrow parallel to the principal axis and striking the mirror. What is the path of the corresponding reflected ray? The rays from an object at infinity are also parallel to the principal axis. We know from Eq. (3.3) that such rays are reflected through the focal point F and the ray corresponding to the incident ray in Fig. 3.8*a* must be so reflected, as illustrated.

Similarly, if we have a point object located at the focal point of the mirror, we see from Eq. (3.3) that all incident rays are reflected parallel to the principal axis so as to form an image at infinity. This means that a ray diverging from the tip of the arrow and passing through the focal point F must be reflected from the mirror as shown in Fig. 3.8*b*.

These two reflected rays intersect at the position of the image of the tip of the arrow (since the incident rays originated from the tip of the object). All other rays that diverged from the tip of the object will also be reflected through the image position for the tip of the arrow. In fact, for each point on the object, rays diverging from that point will be reflected through the image position of the point and the complete image of the object will be produced. Thus, we construct the image of this object as shown in Fig. 3.8*c*. We can see from this figure that the image is real (the reflected light actually passes through the image position), it is inverted, and it is in front of the mirror. By choosing a convenient scale, we can also determine from the drawing the distance of the image from the mirror in the laboratory and

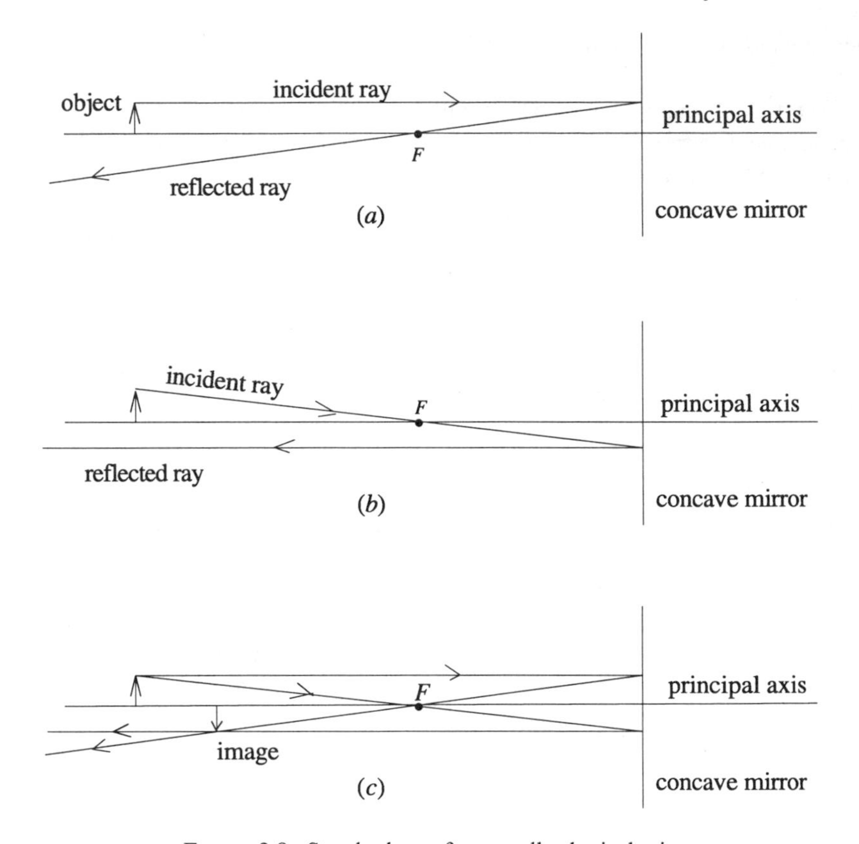

FIGURE 3.8. Standard rays for a small spherical mirror.

the height of the image in the laboratory. The vertical and horizontal scales need not be the same and generally we choose them differently.

This defines two sets of *standard rays* that we can use to construct images formed by small mirrors.

Standard Rays for Small Mirrors

1. Incident ray from tip of object parallel to the principal axis.
   Reflected ray through the focal point of the mirror.
2. Incident ray from tip of object through the focal point of the mirror.
   Reflected ray parallel to the principal axis.

The image of the tip of the object is formed at the intersection of the two reflected rays.

3.2.2 Dispersion by Refraction

In Fig. 3.3, we see that the light changes its direction of propagation as it passes through the surface of the glass. This phenomenon, called *refraction*, is due to the fact that the light travels slower in glass than it does in air. The angle r that the transmitted ray makes with the line perpendicular to the boundary is the *angle of refraction*. For light entering a slower medium (e.g., from air into glass), the angle of refraction is smaller than the angle of incidence ($r < i$) as in Fig. 3.3. In passing from a slower medium into a faster one (glass to air), $r > i$. The quantitative expression of this relationship is called the *law of refraction*.

The speed with which light propagates in a medium is determined by the interaction of the electromagnetic field with the electrically charged particles that make up the medium. In general, this interaction depends on the wavelength of the light. For example, in glass blue light (short wavelength) travels slower than red light (long wavelength).

This can be demonstrated by shining a beam of white (multicolored) light on a glass prism as illustrated in Fig. 3.9. The dashed lines are perpendicular to the air-glass boundaries at the points of incidence. The law of refraction is applied qualitatively to the red and blue components of the white light at each boundary.

As a result, the white light is fanned out by the prism into a spectrum with the different colors physically separated. The spreading of multicolored light into its component colors is known as *dispersion*. In astronomy, this is a very important tool for analyzing the light from celestial objects.

3.2.3 LENSES

Another important application of refraction in astronomy is in the making of lenses. Suppose we polish the ends of a cylindrical piece of glass so that each end has a smooth, spherical shape. This is shown in Fig. 3.10 where the centers of curvature of the left and right ends are indicated by C_1 and C_2, respectively.

Starting from a point object in front of the left surface, we have used the law of refraction to trace rays through the two spherical surfaces. The first surface forms an image labeled *first image* in Fig. 3.10. This serves as an object for the second surface, which produces the final image for the system. An optical system of this type is called a *lens*.

A *thin lens* is a lens with the refracting surfaces separated by a distance that is very small compared to the other relevant lengths (i.e, object distance, image distance, and focal length). For a small thin lens, these quantities are also related by Eq. (3.3), but with a different sign convention:

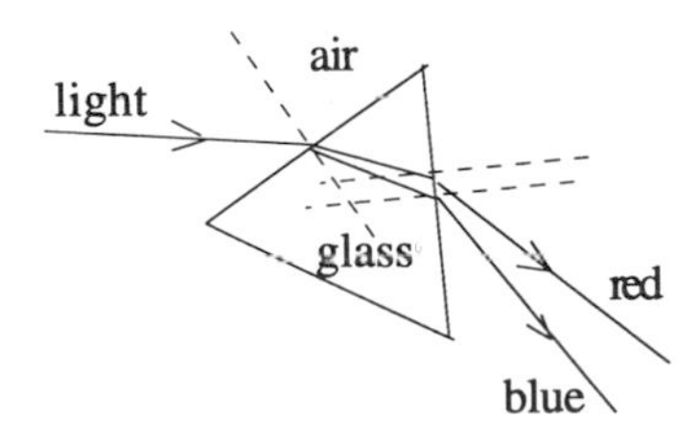

FIGURE 3.9. Dispersion of light in glass.

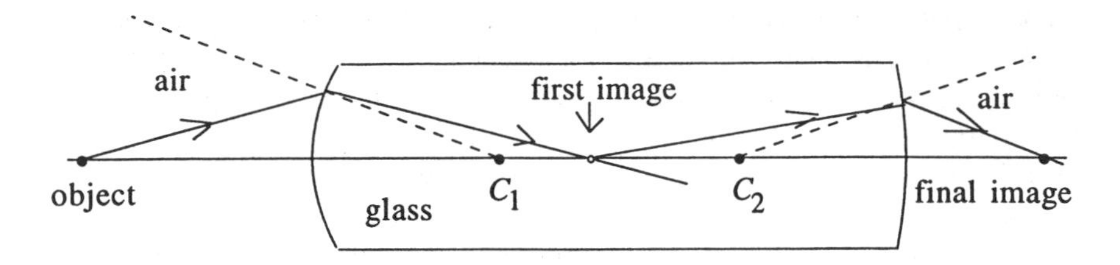

FIGURE 3.10. Ray tracing through a thick lens.

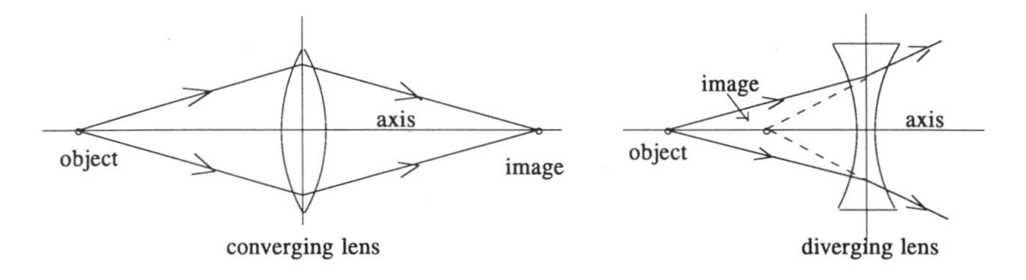

FIGURE 3.11. Profiles of typical lenses.

- $p > 0$ for an object *in front* (incident side) of the lens;
- $q > 0$ for an image formed *behind* the lens;
- $f > 0$ for a *converging* lens.

Light incident on a *converging* lens is refracted *toward* the axis of the lens. The light is refracted *away from* the axis of a *diverging* lens. Some typical lens shapes are illustrated in Fig. 3.11.

The formula, Eq. (3.4), for the ratio of image height to object height is also applicable for a thin lens.

EXAMPLE. A small object 2.1 mm tall stands 38 cm to the left of a small, converging lens with a focal length of 26 cm. Find the position, height, and character of the image formed by this lens.

SOLUTION. Using the data according to the sign convention for a thin lens, Eq. (3.3) becomes

$$\frac{1}{q} = -\frac{1}{38.0 \text{ cm}} + \frac{1}{26.0 \text{ cm}},$$

from which we obtain $q = 82.3$ cm. This is a real image 82.3 cm to the right of the lens. For Eq. (3.4), we have

$$-\frac{h'}{2.1 \text{ mm}} = \frac{83.3 \text{ cm}}{38.0 \text{ cm}}.$$

This implies an inverted image that is 4.6 mm tall.

A geometrical construction similar to that for a small mirror can be used to obtain the position, height, and character of the image formed by a small, thin lens. First, we define two sets of standard rays in analogy with our treatment of the small mirror.

Standard Rays for Small, Thin Lenses

1. Incident ray from tip of object parallel to the principal axis.
   Refracted ray through one focal point of lens.
2. Incident ray from tip of object through *other* focal point of lens.
   Refracted ray parallel to the principal axis.

   Unlike a mirror, light can be incident on either side of a lens. Therefore, a lens has two focal points, whereas a mirror has only one. These focal points are located on the principal axis symmetrically on either side of the lens. In constructing the first refracted ray, the focal point we use is determined by the nature of the lens. If it is a *converging* lens, the first refracted ray must be refracted *toward* the principal axis through the corresponding focal point. If it is a *diverging* lens, this ray must be refracted *away* from the principal axis such that an *extension* of this ray will pass through the corresponding focal point. For the second pair of standard rays, the refracted ray passes through the focal point that was *not* used in constructing the first refracted ray.

The image of the tip of the object is formed at the intersection of the two reflected rays. We illustrate the method for a small object in front of a thin, converging lens.

EXAMPLE. A small object 1.3 mm tall stands 41 cm to the left of a thin, converging lens with a focal length of 18 cm. By means of a scale drawing, determine the position, height, and character of the image of this object.

SOLUTION. The two sets of standard rays are shown in the scale drawing in Fig. 3.12. The two transmitted rays converge at the image position of the tip of the vertical arrow representing the object. We can see directly from the drawing that the image is inverted, it stands to the right of the lens, and it is real since the transmitted rays actually converge at the image position. By measuring the image

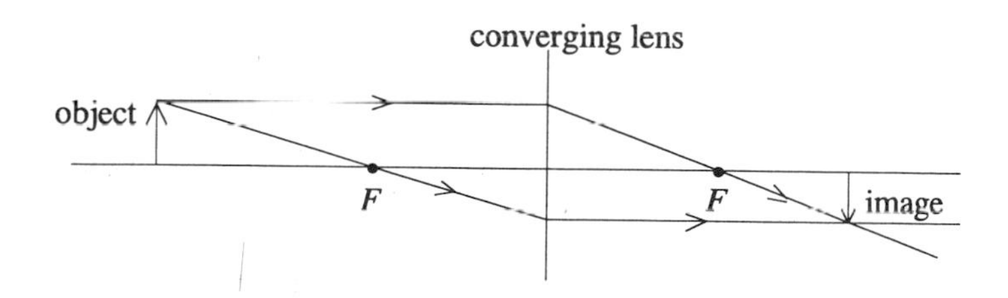

FIGURE 3.12. Construction of standard rays for a thin lens.

distance and height on the drawing and using the horizontal and vertical scales, we find that the image (in the laboratory) is about 1.0 mm tall and stands 32 cm to the right of the lens.

3.2.4 Astronomical Telescopes

In the simplest configuration, a *refracting telescope* consists of a large converging lens (the objective) to collect the light from an object in the sky and a second, smaller lens (the eyepiece). The arrangement is illustrated in Fig. 3.13.

The objective produces at its focal point a real image of the distant object. This image serves as an object for the eyepiece, which produces a magnified final image for the telescope.

The objective in a *reflecting telescope* is a large, concave mirror. Since the mirror produces an image on the *incident* side, it is necessary with smaller telescopes to devise some means of producing a final image so that the observer does not block the incoming light. This is often accomplished through the use of auxiliary mirrors. Two common configurations are illustrated in Fig. 3.14. In each case, light is shown entering the telescope from a distant object, reflected by the primary mirror toward the prime focus where it is intercepted by a secondary mirror which reflects the light to the eyepiece.

The principal function of an astronomical telescope is to gather light. The quality of the image produced depends on two characteristics of the telescope:

1. Light-gathering power;
2. Resolving power.

Both of these are related to the size of the objective. Clearly, the larger the objective, the more light the telescope can collect in a given time interval. The resolving

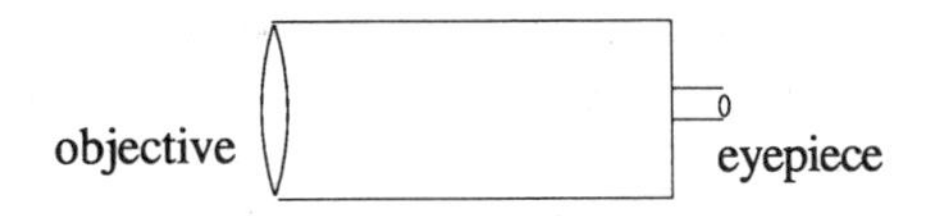

FIGURE 3.13. Principal components of a refracting telescope.

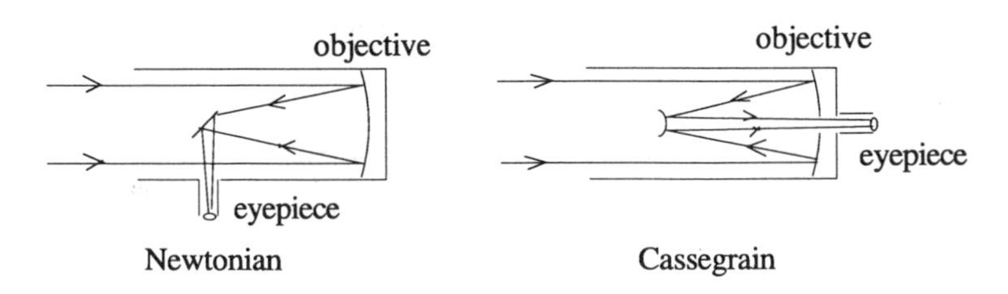

FIGURE 3.14. Common configurations for small reflecting telescopes.

power is a measure of the effectiveness of the telescope to reveal fine details in the image. The sharpness of the image is degraded by a phenomenon known as *diffraction*—a characteristic exhibited by all waves including light waves. This is the ability of a wave to bend around an obstacle, making the shadow of the object fuzzy at the edge. As light enters a telescope, diffraction occurs at the edge of the opening, tending to blur the image. The smaller the aperture, the more noticeable the diffraction and the fuzzier the image. So, in general, a telescope with a large objective will have greater light-gathering power and better resolution than one with a small objective.

Most modern research telescopes are reflectors, because of some disadvantages of large lenses compared to large mirrors.

Problems with Large Lenses

1. A lens has two surfaces to grind. (A mirror has only one.)
2. A heavy, solid glass lens tends to sag, distorting its shape. (A mirror need not be solid glass, but may have a strong, lightweight, honeycomb construction.)
3. The lens must be mounted on its rim so as not to obstruct light passing through it. (A mirror may be supported on the back, since light does not pass through it.)
4. The lens must be free of bubbles or other imperfections that can interfere with the light-gathering or resolving function of the telescope. (This is not a problem with the mirror, since light does not pass through the mirror.)
5. A lens must be corrected for *chromatic aberration*. (This is not a problem with a mirror, since light is reflected off the *front surface* of the mirror.)

Chromatic aberration arises from dispersion in the lens. This is illustrated in Fig. 3.15. As multicolored light from a distant object passes through the lens, the different colored components are refracted by different amounts. The edge of the lens behaves like a small prism with blue light being deviated more than red light. As a result, multiple images of different colors are produced by the lens at different positions as shown. This reduces the sharpness of the image and a correction is required in refracting telescopes. The phenomenon is not a factor in reflectors, because the light is reflected off the front surface of the mirror and does not pass through the glass.

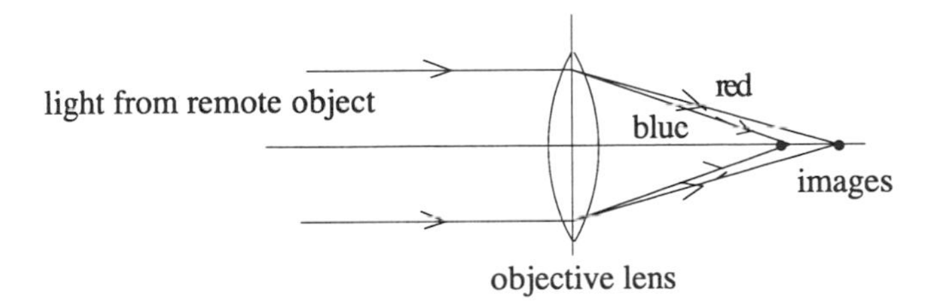

FIGURE 3.15. Chromatic aberration.

3.2.5 Pairs of Lenses

Optical instruments often comprise combinations of lenses arranged with a common principal axis. As light passes through such a system, the image formed by each lens serves as an object for the next lens with the last lens through which the light passes producing the final image for the system.

Example. A converging lens with focal length 17.0 cm stands 15.6 cm to the left of a diverging lens. These lenses have a common principal axis. The focal points of the diverging lens are located 22.8 cm from the lens. A small object 1.38 cm tall is placed on the principal axis of the system 41.3 cm to the left of the converging lens. Find the position, height, and character of the image formed by the lens system.

Solution. First, find the position and height of the image formed by the first (converging) lens the light from the object encounters in passing through the system.

$$\frac{1}{q} = -\frac{1}{p} + \frac{1}{f} = -\frac{1}{41.3\text{ cm}} + \frac{1}{17.0\text{ cm}} = 0.03461\text{ cm}^{-1},$$

$$q = +28.89\text{ cm}.$$

The height of the image is given by Eq. (3.4),

$$\frac{-h'}{h} = \frac{-h'}{1.38\text{ mm}} = \frac{q}{p} = \frac{28.89\text{ cm}}{41.3\text{ cm}},$$

$$h' = -0.965\text{ mm}.$$

Thus, the converging lens produces a real image 28.89 cm to the right of it. (Note that this image is *beyond* the position of the second lens.) This image is 0.965 mm tall and inverted relative to the object.

Next, we use this image as an object for the second (diverging) lens. Since the first image is formed 13.29 cm *behind* (to the right of) the diverging lens, the object distance for this lens is −13.29 cm. For the diverging lens, we have

$$\frac{1}{q} = -\frac{1}{p} + \frac{1}{f} = -\frac{1}{-13.29\text{ cm}} + \frac{1}{-22.8\text{ cm}} = 0.03137\text{ cm}^{-1},$$

$$q = +31.88\text{ cm}.$$

For the height of this image, we find

$$\frac{-h'}{h} = \frac{-h'}{-0.965\text{ mm}} = \frac{q}{p} = \frac{31.88\text{ cm}}{-13.29\text{ cm}},$$

$$h' = -2.32\text{ mm}.$$

From these results, we conclude that the final image formed by this lens system is real, inverted, 2.32 mm tall, and stands 31.9 cm to the right of the diverging lens.

3.3 The Doppler Effect

A useful and interesing phenomenon occurs when waves from a moving source are observed. In Fig. 3.16, a stationary point source emits a spherical wave of a single wavelength and frequency. At any point in the vicinty of the source, an observer receives pulses from the source at a rate that is the same as at any other point. The frequency at which wave crests arrive at the receiver is the same in all directions from the source. Suppose the source moves to the right with a constant velocity. How does this affect the rate at which pulses arrive at the receiver? The situation is depicted in Fig. 3.17

As the source moves to the right, each successive pulse is emitted from a point a little farther to the right than the previous one. Therefore, the centers of the spherical wavefronts are displaced toward the right. As a result, a receiver placed to the right of the source will receive pulses more frequently than with the stationary source. It detects a higher frequency for the wave. An observer on the left receiving the waves as the source moves away from him, detects a longer time interval between pulses—a lower frequency. The effect is reflected in the apparent change in wavelength as can be seen directly from Fig. 3.17. The pulses are bunched closer together (shorter wavelength) to the right and are stretched farther apart (longer wavelength) to the left as the source moves away.

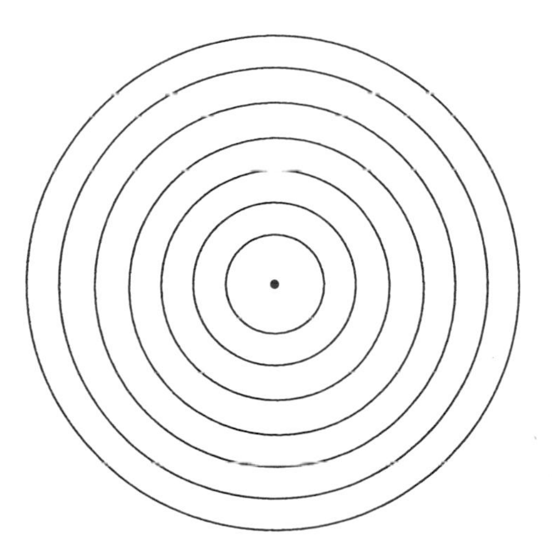

FIGURE 3.16. Stationary source of waves.

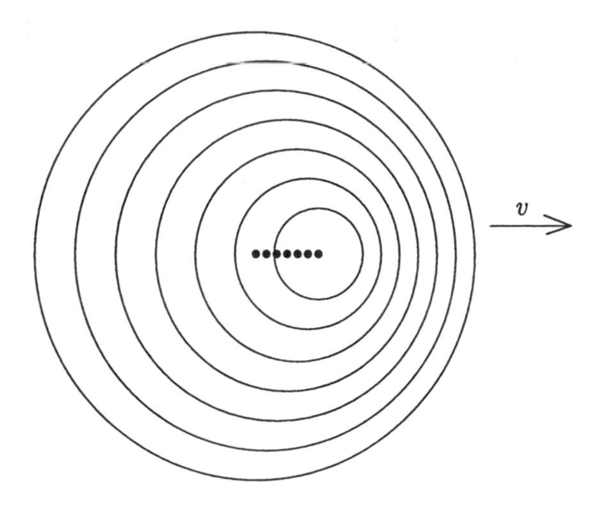

FIGURE 3.17. Source moving to the right with a constant velocity.

Light is a wave phenomenon, hence it also exhibits this effect. For example, the shift in wavelength of starlight from a given star will tell us whether the star is moving toward or away from us. More than that, the amount of shift allows us to determine how fast the star is moving along our line of sight.

This apparent shift in the wavelength resulting from relative motion between source and observer is known as the *Doppler effect*. It will play an important role in our subsequent work.

Exercises

3.1. If light is incident on a mirror from the left, on which side (left or right) of the mirror is a real image formed? From a physical point of view, why is this so?

3.2. An object 1.8 mm tall is placed 58.0 cm to the left of a small, concave mirror with a focal length of 26.0 cm. Find the position, height, and character of the image.

Answer: (47.1 cm to the left of the mirror; 1.5 mm; real; inverted)

3.3. An object 1.1 mm tall stands 31.3 cm in front of a concave mirror with a focal length of 56.2 cm. By means of a scale drawing, find the position, height, and character of the image.

Answer: (70.6 cm behind the mirror; 2.5 mm; virtual; erect)

3.4. The focal point of a small, convex mirror is 36.2 cm from the mirror. What is the location of the image of a small object placed 41.7 cm in front of this mirror?

Answer: (19.4 cm behind the mirror)

3.5. Find the position, height, and character of the image of an object standing in front of a small, convex mirror. The focal point is 19.3 cm from the mirror. The object is 1.3 mm tall and is 40.7 cm from the mirror.

Answer: (13.1 cm behind the mirror; 0.4 mm; virtual; upright)

3.6. A small object stands 49.1 cm to the left of a small, spherical mirror. The image formed by the mirror is 29.1 cm to the right of the mirror. By means of a calculation, determine what kind of mirror this is.

3.7. An object 2.9 mm tall stands 34.2 cm to the right of a convex mirror with the focal point 23.7 cm from the mirror. By means of a scale drawing, find the position, height, and character of the image.

Answer: (14.0 cm to the right of the mirror; 1.2 mm; virtual; erect)

3.8. A small, spherical mirror produces a virtual image of a very small object that stands 23.7 cm in front of the mirror. The image is 2.7 times as tall as the object. In one sentence, give as complete a description as possible of this mirror.

3.9. A small, spherical mirror produces a real image of a very small object located 21.9 cm in front of the mirror. The image is 3.5 times as tall as the object. As completely as you can, describe this mirror.

3.10. A small, spherical mirror produces a virtual image of a very small object located 21.9 cm in front of the mirror. The image is 3.5 times as tall as the object. As completely as you can, describe this mirror.

3.11. A small object stands 26.4 cm in front of a certain mirror. An image of the object is formed 38.9 cm behind the mirror. By means of a calculation, determine whether this is a concave mirror or a convex mirror.

3.12. An object placed 39 cm in front of a small, spherical mirror forms an image 51 cm behind the mirror. Give a quantitative description of the mirror.

3.13. Explain clearly what is meant by "refraction." What is the physical reason refraction occurs?

3.14. Why does a lens have two focal points whereas a mirror has only one? Give a clear and complete physical explanation.

3.15. Explain clearly and completely why we say that a real image is found *in front of* a mirror, but a real image is formed *behind* a lens.

3.16. A small object 2.80 mm tall stands 43.0 cm to the left of a thin, diverging lens with a focal length of 35.0 cm. Use the lens formula to find the position, height, and character of the image formed by this lens. Write your answers in a complete sentence.

Answer: (19.3 cm to the left of the lens; 1.26 mm; virtual; upright)

3.17. A thin, converging lens has a focal length of 27 cm. A small object is placed 35 cm to the right of this lens. Use the lens formula to determine the location of the image relative to the lens.

Answer: (118 cm to the left of the lens)

3.18. The focal points of a thin, diverging lens are located on either side of the lens on its principal axis at a distance of 51 cm from the lens. Use the lens formulas to find the position, height, and character of the image the lens forms of an object 4.20 mm tall standing 22 cm to the left of the lens.

Answer: (15.4 cm to the left of the lens; 2.93 mm; virtual; upright)

3.19. A small object 2.8 mm tall stands 39.0 cm to the left of a thin, diverging lens with a focal length of 75.0 cm. Use the lens formulas to find the position, height, and character of the image formed by this lens. Write your answers in a complete sentence.

Answer: (25.6 cm to the left of the lens; 1.8 mm; virtual; erect)

3.20. Find the position, height, and character of the image of a small object 4.8 mm tall that stands 41.0 cm to the right of a thin, converging lens with a focal length of 52.8 cm. Solve this problem first by using the lens formulas, then choose appropriate scales (vertical and horizontal) and solve it by means of a scale drawing.

Answer: (183 cm to right of lens; 2.15 cm; virtual; erect)

3.21. An object 1.12 mm high is located 30.0 cm in front of a converging lens, which has a focal length of 40.0 cm. Calculate the position and height of the image. Is the image real or virtual? Is it upright or inverted?

Answer: (120 cm in front of the lens; 4.48 mm; virtual; upright)

3.22. The focal points of a diverging lens are located on the principal axis 37.0 cm on either side of the lens. An object 0.75 cm tall is placed 60.0 cm to the left of this lens. By means of a scale drawing, find the position and height of the image in the laboratory. Is the image real or virtual? Is it erect or inverted?

Answer: (22.9 cm to the left of the lens; 0.29 cm; virtual; erect)

3.23. A small object 2.1 mm tall stands 33.0 cm to the left of a thin, converging lens with a focal length of 68.0 cm. By means of a carefully drawn scale drawing, find the position, height, and character of the image formed by this lens.

Answer: (64 cm to the left of the lens; 4.1 mm; virtual; erect)

3.24. Very large telescopes are expensive to build and operate. Explain clearly and completely why it is desirable to build large astronomical telescopes.

3.25. Why is an auxiliary mirror needed in small, reflecting telescopes?

3.26. What are the disadvantages of building large refracting astronomical telescopes compared to building large reflectors?

3.27. Two converging lenses are separated by a distance of 63.0 cm measured along the principal axis of these lenses. Each has a focal length of 21.0 cm. A small object 0.370 cm tall stands 57.0 cm to the left of this system. Find the position, height, and character of the image formed by this system. Solve this problem first by means of a scale drawing, then by means of the thin lens formulas.

Answer: (71.4 cm to right of second lens; 0.518 cm; real; erect)

3.28. A converging lens with a focal length of 42 cm stands 15 cm to the left of a second converging lens, which has a focal length of 27 cm. An object 0.28 cm high stands 17.8 cm to the left of this system. How tall is the image produced by this system? Where is the image located? Describe the character of the image. Solve this problem by using the formulas and then by using a scale drawing.

Answer: (0.694 cm; 65.6 cm to right of second lens; real; inverted)

3.29. An object 0.44 cm tall is placed 72.0 cm to the left of a thin, converging lens, which has a focal length of 17.2 cm. A diverging lens with focal points 22.7 cm from the lens stands 33.3 cm to the right of the converging lens. First, use a scale drawing, then use the lens formulas to determine the location, the height, and the character of the final image produced by this optical system.

Answer: (7.27 cm to the left of the second lens; 0.939 mm; virtual; inverted)

3.30. A thin, converging lens is placed 23.6 cm to the left of a thin, diverging lens. The magnitudes of the focal lengths of these lenses are 52.3 cm and 19.9 cm, respectively. A small object stands 5.2 mm tall at a distance of 176.0 cm to the left of the converging lens. Use the thin lens formulas to determine the location, height, and character of the image of this object. Do not attempt to make a scale drawing for this problem.

Answer: (32.7 cm to the left of the diverging lens; 1.4 mm; virtual; erect)

3.31. A telescope consists of two converging lenses—an objective with a focal length of 0.83 m and an eyepiece with a focal length of 27 mm. The instrument is used to observe a small object located 17.0 m in front of the objective. Find the distance between these lenses if the final image is real and is formed at infinity. Solve this problem by means of the formulas only. Do not make a scale drawing.

Answer: (89.96 cm)

3.32. A small, concave mirror and a thin, diverging lens are placed 25.7 cm apart as measured along their principal axes. The magnitudes of the focal lengths of these elements are 16.2 cm and 36.5 cm, respectively. A small object 2.3 mm tall is placed between the two elements at a distance of 24.3 cm from the mirror. Light from the object reflects off the mirror and passes through the lens to form an image. Use the formulas to find the position, height, and character of the image formed by this optical system.

Answer: (61.4 cm behind the lens; 12.3 mm; real; inverted)

3.33. A thin, converging lens with a focal length of 21.5 cm stands 25.0 cm to the left of a thin, diverging lens of focal length −36.4 cm. A small object 0.37 mm tall is placed 38.7 cm to the left of the converging lens. Use the lens formulas to find the position, height, and character of the image formed by this system.

Answer: (65.3 cm to the right of the diverging lens; 1.29 mm; real; inverted)

3.34. A thin, diverging lens stands to the left of a thin converging lens. A small object 2.2 mm tall is placed to the left of the diverging lens as illustrated in the scale drawing below. By carefully completing the scale drawing, find the position, height, and character of the image (in the laboratory) formed by this lens system. Give your answer in a complete sentence.

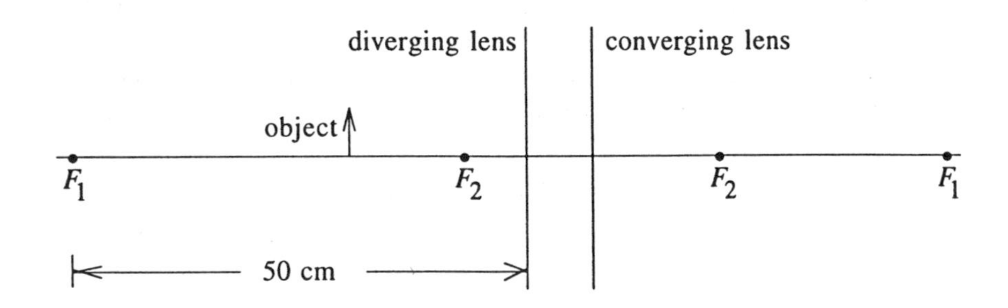

Answer: (46 cm to the right of the converging lens; 3 mm; real; inverted)

3.35. A small object is placed at a point A on the principal axis of a small, thin lens. The image formed by the lens is four times as tall as the object. The object is moved 10 cm farther away from the lens to a point B on the principal axis and again the image is four times as tall as the object. How far from the lens is the object when it is at point B?

Answer: (25 cm)

3.36. By means of words and a drawing, explain clearly why chromatic aberration is objectionable in a refracting telescope.

3.37. Light falls on a glass prism as shown in the drawing. *Based on the drawing*, which color light, if either, travels faster in the glass, yellow or green? Explain clearly and completely how you can tell.

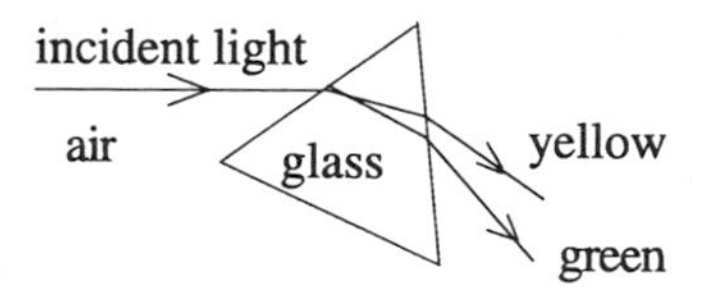

4

Newton Puts It All Together

4.1 Newton's Laws of Motion

In 1686, Isaac Newton published, in Latin, his *PhilosophiæNaturalis Principia Mathematica*, commonly called the *Principia*. In this landmark effort, he was able to synthesize the work of his predecessors, Copernicus, Brahe, Kepler, and Galileo, and produce a "system of the world" that described not only the celestial motions with remarkable precision, but also the behavior of matter here on the earth. At the outset, he wrote down three "laws"—more accurately, axioms—concerning the motion of matter. These he considered to be *given*. According to a translation from the Latin by A. Motte, they are

Law I

Every body continues in its state of rest, or of uniform motion in a right line, unless it is compelled to change that state by forces impressed upon it.

Law II

The change of motion is proportional to the motive force impressed; and is made in the direction of the right line in which that force is impressed.

Law III

To every action there is always opposed an equal reaction: or, the mutual actions of two bodies upon each other are always equal, and directed to contrary parts.

The first law provides us with a *qualitative* definition of *force*, namely, whatever is capable of changing the state of motion of a material body. It is often referred to as "the law of inertia." It is because of inertia that matter tends to retain whatever state of motion it has. A *quantitative* definition of force is given by the second law. These first two laws are concerned with the motion of a *single* body. The third law tells us something about how two different bodies interact with each other.

First, let's understand clearly the distinction between the first and third laws. A book is lying on the table in Fig. 4.1*a*. The book doesn't move relative to its

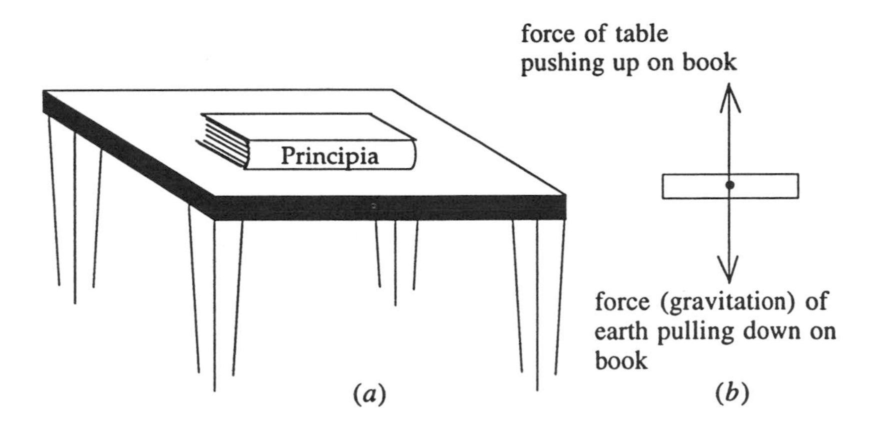

FIGURE 4.1. Newton's first and third laws of motion.

surroundings. Why not? This is a manifestation of Newton's first law. We can see this more clearly in Fig. 4.1*b* where the forces exerted on the book are represented by vectors. The two forces have equal strength, but are exerted in opposite directions on the book. Therefore, they cancel. The net force on the book is zero, so according to Newton's first law the motion of the book doesn't change. What role does the third law play? Take the force of *the table pushing up on the book* in Fig. 4.1*b*. In the sense of Newton's third law, what is the "reaction" force here? It is simply the force that *the book exerts down on the table*. The book pushes down on the table just as hard as the table pushes up on the book—Newton's third law. What is the reaction force to the gravitational force of the earth acting on the book? Be careful! Identify clearly the two bodies involved.

For an examination of the second law, we return to the glider and air track introduced in Chapter 2. Suppose we have a constant *reproducible* force that we can apply horizontally to the glider. Never mind what the source of the force is, we only require that we be able to *reproduce* it—apply it to the glider in exactly the same way over and over again. We also have a set of metal blocks that we can place on top of the glider to change the mass of the system. This is illustrated in Fig. 4.2.

First, we apply the force to the unloaded glider and observe the acceleration. Next, we place one of the metal blocks on top of the glider, apply the force, and observe the acceleration. Block by block we increase the mass of the glider-block

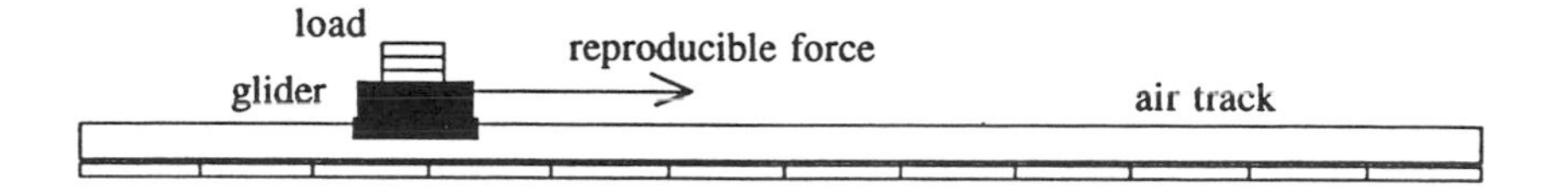

FIGURE 4.2. Frictionless glider on an air cushion.

system, apply the force, and note the resulting acceleration. The only thing that is changing here is the mass of the system. We find that each time we add a block, the acceleration is smaller than for the previous run. The experiment shows that there is an *inverse* relationship between mass and accelertion—as the mass increases, the acceleration decreases.

4.1.1 MASS

So far, we have not given a precise definition of mass—only that it gives a measure of the inertia, matter's ability to resist changes in its motion. The experiment above shows that the rate at which the motion changes (acceleration) and inertia (mass) are inversely related. Let us define mass by making this inverse relationship a *proportionality*. Since we know how to measure time intervals and lengths, we can measure the acceleration of matter. Let's take some piece of matter that we shall call our *standard*. We apply our reproducible force to this standard and measure the acceleration a_s. Now we apply the same force to a different piece of matter and observe its acceleration a_1. We define the ratio of the masses to be equal to the reciprocal ratio of the corresponding accelerations,

$$\frac{m_1}{m_s} = \frac{a_s}{a_1}.$$

In principle, we can continue this procedure and apply our reproducible force to every object in the universe measuring the acceleration and comparing it to a_s. By definition, an object with half the acceleration of the standard has twice the mass m_s. In effect, this is what has been done. The world's standard is a platinum-iridium alloy cylinder that resides in the International Bureau of Weights and Measures at Sèvres in France. It is the standard kilogram. Directly or indirectly, the mass of everything else in the universe is measured relative to this standard.

4.1.2 DENSITY

Quite often, it is needful to know how mass is distributed over an extended body such as a planet, or a star, or even a galaxy. The quantity we use to represent such a distribution is the *mass density*. In general, a density expresses the amount of some quantity per given amount of some other quantity. For example, a *population* density gives the number of people per unit area. Mass density ρ is defined as mass per unit volume. For convenience, when we use the word "density" alone, we shall take it to mean "mass density."

In general, the density may vary from point to point in matter. For a body in which the matter is uniformly distributed, the density ρ is given by

$$\rho = \frac{m}{V},$$

where m is the mass of the body and V is the volume it occupies.

EXAMPLE. The length of a solid metal cylinder is 27.3 cm and the diameter is 5.72 cm. The cylinder has a mass of five kilograms. Calculate the density of this metal.

SOLUTION. First, we calculate the volume V of the cylinder,

$$V = \pi r^2 h = \pi \left(\frac{0.0572 \text{ m}}{2} \right)^2 (0.273 \text{ m}) = 7.015 \times 10^{-4} \text{ m}^3,$$

from which we obtain the density,

$$\rho = \frac{m}{V} = \frac{5.00 \text{ kg}}{7.015 \times 10^{-4} \text{ m}^3} = 7{,}130 \text{ kg/m}^3.$$

4.1.3 Quantitative Definition of Force

Now let's return to our glider experiment to see what else affects the acceleration. This time we assume that we have several *different* reproducible forces. We remove all of the metal blocks and one by one we apply the reproducible forces to the glider and measure the acceleration. We find that the harder we pull, the greater the acceleration of the glider. The mass doesn't change in this series of measurements. Clearly, there is some direct relationship between the force and the acceleration. We can define this relationship to be a direct *proportionality*.

At this point we have defined mass and force such that the acceleration is given by

$$\mathbf{a} = K \frac{\mathbf{F}}{m}, \tag{4.1}$$

where K is a proportionality constant. Note that we have written the acceleration and force as vectors which have the same direction.

What else does the acceleration depend on? The answer is, nothing else. The acceleration or rate of change of motion of a body depends on two things,

1. How hard we try to change its motion (force).
2. The body's own ability to resist that change (mass). That is all—as Newton himself realized.

So to complete Eq. (4.1), all we need is to choose the units with which we measure force. We do this by setting $K = 1$ to obtain

$$\mathbf{F} = m\mathbf{a}, \tag{4.2}$$

which is exactly Newton's second law of motion in more symbolic language. Because both sides of Eq. (4.2) must have the same units, it is clear that in the mks system of units the force is measured in kg·m/s^2, which is defined to be one *newton*, abbreviated N. In the English system, the force unit is the *pound*, abbreviated lb.

Although we shall most often write the second law in this form, we should keep in mind that, in general, the acceleration $\mathbf{a}$ of a body of mass m is determined by

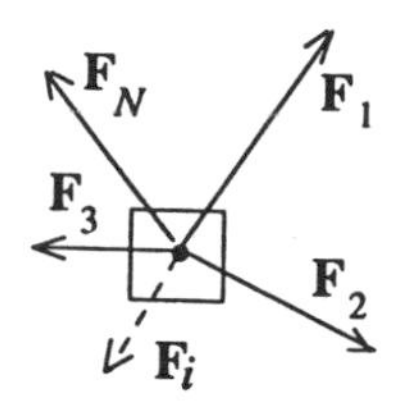

FIGURE 4.3. Vector sum of forces on a body.

the net effect of a number of forces on the body as illustrated in Fig. 4.3. The left-hand side of Eq. (4.2) stands for the vector sum of all of these forces:

$$\mathbf{F}_1 + \mathbf{F}_2 + \cdots + \mathbf{F}_N \equiv \sum_{i=1}^{N} \mathbf{F}_i = m\mathbf{a}.$$

We now see that the first law is a special case of the second law with $\mathbf{a} = 0$,

$$\sum_i \mathbf{F}_i = 0. \tag{4.3}$$

EXAMPLE. Starting from rest, the dock worker in Fig. 4.4*a* pushes a heavy 310.0 kg crate along a rough, horizontal floor until it reaches a speed of 0.44 m/s in fourteen seconds. The floor exerts a steady frictional force of 575.5 N. Find the constant, horizontal force with which the dock worker pushes the crate.

SOLUTION. The diagram in Fig. 4.4*b* shows all of the forces exerted on the crate. From the diagram, we can get the left-hand side of Eq. (4.2). For the horizontal component, we have

$$F_{\text{hand}} - F_{\text{friction}} = ma. \tag{4.4}$$

We wish to find $\mathbf{F}_{\text{hand}}$, which is the force that the worker's hands exert on the crate. We are given $m = 310.0$ kg and $\mathbf{F}_{\text{friction}} = 575.5$ N. From Eq. (2.2), we get

FIGURE 4.4. Dock worker pushing a heavy crate.

$a = 0.0314$ m/s^2. Putting all of this into Eq. (4.4), we find that the force the dock worker exerts on the crate is 585.2 N.

4.2 Linear Momentum

By rewriting Newton's second law, we are led to another very useful physical quantity that involves both the inertia of a body and its motion. This quantity is called the *momentum* of the body. Let's see how it arises.

We apply Newton's second law, Eq. (4.2), to the motion of a body with constant mass m and use Eq. (2.10) for the acceleration to get

$$\mathbf{F} = m\frac{\Delta \mathbf{v}}{\Delta t} = \frac{m\mathbf{v}_{\text{final}} - m\mathbf{v}_{\text{initial}}}{\Delta t} = \frac{\Delta(m\mathbf{v})}{\Delta t}.$$

We rewrite this equation as

$$\mathbf{F} = \frac{\Delta \mathbf{p}}{\Delta t}, \tag{4.5}$$

where the quantity $\mathbf{p} \equiv m\mathbf{v}$ is the *linear momentum* (frequently called simply the "momentum") of the body. The momentum is the product of the mass (inertia) and the velocity (motion) of the body.

This brings us to a very important conservation law. Suppose we have a system that consists of two parts, say two gliders with different masses, on a horizontal air track as in Fig 4.5—two gliders, one system. To a good approximation, we have no net external force on this *system*. The horizontal friction is negligible and the vertical forces of the track and the earth's gravity acting on the gliders cancel. Therefore, the left-hand side of Eq. (4.5) is zero, which tells us that the momentum of the *system* does not change with time as long as there is no net *external* force on the system. The momentum of the system remains constant or is *conserved*. Forces between parts of the system (e.g., the two gliders themselves, two molecules in the same glider, etc.) are *internal* to the system and not relevant here.

Suppose we place one glider at rest on the track. With a brief push, we set the second glider in motion toward the first one. After the second glider is set in motion, there is no longer any external force on the system and momentum is conserved. The gliders collide and separate with different speeds. The motions of both gliders change in the collision, but only in such a way as to keep the total momentum constant.

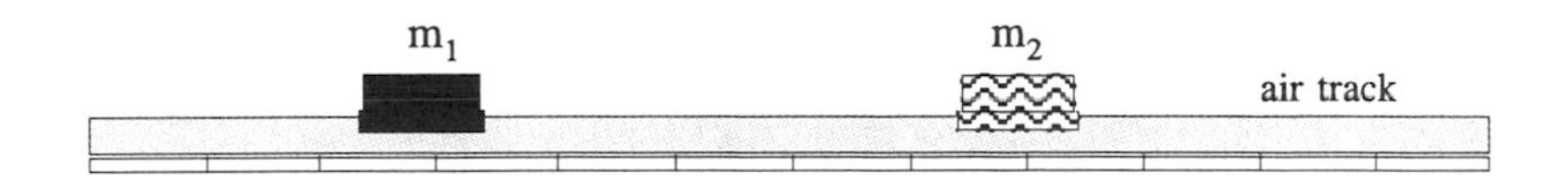

FIGURE 4.5. Colliding gliders.

Momentum is a vector quantity. The total momentum of the system is the vector sum of the momenta of the two gliders.

$$\text{total momentum before collision} = \text{total momentum after collision}$$

$$\mathbf{p}_1 + \mathbf{p}_2 = \mathbf{p}_1' + \mathbf{p}_2'$$

where we have used primes to denote values the quantities have *after* the collision. We can also write this equation explicitly in terms of masses and velocities,

$$m_1\mathbf{v}_1 + m_2\mathbf{v}_2 = m_1\mathbf{v}_1' + m_2\mathbf{v}_2'.$$

These results can be extended to a system with any number of components. If the net external force on the system is zero, the total momentum of the system remains constant even though the momenta of the separate parts may change.

4.3 Weight

Suppose we hold an object, a piece of chalk, for example, above the floor of the laboratory. If we release the chalk, it falls to the floor, picking up speed as it falls. It accelerates. According to Newton's second law, some force must cause the acceleration. As we have already noted, there is a force due to some property of the earth called gravity that pulls objects toward the center of the earth. The gravitational force exerted on a body is called the *weight* of the body. In the case of the falling chalk, other forces such as air friction contribute to the acceleration. For the moment, let us ignore these other effects and consider the gravity only.

We define a *freely falling body* to be a body moving under the influence of gravitational force *only*. This is illustrated in Fig. 4.6 where, by definition, the gravitational force w is the weight of the body.

From Newton's second law, we have

$$F_{\text{gravity}} = ma_{\text{gravity}}$$
$$w = ma_{\text{g}}.$$

From this expression, we see that the weight w of a body depends on two things:

1. an intrinsic property of the body itself—mass (m),
2. the strength of the gravitational field it is in—represented by a_{g}.

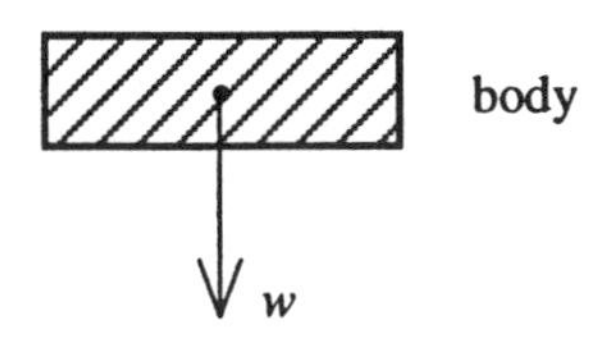

FIGURE 4.6. A freely falling body.

Near the surface of the earth, the acceleration due to gravity is nearly constant and we denote this quantity by g. To a good approximation, $g = 9.8$ m/s$^2 = 32$ ft/s^2. Hence, for objects near the earth's surface, mass and weight are related by $w = mg$.

The mass and weight of a body are related, but distinct, properties. The mass of a body is intrinsic to the body and does not depend on where the body is located. By contrast, the weight depends on where the body is. On the earth, a mass of one kilogram has a weight of 9.8 N. On the moon, where gravity is weaker than on the earth, the weight of this one kilogram body is about 1.6 N. In free space, far from any large gravitating bodies (stars, planets, etc.), its weight is approximately zero. The mass in all three locations is the same—one kilogram.

In the English system, the unit of weight (force) is the *pound* (lb). A body weighing one pound on earth has a mass of about 0.031 lb·s^2/ft.

EXAMPLE. A 76.2 kg rocket is launched vertically upward. The engine supplies a constant thrust of 765 N until it is cut off when the rocket reaches a speed of 233 km/h. What is the altitude of the rocket (in kilometers) when the engine is cut off? Neglect the effect of air friction. The forces exerted on the rocket are shown in Fig. 4.7*b*.

SOLUTION. Applying Newton's second law, we have

$$
\begin{aligned}
F &= ma \\
F_{\text{thrust}} - mg &= ma \\
765\ \text{N} - (76.2\ \text{kg})(9.8\ \text{m/s}^2) &= (76.2\ \text{kg})a,
\end{aligned}
$$

from which we get for the acceleration, $a = 0.239$ m/s^2. We obtain an expression for the distance d traveled (altitude) from Eqs. (2.4) and (2.8),

$$
d = \frac{v^2}{2a}, \tag{4.6}
$$

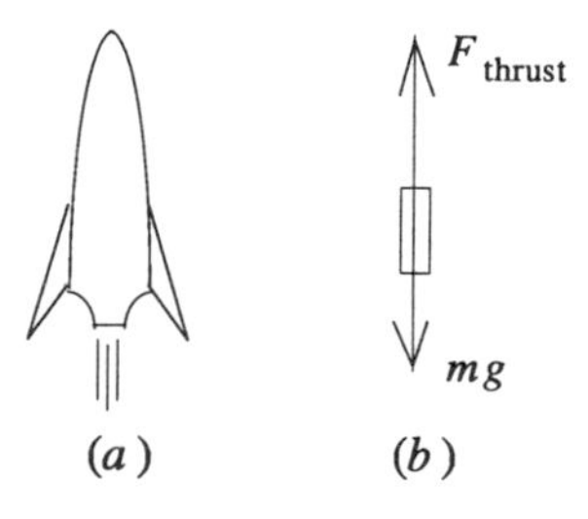

FIGURE 4.7. Rocket propulsion.

where v is the speed at this altitude. The cutoff speed, given in kilometers per hour, is converted to meters per second according to,

$$v = 233\ \frac{\text{km}}{\text{h}} \cdot \frac{1\ \text{h}}{3{,}600\ \text{s}} \cdot \frac{1{,}000\ \text{m}}{1\ \text{km}} = 64.7\ \text{m/s}.$$

Using these values for a and v in Eq. (4.6), we get $d = 8.75$ km for the altitude at which engine cutoff occurs.

Exercises

4.1. A force of 12 N is applied to each of two bodies A and B. The resulting acceleration of A is 15 m/s^2 while B accelerates at 19 m/s^2. Which body, if either, has the greater inertia?

4.2. A little boy sitting in a toy wagon tumbles backward out of the wagon when the wagon is suddenly jerked forward. In terms of inertia, explain clearly why the boy falls backward.

4.3. A horizontal force of thirteen newtons is exerted on a thirty-kilogram crate in pushing it across a smooth, horizontal floor. How far does the crate move in four seconds if it starts from rest?

Answer: (3.47 m)

4.4. Lead is more dense than aluminum. Each of two balls—one aluminum, the other lead—has a mass of 1.32 kg. Which, if either, has the greater inertia? Explain.

4.5. A 1.847-kg metal cylinder is 13.73 cm long and has a diameter of 4.38 cm. Calculate the density of the cylinder and identify the metal from the list in the table.

metal	density (kg/m^3)
aluminum	2,650
brass	8,600
copper	8,930
gold	19,320
lead	11,370
steel	7,380
zinc	7,150

4.6. A rectangular block of gold measures 16.5 mm $\times$ 11.3 mm $\times$ 31.9 mm and a rectangular block of lead 14.5 mm $\times$ 13.6 mm $\times$ 64.2 mm. Which of these, if either, has the greater inertia. (See table in previous exercise.) Explain clearly your reasoning

4.7. Assume that the earth, the sun, and the moon are spheres, and calculate the mean density of each.

Answer: (5,520 kg/m^3; 1,410 kg/m^3; 3,340 kg/m^3)

4.8. It is expected that late in its life the sun will evolve to a very dense, compact object called a *white dwarf*. Typically, a white dwarf star has a mass of the order of that of the sun packed into a volume the size of the earth. Use the data in the appendix to estimate the density of a typical white dwarf. From your result, estimate the mass of a teaspoonful of this matter. Approximately how much would a teaspoonful of this matter weigh on earth?

Answer: (2×10^9 kg/m^3; 7,000 kg; 8 tons)

4.9. A three-engine, 196,000-kg DC-10 requires 3,320 meters of runway to reach its takeoff speed of 310 km/h. Find the average net thrust provided by each engine during the takeoff roll.

Answer: (73,000 N)

4.10. Starting from rest, a man pushes a 380-lb crate a distance of twelve feet in 4.7 seconds over a rough, horizontal floor. Assuming constant acceleration and a frictional force of 95.0 lb, find the horizontal force that the man exerts on the crate.

Answer: (107.9 lb)

4.11. A constant total force of 1.79 N is applied to a certain object. The position s of the object is observed at various times t. The results are displayed in the graph. From the data given, calculate the mass of the object.

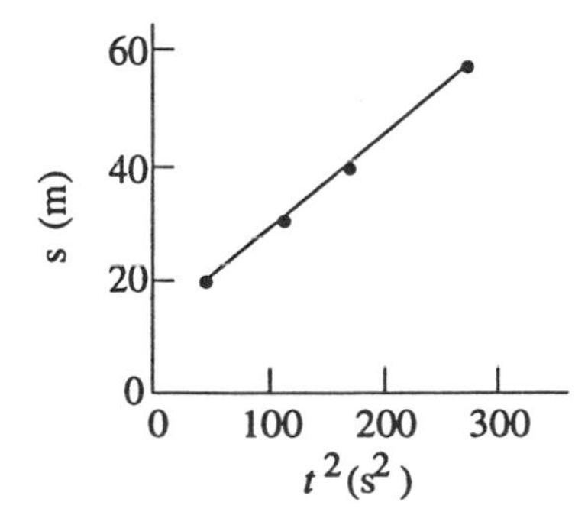

4.12. The brakes of a 1,200-kg truck supply an average force of 4,200 N in stopping the truck. How far does the truck travel before stopping if its speed is 90 km/h when the brakes are applied?

Answer: (89.3 m)

4.13. A small cart is pulled by a constant force along a flat, horizontal track. After the cart is set in motion, its position relative to the left end of the track is observed at various times according to an electronic timer. From the graph, determine the acceleration of the cart. Obtain an equation for the position s of the cart at time t. Assuming that the motion continues with this same acceleration, use your equation to find the position of the cart (relative to the left end of the track) when the timer reads 0.875 second.

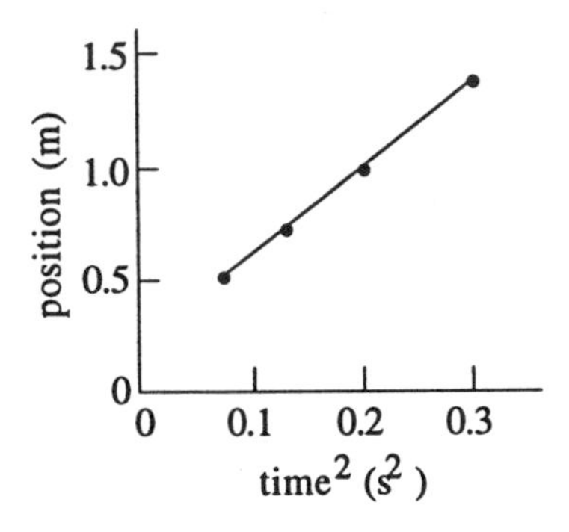

4.14. What force is required to give a 0.49-kg football an acceleration of 400 m/s^2? If this force is applied over a distance of 1.70 m, what is the speed with which the ball leaves the passer's hand?

Answer: (196 N; 36.9 m/s)

4.15. A catapult for launching aircraft from an aircraft carrier exerts a constant horizontal force of 328,000 N on a 11,300-kg aircraft. Through what distance must this force act to give the plane a speed of 220 km/h?

Answer: (64.3 m)

4.16. A girl is lying on a sled which is gliding with a constant velocity of 4.53 m/s over perfectly smooth ice. The combined mass of the girl and the sled is 38.7 kg. The sled enters a patch of rough ice over which the speed is slowed to 2.19 m/s. On the rough ice, the girl and sled experience a constant frictional force of 12.8 N. How far did the sled travel on the rough ice?

Answer: (23.8 m)

4.17. A small, 437-g block is set in motion with a speed of 7.12 m/s on a rough, horizontal surface over which it experiences a constant frictional force of

1.68 N. How fast is the block sliding 0.78 s after it is released? How far does it slide in this time interval?

Answer: (4.12 m/s; 4.38 m)

4.18. An artillery shell explodes in flight. Neglect air friction and discuss the motions of the pieces of the shell before and after the collision.

4.19. Two gliders are free to move on a horizontal air track as in Fig. 4.5. The masses of these gliders are $m_1 = 0.311$ kg and $m_2 = 0.439$ kg. The 0.311-kg glider, moving to the right with a speed of 0.532 m/s, collides with the second glider, which is at rest. The two gliders stick together. What is the velocity of the 0.439-kg glider after the collision?

Answer: (0.221 m/s to the right)

4.20. Suppose the two gliders in the previous exercise do *not* stick together and after the collision the 0.439-kg glider moves to the right with a speed of 0.427 m/s. Calculate the velocity of the 0.311-kg glider after the collision.

Answer: (0.071 m/s to the left)

4.21. A small spacecraft of mass 267 kg on an interplanetary flight is traveling with a speed of 67,317 km/h. Its engine is fired briefly for 2.3 seconds. During this time interval, the speed of the craft is increased uniformly by ejecting the hot gas of the burned fuel. During the burn, 1.70 kg of hot gas is ejected with a speed of 2,150 m/s relative to the spacecraft. Calculate the speed of the spacecraft at the end of the burn. Calculate the thrust supplied by the engine during the burn. Assume that the mass of the spacecraft remains constant.

Answer: (67,366 km/h; 1,589 N)

4.22. A 79-kg lifeguard dives from a 160-kg rowboat initially at rest. The lifeguard leaves the boat with a horizontal speed of 3.44 m/s. Calculate the recoil speed of the boat. Neglect the transfer of momentum to the water.

4.23. A railroad car weighing 38 tons is loaded with nine tons of logs. It rolls along a straight, level track at 8.37 ft/s until it collides with an identical car at rest on the same track with a 23-ton load of logs. The two cars couple together in the collision. What is the velocity of the first car after the collision? Explain clearly your reasoning in arriving at your answer.

4.24. If you drop an object from rest and allow it to fall freely, what is its acceleration at the end of two seconds?

4.25. Suppose you throw a ball vertically upward. The ball will rise to some maximum height, then fall back to the ground. What is the velocity of the ball when it reaches maximum height? Explain. What is the acceleration of the ball at maximum height? Explain.

4.26. If you drop a steel ball, its acceleration toward the ground is about 32 ft/s^2. Suppose you threw the ball toward the ground. Would its acceleration after you released it be less than, greater than, or about equal to 32 ft/s^2? Explain.

4.27. A book rests on a horizontal table. Make a free-body diagram showing all of the forces acting on the book. Beside each vector, write the name of the *physical object* exerting the force. Give the reaction force (in the sense of Newton's third law) to each of these forces.

4.28. The acceleration due to gravity on the moon is about one-sixth of the value on the earth. On the earth, the mass of a 114-lb woman is about 52 kg. Estimate her mass if she were on the moon. Estimate her weight if she were on the moon. Explain clearly and completely your reasoning in each case.

4.29. On the earth, the mass of a 176-lb man is 80 kg. If the man were floating freely in space, far from any large gravitating bodies (planets, stars, etc.), what would be his mass out there? What would be his weight out there? Explain clearly your reasoning.

4.30. A woman stands in an elevator that is moving with a downward acceleration of 4 ft/s^2. Draw a free-body diagram showing all of the forces exerted on the woman. Which, if either, of these forces is greater? Explain. In the sense of Newton's third law of motion, what is the reaction force to each of these forces? From the information given, can you tell in which direction the elevator is moving? Explain.

4.31. A 1,320-kg rocket is launched vertically upward. The engine supplies a constant thrust of 18,000 N until it is cut off when the rocket reaches a speed of 2,300 km/h. What is the altitude of the rocket when the engine is cut off? Neglect the effects of air friction and assume that the mass of the rocket remains constant.

Answer: (53.2 km)

4.32. Starting from rest and accelerating uniformly, a 68-kg man slides down the full length of a four-meter rope in 1.10 seconds. How much force does he exert on the rope while he is sliding down?

Answer: (217 N)

4.33. A string is tied to a four-kilogram metal slug that rests on a workbench. The tension in the string must not be greater than 44 N or the string will break. By jerking upward on the string, what is the largest acceleration that can be given to the slug without breaking the string?

Answer: (1.2 m/s^2)

4.34. A bell boy carries a 35-lb piece of luggage into a hotel elevator. How much force must he exert on the luggage handle to support it when the car moves with an upward acceleration of 9.8 ft/s^2?

Answer: (45.7 lb)

4.35. By means of a light string, a ball hangs suspended from the roof inside a van moving on a straight level road. The string is observed to be inclined at a constant angle with respect to the vertical as shown in the drawing. Draw a carefully labeled free-body diagram showing all of the forces exerted on the ball. State clearly the reaction force (in the sense of Newton's third law of motion) to each of these forces. Describe clearly the motion of the van. Can you tell in which direction the van is moving? If so, give the direction and explain clearly your reasoning. If not, explain why not.

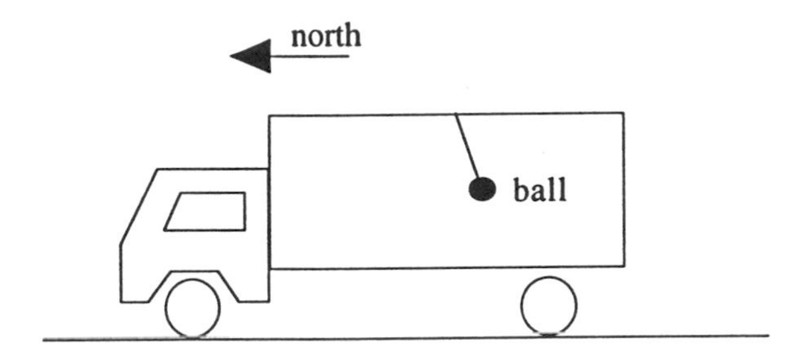

4.36. A 6.43-kg rocket launched vertically upward from a stationary pad accelerates uniformly to a speed of 187 km/h, which it reaches at an altitude of 621 m. Calculate the thrust of the engine during the rise. Neglect frictional effects.

Answer: (77.0 N)

4.37. A 146-lb weight watcher stands on a spring scale in an elevator that is accelerating vertically downward at 2.38 ft/s^2. What is the reading on the scale? Explain clearly.

Answer: (135 lb)

4.38. A 117-lb woman stands in a hotel elevator, which is slowing down at a rate of 1.98 ft/s^2. Draw a clearly labeled free-body diagram showing all of the forces exerted on the woman. Identify clearly the material body exerting each of these forces. Can you tell which force is greater? If so, which? If not, why not? In the sense of Newton's third law of motion, identify clearly the reaction force to each of these forces.

4.39. A ball is seen to be moving vertically upward with a speed of 15.7 m/s. Some time later, it is 22.6 m above the ground and moving straight down toward the earth's surface with a speed of 11.4 m/s. How high above the ground was the ball when it was first observed? Neglect the effects of air friction.

Answer: (16.6 m)

4.40. A small, two-stage rocket is launched vertically upward from its pad. When the rocket reaches an altitude of 3.27 km—where its speed is 1,070 km/h—the burned-out first stage is released and falls back to the earth. At this altitude, the second-stage engine is fired, giving the rocket a constant upward thrust of 681 N. The total mass of this second stage is 25.7 kg. When the second-stage speed reaches 2,250 km/h, the engine cuts off. Find the altitude at which the engine of second stage cuts off. Neglect effects due to friction.

Answer: (12.3 km)

4.41. Two identical 2-kg balls are connected by a light string. With another light string, the system is suspended from the ceiling of an elevator that is accelerating upward at 1.3 m/s^2. The arrangement is illustrated in the drawing. Calculate the tension in each string.

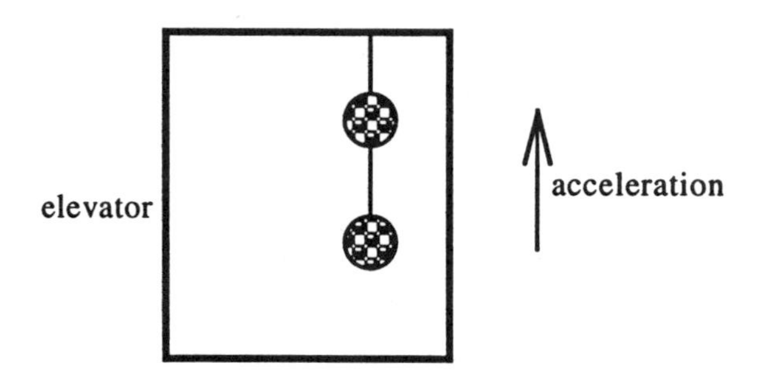

5

Running the Machine

5.1 Mechanical Work

Let's return to the analysis of the book on the table in Fig. 4.1. With the book at rest, we identified two forces on the book: $\mathbf{F}_{\text{table}}$, the table pushing up on the book and $\mathbf{F}_{\text{earth}}$, the earth pulling down. Now suppose we push the book so that it moves to a new position on the table top—there is a displacement $\Delta\mathbf{s}$ of the book. Two more forces enter here as shown in Fig. 5.1: the force of the hand pushing the book, $\mathbf{F}_{\text{hand}}$, and the frictional force that the rough tabletop exerts on the sliding book, $\mathbf{F}_{\text{friction}}$.

Now contrast the two forces $\mathbf{F}_{\text{hand}}$ and $\mathbf{F}_{\text{table}}$ in Fig. 5.1. The book moves horizontally on the table. So for $\mathbf{F}_{\text{hand}}$ there is some component of the force that is parallel (or antiparallel) to the displacement. There is *no* component of $\mathbf{F}_{\text{table}}$ that is parallel or antiparallel to the displacement—$\mathbf{F}_{\text{table}}$ is perpendicular to $\Delta\mathbf{s}$. To distinguish these two, we say that *work* is done by $\mathbf{F}_{\text{hand}}$, since a component of this force lies along the displacement. By constast, $\mathbf{F}_{\text{table}}$ does *no* work, because it has no component along the displacement.

We have introduced a new quality, work, which involves not only a force $\mathbf{F}$, but some displacement $\Delta\mathbf{s}$ along the force as illustrated in Fig. 5.2. To make this notion

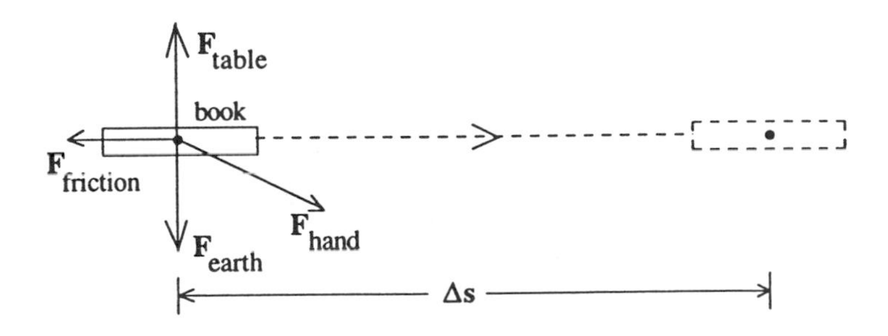

FIGURE 5.1. Horizontal displacement of a book on a table.

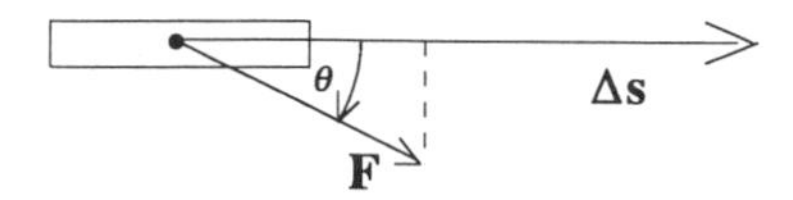

FIGURE 5.2. A component of the force along the displacement.

quantitative, work is defined by

$$\begin{aligned} \text{work} &\equiv \text{displacement} \times (\text{component of force along displacement}) \\ &= \Delta s(F\cos\theta) = F\Delta s\cos\theta. \end{aligned}$$

Clearly, work is a scalar quantity. From the definition, we see that in the mks system work is measured in N·m. This combination of units has a special name—the *joule*, abbreviated J. One joule is equal to one newton·meter.

5.2 Energy

Another quality is the ability that matter has to do work. The quantity we introduce to measure this ability is called *energy*. The energy a body possesses is simply defined to be *the work the body is capable of doing.*

Now let us look at ways in which matter has the ability to do work. As an example, we consider the simple arrangement illustrated in Fig. 5.3. This apparatus consists of a heavy hammer which is free to move vertically, guided by smooth metal rods. A wooden block with a nail driven a short distance into it is placed on the base of the apparatus with the nail directly underneath the hammer. If we

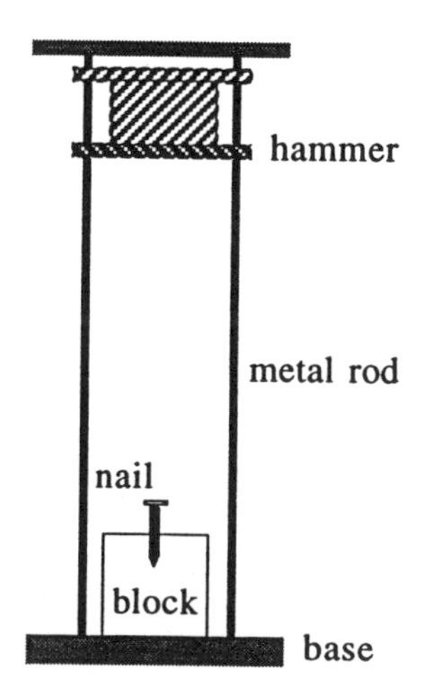

FIGURE 5.3. A falling hammer drives a nail in a wooden block.

release the hammer, it will fall onto the nail, driving it deeper into the block. In the impact, the hammer exerts a force on the nail and displaces it—the hammer does work on the nail. When the hammer is at the top of the apparatus, before we release it, it has the ability to do work on the nail. It hasn't done it yet, but it has the ability to do it. It has energy.

5.2.1 Gravitational Potential Energy

What is it about the hammer that gives it this ability to do work? When the hammer is held at rest at the top of the apparatus, it is *capable* of doing work on the nail. All we have to do is release it. The hammer falls, strikes the nail, and does work on the nail driving it deeper into the block. So at the top, the hammer has energy. Suppose we slowly let the hammer down to rest on top of the nail. Now the hammer is no longer capable of doing the work that it was capable of doing before. It no longer has the energy that it had at the top. What is the difference? From the top, it can fall and hit the nail. At the bottom, it cannot fall onto the nail. Clearly, falling has something to do with it. But what makes the hammer fall? It is the earth's gravity acting on it. If the earth didn't attract the hammer, the hammer would simply stay at the top (Newton's first law).

Now we have identified the source of energy for the hammer. The hammer derives its ability to do work on the nail from its *position* in the earth's *gravitational field*. The higher the hammer is (further from the center of the earth), the more work it is capable of doing when released. Because of its source, we call this kind of energy *gravitational potential energy*, PE_{grav}. How did the hammer acquire its PE_{grav} in the first place? We gave it this energy by doing work on it—pushing it by hand to its position at the top. The gravitational potential energy of the hammer at the top is equal to the minimum amount of work we have to do in pushing it up there. The minimum force F we can apply to move the hammer up is just equal to the weight of the hammer,

$$PE_{\text{grav}} = \text{work on hammer} = Fh = mgh,$$

where h is the height above the top of the nail.*

5.2.2 Kinetic Energy

Let us carry the analysis a little further. After we release the hammer, it falls. Now, just at the instant before the hammer strikes the nail, it is still capable of doing the work on the nail—it hasn't done the work yet so it still has the same energy as before. However, it is no longer its position in the gravitational field that gives it this energy. It now has the same position it had when it was sitting at rest on top of the nail and had no ability to do work on the nail, $PE_{\text{grav}} = 0$. So what kind of

*For simplicity, we neglect here the slight amount the nail is driven deeper into the block. In calculating the work done *on the nail*, of course, this distance is *not* negligible.

energy *does* it have now? Putting it another way, what is the difference between the state of the hammer at the instant before it hits the nail while falling and its state when it is sitting at rest on top of the nail? The answer is motion. The moving hammer is capable of doing work, whereas the hammer at rest is not. We call this kind of energy associated with the motion of matter, *kinetic energy*, *KE*.

At the instant before the hammer hits the nail, it is still capable of doing the same work it was capable of doing when it was at the top of the apparatus (neglecting small frictional effects). Therefore,

$$KE_{\text{bottom}} = PE_{\text{top}} = mgh. \tag{5.1}$$

Because the hammer starts from rest and falls with uniform acceleration g, $h = \frac{1}{2}gt^2$. Its speed just before it hits the nail is $v = gt$. On substituting these results into Eq. (5.1), we find that the kinetic energy of the hammer is

$$KE = \tfrac{1}{2}mv^2. \tag{5.2}$$

So the kinetic energy of a body depends on its inertia (m) and on its motion (v). This relation holds for any material body moving with a speed that is small compared to the speed of light.

5.2.3 Conservation of Energy

Let's consider the energy of the hammer a little more carefully. Suppose we close our eyes while someone else lifts the hammer and releases it. How do we know that the hammer strikes the nail? We *hear* it. That means air molecules get moved around so that the ones near us impinge on our ears. Work is done in vibrating our eardrums. Where does the energy come from to do this work? It was part of the original PE_{grav} the hammer possessed before it was released. In this experiment, the gravitational potential energy of the hammer is transformed into other kinds of energy. During the fall, some of the potential energy is converted to kinetic energy of the hammer. The hammer also does work on air molecules in pushing them out of the way as it falls (air friction). The metal rods that guide the falling hammer rub against the hammer, making the molecules in the rods and the hammer jiggle around faster (increasing the internal molecular kinetic energy). These jiggling molecules quickly come to equilibrium with those in the surrounding air and this energy is dissipated into the environment. So, in falling, the hammer retains some of its initial energy and loses the rest of it to the surroundings.

The energy does not disappear; it simply is transformed from one form to another or transferred from one system to another. This is a manifestation of the principle of *conservation of energy*. For any system, we can state the energy conservation principle* in the form,

$$E_{\text{initial}} + E_{\text{input}} = E_{\text{final}} + E_{\text{output}}. \tag{5.3}$$

---

*In tracking the conservation of energy in Chapter 18, we shall have to include the energy equivalent of mass according to Einstein's relation $E = mc^2$. See also Chapter 13.

It is implied here that the system begins in some initial state with total energy E_{initial} and ends up in some final state with total energy E_{final}. In the process of getting from the initial state to the final state, some energy E_{input} may be added to the system. For example, while the hammer is falling, we could give it a little push downward, increasing its kinetic energy. Also during the transition from initial to final state, the system may give up some energy E_{initial} to the environment.

In striking the nail, the hammer comes to rest. After it comes to equilibrium, it no longer has any of the energy it had before it was released at the top. It lost it all to the environment. Some was lost due to friction while falling, some more went out as sound waves in the impact, and some was used to do work on the nail. As the nail was driven deeper into the block, the molecules of the nail and block were made to jiggle faster, increasing the internal kinetic energies of these objects. As they came to equilibrium with their surroundings, the last of the original gravitational potential energy of the block was transferred to the environment.

EXAMPLE. A 0.1-kg ball is given an upward velocity of 20 m/s. If eight percent of its energy is lost to air resistance during its rise, how high will the ball go?

SOLUTION. We choose the zero of potential energy to be at the level where the ball was released. At that point, it has only kinetic energy. When it reaches maximum height, it is momentarily at rest with only gravitational potential energy and no kinetic energy. No energy is added to the ball during its flight, therefore $E_{\text{input}} = 0$. For Eq. (5.3), we write

$$\tfrac{1}{2}(0.1\text{ kg})(20\text{ m/s})^2 + 0 = (0.1\text{ kg})(9.8\text{ m/s}^2)h + (0.08)\left(\tfrac{1}{2}(0.1\text{ kg})(20\text{ m/s})^2\right),$$

from which we find that $h = 18.8$ m. The ball rises to a height of 18.8 m above the point at which it was released.

EXAMPLE. A 41.2-kg boy, starting from rest, slides down a hill 9.2 m high on a 3.42-kg sled. He experiences a constant frictional force of 28.3 N over the full length of the 138.4-m slope. Find the speed of the sled when it reaches the bottom of the slope.

SOLUTION. The system consists of the boy and the sled with a total mass of 44.62 kg. They begin with gravitational potential energy only. At the bottom of the slope where the potential energy is zero, they have only kinetic energy. During the slide down, no energy is added to the system, but it loses energy to the surroundings because of the friction of the slope. In this case, for Eq. (5.3) we have

$$\underbrace{(44.62\text{ kg})(9.8\text{ m/s}^2)(9.2\text{ m})}_{E_{\text{initial}}} + \underbrace{0}_{E_{\text{input}}} = \underbrace{\frac{1}{2}(44.62\text{ kg})v^2}_{E_{\text{final}}} + \underbrace{(28.3\text{ N})(138.4\text{ m})}_{E_{\text{output}}}.$$

On solving this equation, we get $v = 2.18$ m/s. At the bottom of the slope, the speed of the sled is 2.18 m/s.

5.3 Collisions

We saw in Chapter 4 that if there is no net external force on a many-body system, the momentum of the system is conserved. This is particularly useful in analyzing collisions between bodies.

Let us return to the specific case of the two colliding gliders on the air track in Fig. 4.5. The track is perfectly level and the air cushion on which the gliders ride renders friction negligible. So, once the gliders are set in motion, the net external force on the system is zero. Therefore, the total momentum of the two-body system is the same before and after the gliders collide. We express this conservation of momentum as

$$m_1\mathbf{v}_1 + m_2\mathbf{v}_2 = m_1\mathbf{v}_1' + m_2\mathbf{v}_2', \tag{5.4}$$

where the primes denote velocities *after* the collision.

Now let us apply the energy conservation principle, Eq. (5.3). It is convenient to choose the zero of gravitational potential energy at the level of the air track. Since the air track is horizontal, E_{initial} is just the sum of the kinetic energies of the two gliders *before* the collision. Similarly, E_{final} is the sum of their kinetic energies *after* the collision. No energy is added to the system in the collision, so $E_{\text{input}} = 0$. However, some energy may be lost by the two-body system. For example, some of the energy may be transferred to the environment as sound or as thermal energy (heat). So, in general, $E_{\text{output}} \neq 0$ and Eq. (5.3) reduces to

$$\tfrac{1}{2}m_1\mathbf{v}_1^2 + \tfrac{1}{2}m_2\mathbf{v}_2^2 = \tfrac{1}{2}m_1\mathbf{v}_1'^2 + \tfrac{1}{2}m_2\mathbf{v}_2'^2 + E_{\text{output}}. \tag{5.5}$$

In the special case when $E_{\text{output}} = 0$, we say that the collision is *elastic*. If $E_{\text{output}} \neq 0$, then the collision is *inelastic*.

EXAMPLE. Suppose the gliders in Fig. 4.5 have masses $m_1 = 0.383$ kg and $m_2 = 0.249$ kg. The heavier glider, moving to the right with speed 0.293 m/s, overtakes the lighter one, which is moving to the right with speed 0.185 m/s. The gliders collide and separate. After the collision, the lighter glider is moving to the right with speed 0.254 m/s. Find the velocity of the heavier glider after the collision. Is the collision elastic or inelastic?

SOLUTION. First, we invoke Eq. (5.4) to find the velocity of the heavier glider after the collision,

$$\begin{aligned}(0.383\ \text{kg})(0.293\ \text{m/s}) &+ (0.249\ \text{kg})(0.185\ \text{m/s}) \\ &= (0.383\ \text{kg})\mathbf{v}_1' + (0.249\ \text{kg})(0.254\ \text{m/s}).\end{aligned}$$

From this equation, we find that the velocity of the heavier glider after the collision is 0.248 m/s to the right. Inserting these results into Eq. (5.5),

$$\begin{aligned}&\tfrac{1}{2}(0.383\ \text{kg})(0.293\ \text{m/s})^2 + \tfrac{1}{2}(0.249\ \text{kg})(0.185\ \text{m/s})^2 \\ &\quad = \tfrac{1}{2}(0.383\ \text{kg})(0.248\ \text{m/s})^2 + \tfrac{1}{2}(0.249\ \text{kg})(0.254\ \text{m/s})^2 + E_{\text{output}},\end{aligned}$$

yields $E_{\text{output}} = 8.8 \times 10^{-4}$ J. Since $E_{\text{output}} \neq 0$, the collision is inelastic.

5.4 Power

When an amount of work ΔW is done over a finite time interval Δt, the average power is given by

$$\bar{P} = \frac{\Delta W}{\Delta t}. \tag{5.4}$$

This expression can also be extended to apply to the production or absorption of energy. If energy ΔW is produced (absorbed) in the time interval Δt, Eq. (5.4) gives the average rate at which the energy is produced (absorbed).

Notice that power has the dimensions of energy per unit time. In the metric system, this is joules per second. This combination of units is called the *watt.*

$$\text{one watt} \equiv \frac{\text{one joule}}{\text{one second}}.$$

EXAMPLE. A 68.0-kg woman is lifted by an elevator through a distance of 73.0 m in three-fourths of a minute. What is the increase in her gravitational potential energy? What average power is required to raise her?

SOLUTION. The change in her gravitational potential energy is

$$PE_{\text{grav}} = mgh = (68.0\,\text{kg}) \cdot (9.8\,\text{m/s}^2)(73.0\,\text{m}) = 48{,}647\,\text{J}.$$

We assume that the average work done on the woman by the floor of the elevator is just equal to the change in her potential energy. Thus,

$$\bar{P} = \frac{\Delta W}{\Delta t} = \frac{48{,}647\,\text{J}}{(0.75\,\text{min}) \cdot (60\,\text{s/min})} = 1{,}081\,\text{W}.$$

The woman's gravitational potential energy increases by about 48,650 J and the elevator delivers about 1,080 W of power to raise her.

Exercises

5.1. A little boy pulls with a constant horizontal force of 58 N on the handle of a wagon in which his friend sits. He gives his friend a 38-m ride along a horizontal sidewalk. The combined mass of the rider and the wagon is 39.5 kg. How much work is done on the wagon by the boy pulling on it?

5.2. By applying a horizontal force of 289 N, a dock worker pushes a 1,220-N crate at a constant velocity of 1.17 m/s a distance of 10.3 m over a rough, horizontal floor in 8.80 seconds. Calculate the work done on the crate by the dock worker.

Answer: (2,980 J)

5.3. A frictionless block and tackle is used to raise a 318-kg steel beam at constant speed from the ground to a height of 34.6 m. How much work is done in lifting the beam?

Answer: (108,000 J)

5.4. A 384-kg rocket is launched vertically upward. The rocket accelerates uniformly until it reaches a speed of 1,320 km/h at an altitude of 37 km at which point the engine cuts off. While the engine is running, it supplies a constant thrust of 4,460 N. Calculate the work done by the engine.

Answer: (1.65×10^8 J)

5.5. Show clearly that Eq. (5.2) follows from Eq. (5.1) as outlined in the text.

5.6. A constant vertical force of 9,500 N is used to raise a 950-kg granite block from the ground to a height of twelve meters above the ground. Calculate the work done in lifting the block and the change in gravitational potential energy of the block. Are your two answers equal. Should they be? If so, why? If not, why not?

5.7. A 125-g ball is thrown vertically upward and rises to a maximum height of 6.37 m above the ground. During its rise, the ball experiences an average frictional force of 0.183 N. Use the energy conservation principle to calculate the speed of the ball as it left the thrower's hand 2.13 m above the ground.

Answer: (9.77 m/s)

5.8. A 41.8-kg boy sits on a 3.6-kg sled, which is coasting with a constant velocity of 5.27 m/s over perfectly smooth, level ice. The sled enters a patch of rough ice 17.8 m long over which it experiences a constant frictional force of 15.2 N. Use the energy conservation principle to find the speed at which the sled emerges from the rough stretch. Neglect the effects of air friction. What percent of the original energy of the boy and sled is lost due to friction on the rough patch of ice?

Answer: (3.98 m/s; 42.9%)

5.9. A 39.34-kg girl starts from rest on a 3.29-kg sled and slides down a hill that is 8.62 m high. She experiences a constant frictional force over the full length of the 143.5-m slope. As she reaches the bottom of the slope, her speed is 4.17 m/s. Calculate the combined frictional force she and the sled experience during the slide.

Answer: (22.5 N)

5.10. A small, 1.37-kg rocket is launched vertically upward from rest on the ground. The engine supplies a constant thrust of 24.25 N over a distance of 402 m at which point it runs out of fuel. As the rocket rises, it experiences an average frictional force of 4.54 N. Use the energy conservation principle to calculate the maximum altitude reached by this rocket.

Answer: (543 m)

5.11. A 2.46-kg block is set in motion with an initial velocity of 9.02 m/s on a rough, horizontal surface. The combined effects of air resistance and the rough surface result in a constant frictional force of 6.82 N exerted on the block as it slides. Use the energy conservation principle to find how fast the block is moving when it is 3.19 m from the point at which it is released.

Answer: (7.98 m/s)

5.12. Two balls—one of mass m, the other of mass M—are both at rest. The same constant force F is applied to each ball separately over the same time interval Δt. At the end of the time interval, which ball, if either, has the greater momentum? Which, if either, has the greater kinetic energy?

5.13. A 287-g ball attached to a string is set in motion in a vertical circle of radius 1.09 m. The speed of the ball at the top of the circle is 5.66 m/s. Calculate the tension in the string when the ball is at the top of the circle. Use the energy conservation principle to find the speed of the ball at the bottom of the circle. Neglect frictional effects.

Answer: (5.62 N; 8.65 m/s)

5.14. A 0.189-kg ball is thrown vertically downward from the top of a building that is 22.9 m tall. The speed of the ball as it leaves the thrower's hand is 8.75 m/s. Air resistance results in an average frictional force of 0.652 N on the ball as it travels downward. Use the energy conservation principle to find the speed of the ball at the instant before it hits the ground.

Answer: (19.2 m/s)

5.15. A windsurfer is sailing along at 3.56 m/s when a sudden gust of wind hits his sail with a constant force of 392.0 N and continues over a distance of 42.7 m. During this run, he and his board experience a constant frictional force of 371.9 N. The total mass of the surfer and his board is 82.4 kg. Use the energy conservation principle to calculate the speed of the surfer at the end of the 42.7-m run.

Answer: (5.79 m/s)

5.16. A gun is aimed vertically upward and fired. The 4.22-g bullet leaves the barrel with a speed of 163 m/s. As the bullet rises, it experiences an average frictional force of 0.0179 N. How high above the end of the barrel of the gun does the bullet rise?

Answer: (946 m)

5.17. A 38.15-kg girl sits on a 2.32-kg sled at rest at the top of a straight slope that is 33.5 m long and 9.1 m high. The girl's father pushes the sled downhill with a constant force of 120 N parallel to the slope. He stops pushing after the sled has moved 3.22 m along the slope.

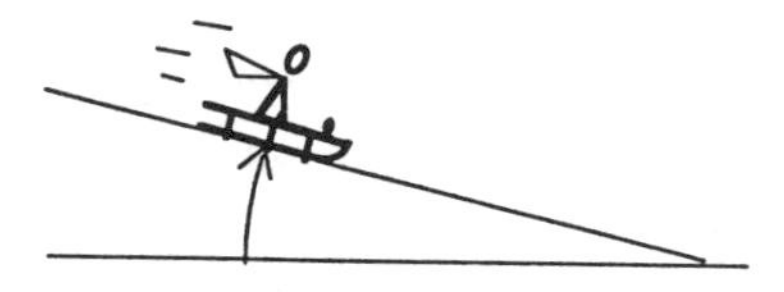

The girl and her sled continue coasting to the bottom of the slope. The frictional force they experience during the slide is 83.6 N. Use the energy conservation principle to find the speed of the sled when it reaches the bottom of the slope.

Answer: (7.68 m/s)

5.18. A small boy sits in a soap box racer at rest at the top of a straight slope that is 87.3 m long and 18.2 m high. The mass of the boy and the racer combined is 42.3 kg. With a constant force applied parallel to the slope, the boy's brother pushes the racer downhill. The brother stops pushing after the racer has moved 2.75 m along the slope. The racer continues coasting downhill until it reaches the bottom of the slope where it has a speed of 4.80 m/s. The constant frictional force the boy and the racer experience during the downhill ride is 84.2 N. Use the conservation of energy principle to calculate the force the brother exerts on the racer while he is pushing it.

Answer: (107 N)

5.19. By blowing with the mouth on the lower end of a straight hollow tube, a small wooden ball is projected from the upper end of the tube as illustrated. The mass of the ball is 28.4 g. The ball leaves the tube with a speed of 5.38 m/s. Calculate the total energy gained by the ball as it travels along the length of the tube.

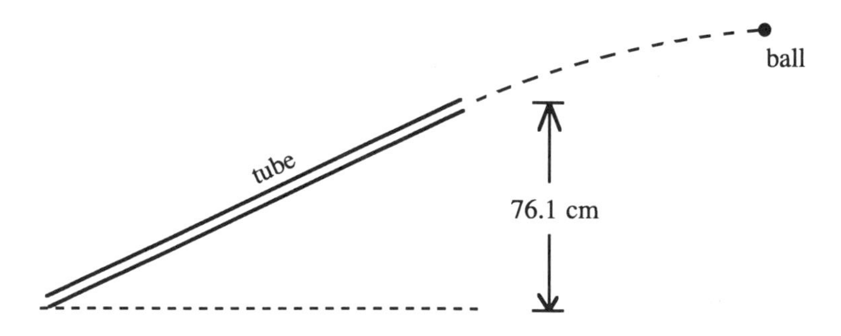

5.20. A rough track consists of a straight, horizontal section and a curved section oriented in a vertical plane. A small, 119-g block is released from rest at the top of the curved section as illustrated. The block reaches the horizontal section with a speed of 3.65 m/s. The upper end of the curved part of the track is 87.3 cm above the horizontal part.

Along the horizontal part of the track, the block experiences a constant frictional force of 1.185 N. Use the conservation of energy principle to find the energy lost by the block due to friction on the curved part of the track and how far the block slides on the horizontal part.

Answer: (0.225 J; 66.9 cm)

5.21. A 227-g glider moves along a straight, horizontal air track with a speed of 0.417 m/s. It strikes a second glider at rest. The two gliders stick together and move with a velocity of 0.238 m/s in the original direction of the first glider. Find the mass of the second glider. How much energy did the gliders lose to the environment in the collision?

Answer: (171 g; 8.47×10^{-3} J)

5.22. A particle of mass 1.67×10^{-27} kg moving to the right with speed 9.380×10^6 m/s collides head on with a second particle of mass 6.680×10^{-27} kg moving to the right with a speed of 0.260×10^6 m/s. After the collision, the second particle is still moving to the right, but with speed 2.345×10^6 m/s. Find the speed and direction of motion of the first particle after the collision. By means of a calculation determine whether this is an elastic collision.

Answer: (1.040×10^6 m/s to the right; inelastic)

5.23. A 0.783-kg mass moving to the right with a speed of 5.21 m/s collides head on with a 1.223-kg mass moving to the left with a speed of 4.07 m/s. After the collision, the 1.223-kg mass is moving to the right with a speed of 1.130 m/s. Find the velocity (magnitude and direction) of the 0.783-kg mass after the collision. What fraction of the original energy of the system is lost in the collision? Explain clearly what happened to this energy.

Answer: (2.91 m/s to the left; 80.2%)

5.24. A 813-g ball moving to the right with a speed of 5.08 m/s collides head on with a 1,228-g ball moving in the opposite direction with a speed of 3.88 m/s. After the collision, the 813-g ball is moving to the left with a speed of 4.17 m/s. From a calculation, determine whether the collision is elastic or inelastic.

Answer: (inelastic)

5.25. A 142-g hockey puck sliding with a speed of 36.4 m/s to the right on smooth ice collides head on with a second puck at rest. After the collision, the 142-g puck is moving to the left with a speed of 1.74 m/s and the other puck is moving to the right with a speed of 29.7 m/s. Find the mass of the second puck. What percent of the total energy was lost by the two-puck system in the collision?

Answer: (182 g; 14.3%)

5.26. Two blocks, equipped with bumpers, are free to slide on a smooth, horizontal surface. The block of mass 1.837 kg slides with a velocity of 9.25 m/s toward the right. It collides head on with the second block, which has mass 0.722 kg, sliding toward the left with a speed of 3.84 m/s. After the collision, the second block is seen to be moving to the right with a speed of 6.54 m/s. Calculate the velocity of the first block after the collision. What percent of the initial kinetic energy of the system is lost in the collision?

Answer: (5.17 m/s to the right; 52.3%)

5.27. Estimate the earth's kinetic energy due to its orbital motion around the sun.

5.28. The Electric Company charges customers for energy at a rate of 8.2 cents per kilowatt·hour. An automatic timer is set to switch on a 250-W outdoor flood lamp at 6:00 p.m. each day. At 11:30 p.m. each night, the timer automatically switches off the lamp. Estimate the monthly cost of operating this lamp.

Answer: ($3.38)

5.29. A small aircraft is cruising in level flight with a constant speed of 180 km/h. The aircraft is experiencing a constant drag of 930 N due to air friction. Calculate the power the engine must deliver to overcome the drag.

Answer: (46.5 kW)

5.30. It takes exactly one minute for a winch to drag a heavy crate at constant speed a distance of 11.7 m over a rough, horizontal floor. The tension in the winch cable as it pulls horizontally on the crate is 2,930 N. Calculate the power delivered to the crate by the winch.

5.31. A 1,290-kg automobile starts from rest and accelerates uniformly until it reaches a speed of 110 km/h in 31.3 seconds. Calculate the power delivered to the wheels of the automobile in this run. Neglect frictional effects.

Answer: (19.2 kW)

5.32. A woman draws water from a deep well as shown in the illustration. The surface of the water is 27.4 m below the top of the well where the woman is standing. She hauls the water up by turning a windlass by means of a crank. The well rope passes over a pulley and winds around the turning windlass. It takes her 21.3 s to draw the filled bucket out of the well. She turns the crank so that the bucket rises at constant speed. The total mass of the water and the bucket is 25.5 kg. Calculate the power the woman delivers to the bucket of water in raising it. Suppose she uses an electric motor to turn the crank instead of turning it by hand. Assume that 70% of the energy the motor consumes actually goes to raising the bucket of water. At 8.2 cents per kW·h, how much would it cost to run the motor in bringing up each bucketful of water?

Answer: (321 W; 0.022 cents)

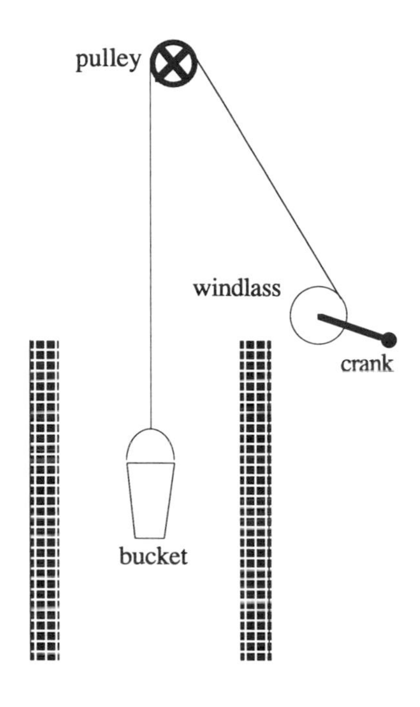

6

Off the Straight and Narrow

So far we have only considered one-dimensional motion—motion along a straight line. In general, the celestial objects—the moon, the planets, comets, stars, and so on—move along *curved* paths—*curvilinear* motion. We must generalize.

6.1 Uniform Circular Motion

We begin with the simplest case—uniform circular motion. A particle of mass m moves with constant speed v in a circular path about a fixed point. A good approximation is realized with a small mass attached to one end of a length of string. The other end of the string is held in the hand and the ball is whirled around overhead in a horizontal circle. The geometry is illustrated in Fig. 6.1*a*.

In the time interval $\Delta t = t_2 - t_1$, the particle moves a distance s along the arc of radius r. At time t_1 its velocity is $\mathbf{v}_1$ and at t_2 the velocity is $\mathbf{v}_2$. Because the speed is constant, $v_1 = v_2 = v$. Only the *direction* of the velocity changes in this case. From Eq. (2.10), the average acceleration is

$$\bar{\mathbf{a}} = \frac{\mathbf{v}_2 - \mathbf{v}_1}{\Delta t}.$$

We can solve this equation for $\mathbf{v}_2$,

$$\mathbf{v}_2 = \mathbf{v}_1 + \bar{\mathbf{a}}\Delta t.$$

Here two vectors are added to obtain a third vector. This addition is shown graphically in Fig. 6.1*b*. Note that as Δt becomes smaller, the angle ϕ between $\mathbf{v}_1$ and $\mathbf{v}_2$ gets smaller and the chord length $\bar{a}\Delta t$ approaches the arc length ϕv in Fig. 6.1*b*,

$$\bar{a}\Delta t \approx \phi v \text{ (for small } \Delta t). \tag{6.1}$$

From Fig. 6.1*a*, we see that θ and ϕ are both acute angles in right triangles, which have a common acute angle β. Since both angles are complementary to β, we have $\theta = \phi = s/r$, where this second equality follows from Fig. 6.1*a* and the

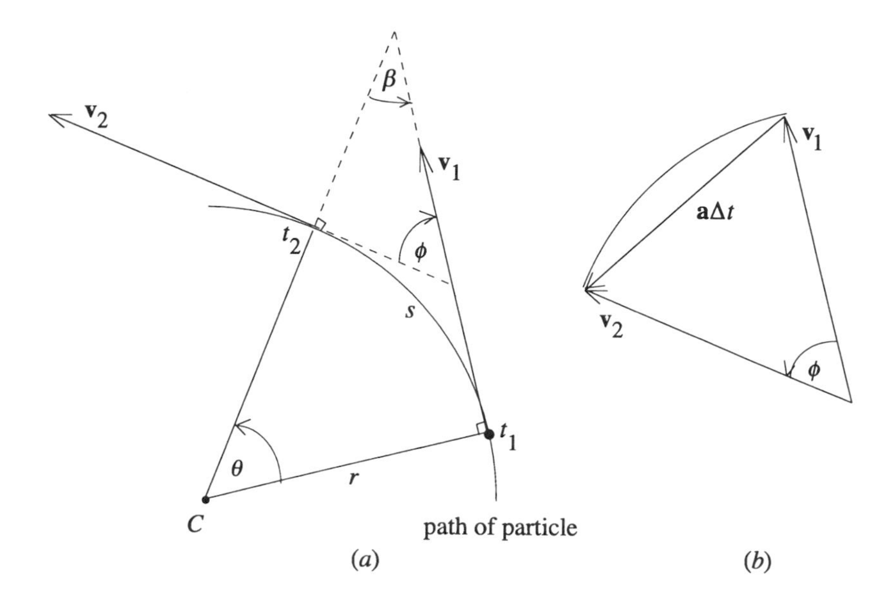

FIGURE 6.1. Particle moving at constant speed along a circular arc.

definition of radian measure of an angle.* The arc length s is just the distance the particle travels at speed v in time Δt. That is, $s = v\Delta t$.

With these substitutions, Eq. (6.1) can be written,

$$\bar{a} \approx \frac{v^2}{r} = \text{constant}.$$

In the limit that Δt becomes vanishingly small, the average acceleration $\bar{a}$ approaches the instantaneous acceleration a and we get,

$$a = \frac{v^2}{r}. \tag{6.2}$$

Since the path of this particle lies in a plane, we can resolve the acceleration at any instant into two components that lie in this plane. We choose one component to be parallel (or antiparallel) to the velocity $\mathbf{v}$ at that instant and the other component to be perpendicular to $\mathbf{v}$. We call the first of these the *tangential* acceleration, denoted by a_t, because its direction is tangent to the circular path of the particle. Clearly, if this component of the acceleration is different from zero, then the velocity must change in this direction, which means that the particle must either speed up or slow down. In the case here, the speed is constant. Therefore, for uniform circular motion, $a_t = 0$ and it is only the perpendicular component of $\mathbf{a}$ that is represented

*According to the definition of radian measure, an angle α subtends a circular arc of length l and radius of curvature r with $\alpha = l/r$.

by Eq. (6.2). This acceleration represents a change in the *direction* of the velocity and not the speed. It corresponds to a change in the direction of the motion of the particle *toward* the center of the circle. For this reason, it is called *centripetal* accleration, denoted by a_c, meaning acceleration "toward the center."

EXAMPLE. A racing car is traveling at a constant speed of 240 km/h around a circular track. The radius of curvature of the track is 760 meters. Calculate the centripetal acceleration of the racer.

SOLUTION. Using Eq. (6.2), we have

$$a = \frac{v^2}{r} = \frac{1}{760\text{ m}} \cdot \left(\frac{240\text{ km}}{\text{h}} \cdot \frac{1000\text{ m}}{\text{km}} \cdot \frac{1\text{ h}}{3600\text{ s}}\right)^2 = 5.85\text{ m/s}^2.$$

The centripetal accleration of the racer is 5.85 m/s^2.

EXAMPLE. The planet Venus travels at nearly constant speed in a circular orbit around the sun. Estimate Venus's orbital acceleration in m/s^2.

SOLUTION. With Eq. (6.2) and data from Table 1.1, we have

$$\begin{aligned} a &= \frac{v^2}{r} = \frac{1}{r}\left(\frac{\text{circumference of orbit}}{\text{time for complete orbit}}\right)^2 = \frac{1}{r}\left(\frac{2\pi r}{T}\right)^2 = \frac{4\pi^2 r}{T^2} \\ &= \frac{4\pi^2(0.72\text{ AU})}{(0.62\text{ yr})^2} \cdot \frac{1.50 \times 10^{11}\text{ m}}{\text{AU}} \cdot \left(\frac{1\text{ yr}}{3.156 \times 10^7\text{s}}\right)^2 = 0.011\text{ m/s}^2. \end{aligned}$$

Thus, the orbital acceleration of Venus is 0.011 m/s^2.

6.2 Centripetal Force

The relation in Eq. (6.2) is more general than our derivation here suggests. Even when the particle is moving along a curved path that is *not* circular and the speed is *not* constant, Eq. (6.2) still gives the centripetal component of the acceleration with v and r the *instantaneous* speed and radius of curvature, respectively.

For the centripetal component of Newton's law, we have

$$F_c = m\frac{v^2}{r}, \tag{6.3}$$

where F_c is called the *centripetal* force. Notice that, in general, F_c is *not* a single force but a combination of forces that collectively cause the centripetal acceleration.

EXAMPLE. A small, 120-g ball is attached to one end of a string of length 40 cm. The other end of the string is fixed at a point so that the ball swings in a vertical

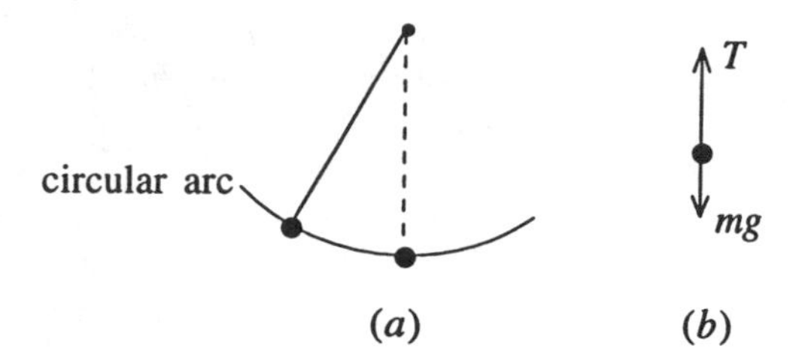

FIGURE 6.2. A ball moving in a vertical circle.

circle about this point as shown in Fig 6.2*a*. When the ball reaches the lowest point on the arc, it has a speed of 7.2 m/s. Calculate the tension T in the string at this point.

SOLUTION. Use the diagram in Fig. 6.2*b* to get the components of the forces contributing to the centripetal force.

$$F_c = m\frac{v^2}{r}$$

$$T - mg = m\frac{v^2}{r}$$

$$T - (0.12\,\text{kg})(9.8\,\text{m/s}^2) = (0.12\,\text{kg})\frac{(7.2\,\text{m/s})^2}{0.40\,\text{m}}.$$

On solving for T, we find that the tension in the string is 16.7 N. Note that the centripetal force F_c is 15.6 N, the resultant of the tension pulling the ball *toward* the center of the circular arc and the earth's gravitational force pulling the ball *away* from the center.

Exercises

6.1. The diameter of an old 78-rpm phonograph record is ten inches (24.5 cm). Calculate the centripetal acceleration of a point on the rim of this record when it is playing on the phonograph.

Answer: (8.17 m/s^2)

6.2. Assume that the planet Mars moves with a constant speed in a circular orbit around the sun. Calculate the orbital acceleration (in m/s^2) of Mars.

6.3. On October 4, 1957, the USSR placed Sputnik, the first artificial satellite, in orbit around the earth. At an altitude of 575 km, Sputnik followed a nearly circular orbit around the earth, completing an orbit in 96 minutes. From the data, calculate the centripetal acceleration of Sputnik.

Answer: (8.26 m/s^2)

6.4. A motorcycle on a circular track has a constant speed of 150 km/h. It takes 38.4 s for the motorcycle to make a complete circuit of the track. What is the centripetal acceleration of the motorcycle?

Answer: (6.82 m/s^2)

6.5. A cyclist is traveling at 24.3 km/h. The wheel of his bicycle has a diameter of 26 inches (66.0 cm). Calculate the centripetal acceleration of a small section of the tread of his front tire. How fast is the wheel turning at this speed?

Answer: (138 m/s^2; 195 rpm)

6.6. A pilot pulls his aircraft out of a dive along the arc of a large vertical circle. At the lowest point (bottom of the dive) what forces are exerted on the pilot? Name the material body that exerts each force. How are these forces related to the centripetal force on the pilot?

6.7. Referring to the previous exercise, find the force that the aircraft exerts on an 80-kg pilot if the speed of the aircraft at the bottom of the circle is 690 km/h and the radius of the circle is 2,030 meters.

Answer: (2,232 N)

6.8. A small ball is attached to a light string and set in motion in such a way that it swings completely around in a vertical circle as shown. At the bottom of the circle the tension in the string is 9.35 N, and when the ball is at the top of the circle the tension is 2.06 N. Find the mass of the ball.

Answer: (0.124 kg)

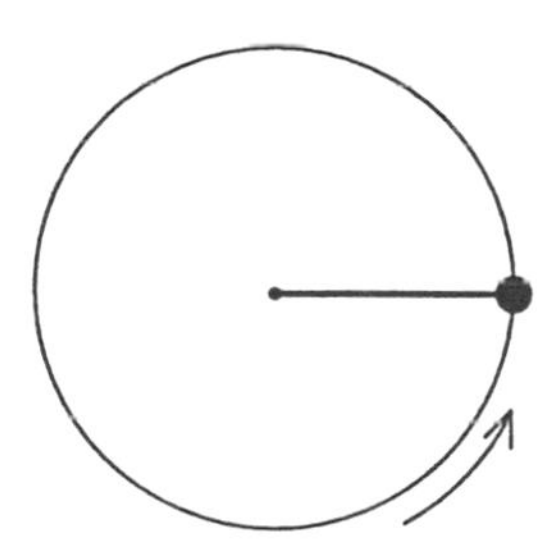

6.9. The earth describes a nearly circular orbit around the sun. The earth's orbital speed is about 1.075×10^5 km/h. Other relevant data are given in Appendix B. From these data, estimate the average centripetal force exerted on the earth as it orbits the sun.

Answer: (3.5×10^{22} N)

6.10. A woman standing on the surface of the rotating earth at the equator experiences a centripetal acceleration. Making use of a clearly labeled free-body diagram explain clearly and completely why the woman expe-

riences a centripetal acceleration. Calculate the centripetal acceleration of the woman.

Answer: (0.034 m/s^2)

6.11. A 352-g ball hangs vertically at one end of a light string. The other end of the string is fixed at a point 73.4 cm directly above the ball. The ball is suddenly given a sharp, horizontal blow which sets it in motion with an initial speed of 5.26 m/s. This will cause the ball to swing along a circular arc as illustrated below. What is the tension in the string at the instant after the ball is struck? When the ball reaches a position at the same height as the point of suspension, the string is stretched out horizontally. What is the tension in the string at this point? Will the ball swing all the way around or only part of the way?

Answer: (16.7 N; 6.37 N)

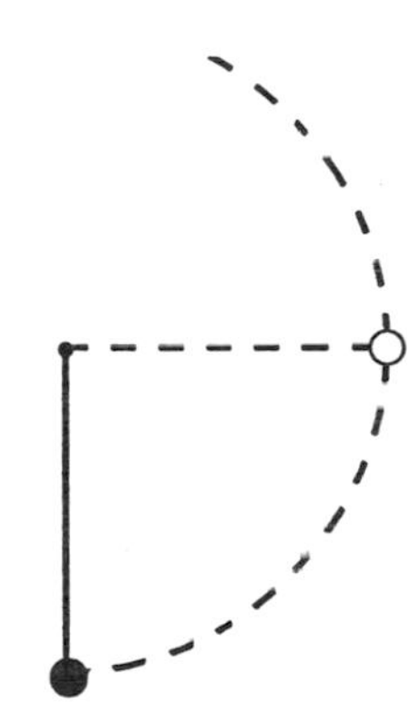

6.12. Assume that the moon describes a circular orbit around the earth. Calculate the corresponding centripetal acceleration of the moon.

6.13. In terms of centripetal force and centripetal acceleration, explain clearly and completely why it is dangerous to drive a car too fast around a highway curve.

6.14. A 235-g steel ball is attached to one end of a light string. The other end of the string is fastened to the ceiling so that the ball hangs 1.332 m directly below the point of suspension. The ball is pulled aside (as indicated by the dashed line in the illustration at the top of the next page) to a point 1.252 m below the ceiling. It is released from rest at that point to swing back and forth as a pendulum. Calculate the tension in the string when the ball is at the lowest point in its swing. Neglect frictional effects.

Answer: (2.58 N)

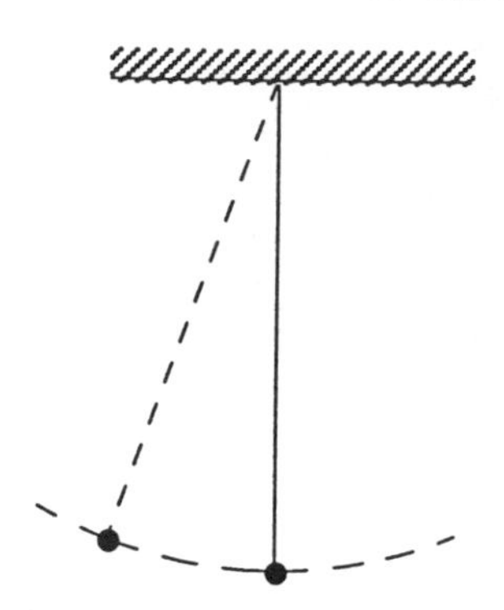

6.15. Use the orbital data for the planets given in Table 1.1 to calculate the centripetal acceleration a_c (in AU/yr^2) of each of the planets as it orbits the sun. Tabulate your results and note that as the period T increases, the centripetal acceleration a_c decreases. Assume that the relationship between a_c and T is a power law. That is,

$$a_c = KT^x,$$

where the constant parameters x and K have the same values for all planets. If we take the logarithm of both sides of this equation, we get

$$\log a_c = x \log T + \log K.$$

Use your results for the planets to make a plot of $\log a_c$ versus $\log T$. You should have a linear graph from which you can obtain values for x and K. Express x as the ratio of two small integers and rewrite the power law for a_c above using your values of x and K.

6.16. A hydrogen atom consists of an electron and a proton. According to a simple model of the atom, the electron travels at constant speed in a circular orbit around the much heavier proton. The separation between the electron and proton is 5.28×10^{-11} m. The electron makes 6.60×10^{15} complete orbits each second. Use these data and the information in Appendix B to calculate the centripetal force experienced by the electron in this model of the hydrogen atom.

Answer: (8.27×10^{-8} N)

6.17. A 1,230-kg automobile travels around a highway curve at a constant speed of 130 km/h. The constant radius of curvature of the curve is 483 m. Draw a free-body diagram showing all of the forces exerted on the vehicle. Calculate the centripetal force on the vehicle as it rounds the curve. Describe clearly in words the two most probable contributions to the centripetal force.

Answer: (3,320 N)

7

The Gravity of It All

7.1 Early Speculations

Early in the seventeenth century, Galileo did extensive studies of bodies falling near the earth's surface. Bodies projected with a component of velocity tangent to the earth's surface—cannonballs, for example—are called *projectiles*.

As projectiles move horizontally, they also fall vertically. Some typical trajectories for projectiles are illustrated in Fig. 7.1. As the horizontal component of the muzzle velocity increases, the range of the cannonball increases. When fired over level ground, the ball travels farther down range before it strikes the ground.

Later in that century, Newton came to realize that the moon behaves the same way cannonballs do. The moon, like the cannonball, is attracted toward the center of the earth by some property of the earth he called *gravity*. He imagined firing cannonballs from a powerful cannon on top of a tall mountain on the earth.

In the *Principia*, Newton published a drawing similar to that in Fig. 7.2, which shows the trajectories of imaginary cannonballs fired with various muzzle velocities. The greater the velocity, the farther around the earth the projectile travels before striking the earth. Given enough velocity, the ball will travel all the way around the earth without hitting it. It will orbit the earth. (We have ignored friction here and assumed that only the earth's gravity is exerted on the ball.)

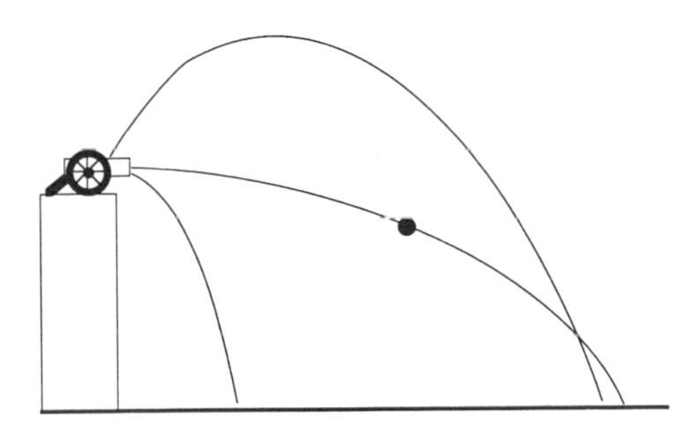

FIGURE 7.1. Cannonball trajectories near the earth's surface.

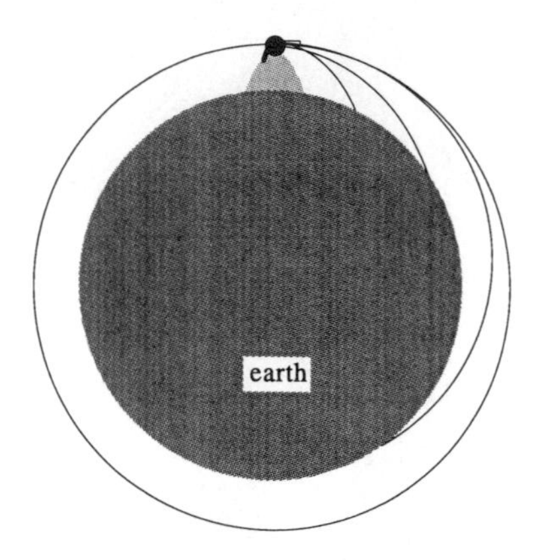

FIGURE 7.2. High-velocity cannonball trajectories.

Newton concluded that this is just what the moon does. It is continuously falling toward the center of the earth, but it has enough tangential velocity to keep it from hitting the earth. The moon—like the cannonball—orbits the earth. He assumed that it is the same property of the earth—gravity—that acts on the cannonball and on the moon.

Still later in the seventeenth century, others around the Royal Society of London—Christopher Wren, Robert Hooke, Edmund Halley—began speculating about the nature of this force of gravity and its extension to other celestial objects—the planets orbiting the sun, for example. In 1686, Newton put an end to their speculations with the publication of the *Principia* including his law of universal gravitation. But all of this had its genesis some twenty years earlier with his thoughts about things like the moon and cannonballs.

The actual orbit of the moon is an ellipse, but not a very eccentric one. It is nearly circular. To get a sense of Newton's reasoning without distracting mathematical complications, it will be completely adequate to assume that the moon follows a circular orbit around the center of the earth. In this approximation, the moon moves around the earth with uniform circular motion.

The centripetal acceleration of any object in a circular orbit around a fixed point is given by Eq. (6.2),

$$a = \frac{v^2}{r}, \tag{7.1}$$

where v is the orbital speed and r is the distance from the center of the orbit to the center of mass of the object.* Since v is constant, it is equal to any distance the object travels divided by the corresponding time. The time required for the object

*The center of mass of a spherically symmetric object is located at the geometrical center of the object. The concept of center of mass is discussed more fully in Chapter 8.

to traverse the circumference of the orbit is the *period* of the orbit, denoted by T. Thus,

$$v = \frac{2\pi r}{T}. \tag{7.2}$$

On substituting this result into Eq. (7.1), we get for the acceleration a of any object moving in a circular orbit of radius r with orbital period T,

$$a = \frac{4\pi^2 r}{T^2}. \tag{7.3}$$

We shall apply this equation to objects orbiting the center of the earth.* The observational data available to Newton included the following:

average distance from earth to moon = 384,500 km = 60 earth radii
period of the moon's orbit about the earth = 27.3 days
acceleration of a falling body at the earth's surface = 9.80 m/s^2

With these data, we can use Eq. (7.3) to calculate the acceleration due to gravity of an object with the same orbital parameters (r and T) as the moon,

$$a = \frac{4\pi^2 r}{T^2} = \frac{4\pi^2(3.845 \times 10^8 \text{ m})}{(27.3 \cdot 24 \cdot 3600 \text{ s})^2} = 0.00273 \text{ m/s}^2.$$

If this object were at the earth's surface instead of out where the moon is, the acceleration due to gravity would be 9.80 m/s^2. Clearly, the acceleration due to the earth's gravity gets smaller with increasing distance from the center of the earth. Let's make a conjecture that the acceleration decreases simply in the inverse proportion of some power of the distance r from the earth's center. That is,

$$a = \frac{C}{r^x}, \tag{7.4}$$

where x is a number to be determined and C is a constant that can depend on properties of the earth. Now let's apply this expression to the two places where we have data—on the earth's surface and out where the moon is. Let R denote the radius of the earth. According to the data, the radius of the moon's orbit is $60R$.

at the moon's orbit:	$0.00273 \text{ m/s}^2 = C/(60R)^x$
at the surface of the earth:	$9.80 \text{ m/s}^2 = C/R^x$

Setting the ratio of the left-hand sides of these equations equal to the ratio of the right-hand sides yields

$$\frac{0.00273 \text{ m/s}^2}{9.80 \text{ m/s}^2} = \left(\frac{1}{60}\right)^x.$$

*This development through Eq. (7.5) follows B. M. Casper and R. J. Noer, *Revolutions in Physics*, Norton, New York, 1972, p. 222.

By taking logarithms of both sides of this equation, we get

$$x = \frac{\log\left(\frac{0.00273}{9.80}\right)}{\log\left(\frac{1}{60}\right)} = 1.9994 \approx 2.$$

From these results and Newton's second law of motion, we find that the gravitational force exerted by the earth on a body of mass m is,

$$F_{\text{grav}} = \frac{Cm}{r^2}, \tag{7.5}$$

where the value of C may depend on properties of the earth.

7.2 Newton's Law of Universal Gravitation

Our results so far suggest that the law of force for bodies under the influence of the earth's gravity is an *inverse square law*, meaning that the gravitational force the earth exerts on a body outside the earth is inversely proportional to the square of the distance of the body from the center of the earth.

On recognizing that it is the earth's gravity that holds the moon in orbit around the earth, Newton assumed that gravity is also responsible for binding the planets in orbit around the sun. The information he had regarding the orbits of the planets was summarized in Kepler's three laws of planetary motion. With these three laws of Kepler and Newton's three laws of motion, let us see how far we can get toward a law of gravity for the sun and a planet. The orbits of the planets that can be seen with the naked eye (the only ones Newton and Kepler knew about) are not very eccentric ellipses. The two foci for a given orbit are very nearly at the same point. For simplicity, we consider them to be coincident. For our purpose here, this is an acceptable assumption. First, we state Kepler's laws again and consider the implications of this approximation for each.

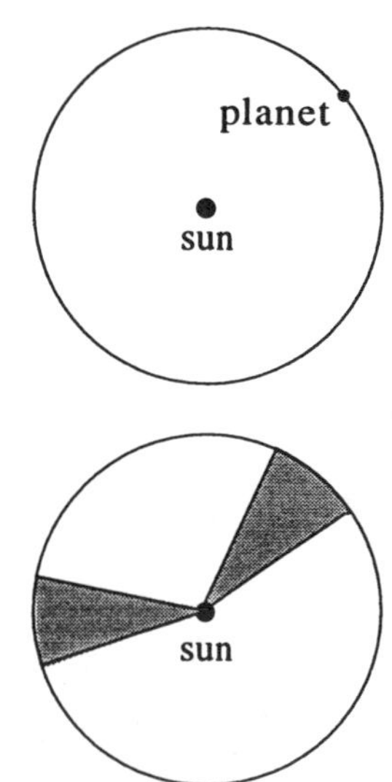

Kepler's First Law

The path of each planet is an ellipse with the sun at one focus.

When the two focal points of an ellipse coincide, the figure is a circle.

Kepler's Second Law

The line from the sun to the planet sweeps out equal areas in equal times.

For a circular orbit, this means that the orbital speed is constant.

Kepler's Third Law

The square of the orbital period of a planet is proportional to the cube of the semimajor axis of the orbit.

For a circle, the semimajor axis is equal to the orbit radius r. Thus,

$$T^2 = K_{\text{sun}} r^3.$$

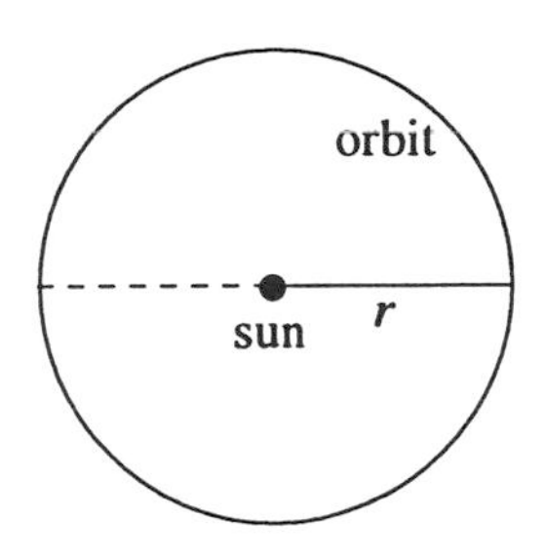

For a circular orbit, Kepler's second law implies that the orbital speed of the planet is constant and given by Eq. (7.2),

$$v = \frac{2\pi r}{T}.$$

In our approximation, the only force the planet experiences is the gravitational force of the sun that draws the planet toward the center of the orbit. Therefore, the acceleration of the planet is centripetal acceleration. According to Newton's second law, the gravitational force the sun exerts on the planet is,

$$F_{\text{sun-planet}} = m_{\text{planet}} \frac{v^2}{r} = \frac{4\pi^2 r}{T^2} m_{\text{planet}}.$$

By Kepler's third law, we can replace T^2 by $K_{\text{sun}} r^3$ to get,

$$F_{\text{sun-planet}} = \frac{4\pi^2}{K_{\text{sun}} r^2} m_{\text{planet}}. \tag{7.6}$$

According to Newton's third law, the planet exerts a force on the sun with

$$F_{\text{sun-planet}} = F_{\text{planet-sun}} = \frac{4\pi^2}{K_{\text{planet}} r^2} m_{\text{sun}}, \tag{7.7}$$

where we have required that the structure of the gravitational force be the same for the planet-on-the-sun as for the sun-on-the-planet. On comparing Eqs. (7.6) and (7.7), we find that

$$m_{\text{sun}} K_{\text{sun}} = m_{\text{planet}} K_{\text{planet}}.$$

Now, the left-hand side of this equation depends on properties of the sun *only* and the right-hand side depends on the properties of the planet—*any* planet—*only*. Therefore, both sides must be equal to some constant that has the same value for the sun and each of the planets. We define this constant to be $4\pi^2/G$, so that

$$G = \frac{4\pi^2}{m_{\text{sun}} K_{\text{sun}}} = \frac{4\pi^2}{m_{\text{planet}} K_{\text{planet}}}.$$

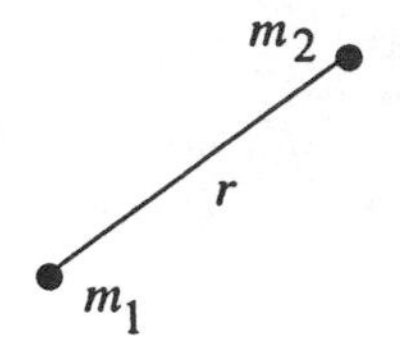

FIGURE 7.3. Gravitational interaction between two masses.

From this equation, we get $K_{\text{planet}} = 4\pi^2/Gm_{\text{planet}}$. On inserting this result into Eq. (7.7), we get

$$F_{\text{sun-planet}} = F_{\text{planet-sun}} = G\,\frac{m_{\text{planet}}m_{\text{sun}}}{r^2}, \tag{7.8}$$

for the gravitational force between the sun and any planet. Clearly, it is an inverse square law, as is the gravitational force we found earlier between the earth and any object as given by Eq. (7.5).

Newton extended this force law to apply to all matter in the universe. That is, any two masses m_1 and m_2 separated by a distance r, as in Fig. 7.3, exert a gravitational force on each other with the magnitude of the force given by

$$F_{\text{grav}} = \frac{Gm_1m_2}{r^2}, \tag{7.9}$$

where G is a universal constant with the same value for *any* pair of masses m_1 and m_2. This is called Newton's *law of universal gravitation*. It plays a key role in our study of the cosmos.

7.3 Measurement of the Force Constant

The gravitational force is extremely weak, the weakest of all of the four known fundamental forces in nature. Generally, it is of practical importance only when at least one of the bodies has a very large mass of the order of that of a planet, star, or other large astronomical body. For this reason, a determination of G in the laboratory requires a very sensitive apparatus. This measurement was first carried out by the British scientist, Henry Cavendish in 1797. Two small, spherical masses m (a few kilograms each) are attached to the ends of a light rod as illustrated in Fig. 7.4. This system is suspended by a fine wire so that the system is balanced with the rod horizontal. Two large, uniform spherical masses M (a few hundred kilograms each) are placed near the smaller masses as shown. The gravitational attraction between the larger and smaller masses will twist the wire. Since the elastic properties of the wire are known, the force of attraction between m and M is easily measured. With these data, the value for G is obtained from Eq. (7.9). The current accepted value is 6.67×10^{-11} N·m^2/kg^2, which is very close to the result Cavendish himself obtained.

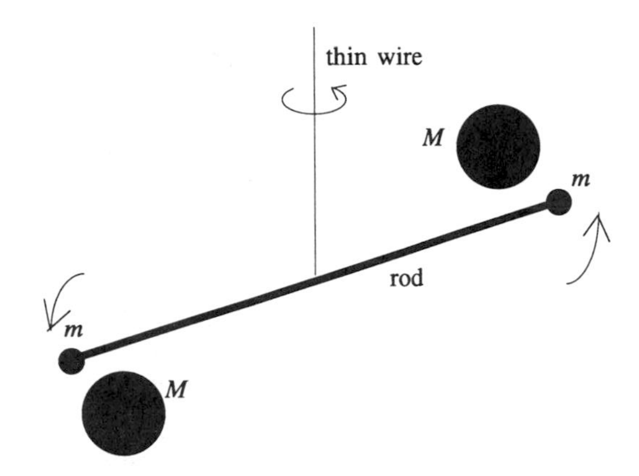

FIGURE 7.4. The Cavendish experiment for measuring G.

Once Cavendish had a laboratory value for G, he could use Eq. (7.9) to obtain an estimate of the mass of the earth. Writing Newton's second law for a freely falling body of mass M near the earth's surface, we have

$$\frac{GmM}{R^2} = mg, \tag{7.10}$$

where M and R are the mass and radius of the earth, respectively. The size of the earth, hence R, was known to Cavendish. Dividing both sides of Eq. (7.10) by m and substituting known values for the parameters, we get

$$\frac{(6.67 \times 10^{-11}\ \mathrm{N \cdot m^2/kg^2})\, M}{(6.38 \times 10^6\ \mathrm{m})^2} = 9.80\ \mathrm{m/s^2},$$

from which we obtain $M = 5.98 \times 10^{24}$ kg for the mass of the earth.

When we calculated gravitational potential energy in Chapter 5, we assumed that the force of gravity on the body was constant. As long as we stay near the earth's surface, that is a good approximation. However, we see from Eq. (7.9) that the gravitational force, hence the gravitational potential energy, depends on distance. The potential energy of a small mass m at a distance r from the center of the earth (mass M) is

$$PE_{\text{grav}} = -\frac{GmM}{r}, \tag{7.11}$$

where we have arbitrarily chosen the zero of potential energy to be at infinity. The minus sign is introduced so that the farther m is from the center of the earth, the greater will be its potential energy.

Table 7.1. Physical Data for the Planets[a]

planet	mass (kg)	radius (km)	mean density (kg/m^3)	surface gravity (m/s^2)
Mercury	3.30×10^{23}	2,440	5,430	3.7
Venus	4.87×10^{24}	6,050	5,250	8.9
Earth	5.98×10^{24}	6,370	5,520	9.8
Mars	6.42×10^{23}	3,400	3,930	3.7
Jupiter	1.90×10^{27}	71,490	1,330	23.1
Saturn	5.69×10^{26}	60,270	710	9.0
Uranus	8.68×10^{25}	25,560	1,240	7.8
Neptune	1.02×10^{26}	24,760	1,670	11.0
Pluto	1.36×10^{22}	1,120	2,000	0.1

[a] K. R. Lang, *Astrophysical Data*, Springer-Verlag, New York, 1991.

7.4 Escape Velocity

If we throw a piece of chalk vertically upward, it rises only so high and then falls back down. Throw it harder with greater velocity and it rises higher before falling back. How fast do we have to throw it so that it never falls back? The smallest velocity that will accomplish this is called the *escape velocity.*

Let us restate the problem in terms of energy. We want to give the chalk just enough kinetic energy so that it will go out to an infinite distance (where the earth's gravitational force on it is zero) and not have any kinetic energy left over. We just give it the minimum.

In applying Eq. (5.3) for energy conservation, E_{initial} consists of the kinetic and potential energy of the chalk when we release it at the earth's surface. We assume that after releasing it, there is no energy input or energy output. The chalk arrives at $r = \infty$ with zero kinetic energy and zero potential energy. Thus, Eq. (5.3) is

$$\frac{1}{2}mv_{\text{e}}^2 - \frac{GmM}{R} + 0 = 0 + 0,$$

which yields

$$v_{\text{e}} = \sqrt{\frac{2GM}{R}}. \tag{7.12}$$

This relation holds for any spherically symmetric, gravitating body of mass M and radius R. Notice that the escape velocity v_{e} depends on the mass and size of the large gravitating body and *not* on the mass of the chalk. Using the parameters given in Table 7.1 for the earth, we find that the escape velocity for the earth is about 11.2 km/s. Escape velocities for the other planets can also be obtained from the data in Table 7.1.

7.5 Planetary Atmospheres

The earth's atmosphere is a mixture of nitrogen (78%), oxygen (21%), and very small amounts of other stuff (argon, water vapor, carbon dioxide, etc.). These substances are in the form of a gas with the molecules in constant random motion flying around with a distribution of kinetic energies. At a given temperature, the average kinetic energy per molecule is the same for all substances.* This means that on the average, the nitrogen molecules, which are lighter than oxygen molecules, are moving a little bit faster than the oxygen molecules.

What holds the atmosphere around the earth? Why doesn't it float off into space? The answer is gravity. The oxygen and nitrogen molecules are flying around rapidly, but on the average, their speeds are well below the escape velocity of the earth, so they don't leave. Hydrogen and helium are the most abundant elements in the universe. Why is there no free hydrogen or helium in the earth's atmosphere? These are the lightest of all the elements. At the temperature of the earth's atmosphere, on the average these particles move at speeds that exceed the escape velocity for the earth. The earth's gravity is not strong enough to hold them in the atmosphere.

Planets closer to the sun receive more intense energy from the sun and are hotter. On these planets, the average kinetic energy per molecule is higher, making it more difficult for the planet to retain an atmosphere. Mercury, for example, is so hot and has so little surface gravity that it is unable to keep any atmosphere at all around it.

The moon is not as close to the sun as Mercury, but it has no atmosphere either. Even the elements that are in the earth's atmosphere have speeds that exceed the moon's escape velocity.

Exercises

7.1. How was the study of the motions of projectiles useful to Newton in obtaining the gravitational force law?

7.2. If the earth's present mass were compressed into a sphere with a diameter only one-half the earth's present diameter, what would be the gravitational acceleration at the surface of this body?

*A precise definition of temperature will be given in Chapter 17.

7.3. Estimate the acceleration due to gravity on the surface of Saturn and on the surface of Titan, Saturn's largest satellite. Calculate the acceleration of Titan in the gravitational field of Saturn.

object	mass (kg)	radius (km)	radius of orbit (km)
Saturn	5.69×10^{26}	60,000	1.427×10^9
Titan	1.37×10^{23}	2,900	1.220×10^6

7.4. In another universe,* the planet Boldan describes a circular orbit about its central star Shinar. The inhabitants of Boldan know that there are three other planets in their system. To these, they have given the names Alfak, Cordol, and Daphos. From their observations, they discover that these planets also have circular orbits around Shinar. The orbital data are given in the table below. A bright, young Boldanese physicist has found that Newton's three laws of motion hold in his universe. He also has discovered experimentally that near the surface of Boldan, all freely falling bodies have the same acceleration g. Follow the procedure in the text to obtain the gravitational force law that operates in Shinar's planetary system.

planet	radius of orbit (AU)	period (years)
Alfak	0.520	0.647
Boldan	1.000	1.000
Cordol	3.201	2.172
Daphos	7.300	3.763

The Boldanese have been able by geometrical means to determine that Boldan is a sphere of radius R. Using their gravitational force law, they have also obtained the value M for the mass of Boldan. Generalize the force law obtained above to give the gravitational force between two point masses m_1 and m_2 separated by a distance r. Express your result in terms of the given parameters, m_1, m_2, M, g, R, and r.

7.5. Suppose a planet is found to revolve around the sun in a circular orbit that lies between the orbits of Mercury and Venus at a distance of 8.15×10^7

*A similar exercise appears in B. M. Casper and R. J. Noer, *op.cit.*, p. 234.

km from the sun. The sun exerts exactly the same gravitational force on this planet that the sun exerts on the earth. What is the mass of this planet?

Answer: (1.76×10^{24} kg)

7.6. The centers of two identical 200-kg brass spheres are separated by 47.3 cm. Compare the gravitational force one of the spheres exerts on the other with the gravitational force the earth exerts on one of the spheres.

Answer: (The earth's force is about 165 million times greater.)

7.7. Suppose you repeat Cavendish's experiment (see Fig. 7.4) using lead balls with the smaller masses equal to 789 g and the larger ones 167 kg. These are comparable to the masses used by Cavendish himself. From your measurements, you find that when the large mass and the small mass are separated by 21.5 cm, the value you obtain for the force between the two is 1.92×10^{-7} N. From your data, obtain a value for the gravitational constant G.

7.8. An astronaut weighs 175 lb on the earth. What would be his weight if he stood on the surface of Mars? On Neptune?

7.9. Consider an object placed directly *between* the earth and the moon. At what distance from the center of the earth must the object be placed such that the combined gravitational force exerted on it by the earth and the moon will be zero?

7.10. Calculate the force that the sun exerts on the earth and the force that the sun exerts on Jupiter. Which is greater? How many times greater?

Answer: (3.53×10^{22} N; 4.14×10^{23} N; 11.7)

7.11. Three small masses are arranged along a straight horizontal line as illustrated in the drawing below. The horizontal gravitational force on the mass m is zero. Find the value of the mass M.

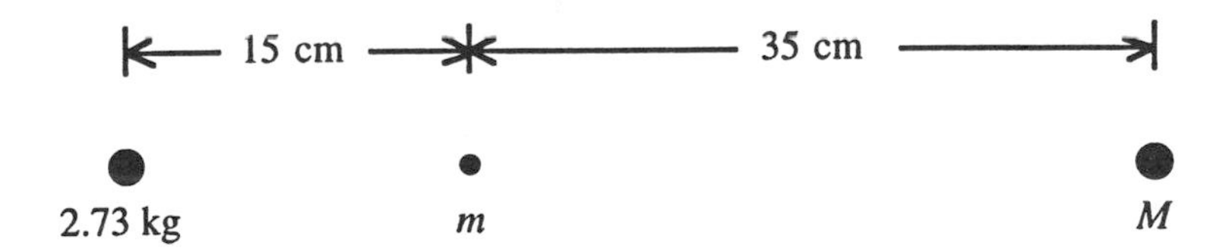

7.12. Uranus and Neptune describe nearly circular orbits around the sun with both orbits lying very close to the ecliptic plane. Obtain an estimate of the gravitational force that each of these planets exerts on the other when they are nearest to each other.

7.13. Estimate the gravitational force that Venus exerts on the earth when the two planets are closest to each other and when they are at their greatest separation.

Answer: (1.1×10^{18} N; 2.9×10^{16} N)

7.14. Compare the gravitational force that Jupiter exerts on the earth at closest approach with the gravitational force that Venus exerts on the earth at closest approach.

Answer: (Jupiter's force is about 1.7 times larger than the force Venus exerts.)

7.15. In a simple model of the hydrogen atom, an electron describes a circular orbit about the proton. The radius of the orbit is about 5.29×10^{-11} m. Calculate the gravitational force that the proton exerts on the electron in this model.

Answer: (3.63×10^{-47} N)

7.16. The average density of a certain planet is found to be 4,730 kg/m^3. The mass of the planet is 7.38×10^{24} kg. Calculate the escape velocity for this planet.

Answer: (11.7 km/s)

7.17. A spherically shaped asteroid with a radius of 187 km has an average density of 2,730 kg/m^3. Calculate the escape velocity at the surface of this body.

7.18. Calculate the escape velocity for the moon and for Mercury.

Answer: (2.4 km/s; 4.2 km/s)

7.19. The average density of a spherically symmetric planet is ρ. Obtain an expression for the escape velocity of this planet in terms of ρ. Does the escape velocity depend on any other physical parameters of this planet? If so, which ones? Explain.

7.20. With a diameter of 940 km, Ceres is the largest of the minor planets. The mass of Ceres is 1.03×10^{21} kg. Calculate the average density of this planet. Estimate the acceleration due to gravity on the surface of Ceres. Suppose that by bending your knees and springing vertically upward, your feet rise to a height of one foot above the ground. If you did this on Ceres, about how high above the surface of Ceres would your feet rise? Explain clearly your reasoning.

Answer: (2,370 kg/m^3; 0.311 m/s^2; 32 ft)

7.21. A 875-kg artificial satellite is in a circular orbit around the earth at an altitude of 545 km. Calculate the total energy acquired by the satellite in taking it from its position at rest on the launch pad at Cape Canaveral and placing it in this orbit.

Answer: (3.0×10^{10} J)

7.22. In Exercise 6.15, a graphical analysis of the orbital data for the planets yields the relationship

$$a_c = \frac{K}{T^{4/3}},$$

between the planet's centripetal acceleration a_c and the period T of its orbit with a_c expressed in AU/yr^2 and T in years. The constant parameter K is found from the graph to have the value $K = 39.8$ AU/yr$^{2/3}$. Show that these results are consistent with the centripetal acceleration a_c of the planets obtained from the gravitational force law

$$a_c = \frac{GM_{sun}}{r^2}.$$

Find the relationship between the parameter K and the mass of the sun M_{sun}. Use this value of K to obtain a value for the solar mass in kilograms.

7.23. Calculate the total energy (kinetic and potential) of the earth as it orbits the sun. (Assume a circular orbit and ignore the effects of the other planets and the earth's rotation.)

Answer: (-2.64×10^{33} J)

7.24. Orbital data for the four Jovian satellites discovered by Galileo in 1609 are given in the table. Assume that the radius r of the orbit and the period T are related by a power law,

$$r = CT^x,$$

where C and x are constants. Taking the logarithms of the two sides of this equation yields

$$\log r = x \log T + \log C.$$

Satellites of Jupiter

satellite	r (km)	T (days)
Io	4.22×10^5	1.77
Europa	6.71×10^5	3.55
Ganymede	10.70×10^5	7.16
Callisto	18.80×10^5	16.69

Plot $\log r$ versus $\log T$ and obtain from the graph the values of x and C. How does this relationship between r and T compare with Kepler's third law for the orbits of the planets around the sun?

7.25. One of the brighter stars in the sky is in the constellation Centaurus. The reason it appears so bright is due partly to the fact that it is at a distance of about 4.3 light years from the sun. (A light year is the distance light travels in free space in one year.) No other known star lies closer to the sun. Actually, this is a system of three stars gravitationally bound together. The brightest of the three components is very similar to the sun. Calculate the distance in meters from the sun to this star system. Assume that the brightest component of the system has the same mass as the sun and calculate the gravitational force that the sun exerts on it. Calculate the ratio of this force to the force that the sun exerts on the earth. Calculate the ratio of the gravitational force this star exerts on the earth to the gravitational force the sun exerts on the earth. How many times greater is the gravitational force that the sun exerts on the earth compared to the gravitational force the next nearest star exerts on the earth?

Answer: (4.07×10^{16} m; 1.59×10^{17} N; 4.52×10^{-6}; 7.36×10^{10} times)

7.26. How far from the surface of the sun would an object have to be for the gravitational force on it to be equal to its weight on the surface of the earth?

Answer: (2.98×10^{6} km)

7.27. We shall attempt to discover a power law* of the form

$$r_n = ab^n, \tag{1}$$

where r_n is the mean orbital radius of the nth planet (counting out from the sun) and a and b are constants to be determined.

n	planet	actual r_n (AU)	$\log r_n$	predicted r_n (AU)
1	Mercury	0.39		
2	Venus	0.72		
3	Earth	1.00		
4	Mars	1.52		
	Jupiter	5.20		
	Saturn	9.54		
	Uranus	19.19		
	Neptune	30.06		
	Pluto	39.53		
	X			
	Y			
	Z			

Calculate $\log r_n$ for the known planets given in the table. Beginning with only the first four planets (out to Mars) make a plot of n versus $\log r_n$.

*L. Basano and D. W. Hughes, *Il Nuovo Cimento*, **C2** (1979), p. 505.

With your ruler, draw a straight line through the data making the best fit you can make to these four points. Keeping in mind that n must be an integer, fit Jupiter and Saturn on this line as well as you can and record the corresponding values of n in the table. Next, make a *new* plot of n versus $\log r_n$ using the six *known* planets (out to Saturn) and the corresponding n-values you recorded in the table. Again, draw through the data the best straight line you can. Fit Uranus, Neptune, and Pluto to this line and record in the table the n-values you obtain for these three known outer planets. Finally, make another *new* plot of n versus $\log r_n$ using the data you now have for all nine known planets. Draw through the data the best straight line you can.

By taking the common logarithm of both sides and rearranging terms, Eq. (1) can be written as

$$n = \frac{1}{\log b} \log r_n - \frac{\log a}{\log b}, \tag{2}$$

which shows a linear relationship between n and $\log r_n$ as does your final graph. From the slope and the vertical intercept of your graph, determine a and b. Using these values of a and b, rewrite Eq. (1).

Your results should predict the existence of three *missing* planets (denoted by X, Y, and Z in the table). Relative to the orbits of the known planets, where are these missing planets? Use the equation you obtained to calculate r_n for all planets (known and missing ones). Enter your predicted values for r_n in the table.

Calculate the semimajor axis of the object described in Exercise 1.13. How does that object fit with the results you obtained in *this* exercise? Explain clearly. What is the implication of the discovery described in Exercise 1.8 with regard to your results? Explain clearly.

8

Round and Round She Goes

8.1 Orbital Motion

The orbital motions of two gravitating bodies can provide information about the mass of the two body system. Let's see how this works. In general, two point masses orbiting under the influence of their mutual gravitational attraction describe orbits about a common point called the *center of mass* of the two-body system.

8.1.1 Center of Mass

The center of mass is a point directly between the two masses on the straight line connecting them. Let m represent the smaller mass and M the larger one. The distance of M from the center of mass is denoted by R and m is located at a distance r from the center of mass as illustrated in Fig. 8.1 To complete the definition of the center of mass, we require that

$$MR = mr. \tag{8.1}$$

The notion of a center of mass can be generalized to a system consisting of any number of mass points. Extended bodies are *many-body* systems consisting of large numbers of particles (molecules, etc.) that can be regarded as point masses. Therefore, every extended body has its own center of mass. To a good approximation, most celestial objects—stars, planets, and so on—have enough symmetry that we can regard the center of mass of any of these objects to be located at the geometrical center of the body.

In our treatment of the orbital motions of gravitating bodies, we shall consider the motions of the centers of mass of the two interacting bodies orbiting about

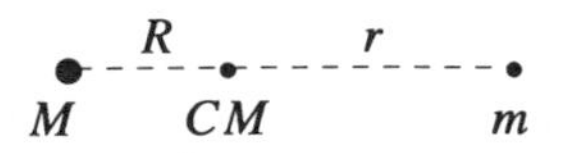

FIGURE 8.1. Center of mass of a two-particle system.

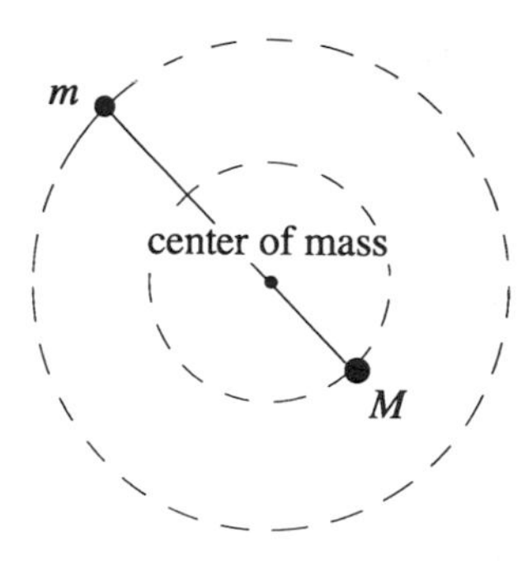

FIGURE 8.2. Orbital motion of two gravitating bodies.

their common center of mass. This is illustrated in Fig. 8.2 for two bodies of different masses in circular orbits. From Eq. (8.1), it is clear that the center of mass is closer to the heavier body (i.e., $R < r$). For very large mass differences, the center of mass of the system may be *inside* the more massive body. For example, the earth's mass is so large compared to the mass of the moon that the center of mass of the earth–moon system is inside the earth.

Strictly speaking, the earth and the moon orbit about a common center of mass, but the orbit of the earth is so small because of its large mass that it makes sense to say that the moon goes around the earth. Actually, they both go around their common center of mass.

Now let's consider in more detail the motion of the lighter mass m in Fig. 8.2. The only force on m is the gravitational attraction of M. Therefore, Newton's second law for the mass m can be written

$$\frac{GmM}{(r+R)^2} = m\frac{v^2}{r}. \tag{8.2}$$

In Eq. (7.2), the constant speed for a circular orbit is given as $v = 2\pi r/T$. With this substitution, Eq. (8.2) becomes

$$\frac{GM}{(r+R)^2} = \frac{4\pi^2 r}{T^2}. \tag{8.3}$$

From Eq. (8.1), we can write

$$R + r = \frac{mr}{M} + r = \frac{r}{M}(m+M),$$

from which we get the expression,

$$r = \frac{M(r+R)}{m+M}.$$

Using this expression for r on the right-hand side of Eq. (8.3) gives, after some rearrangement,

$$\frac{(r+R)^3}{T^2} = \frac{G(m+M)}{4\pi^2}. \tag{8.4}$$

The length $d = r + R$ is the distance between the centers of mass of the two gravitating bodies. Making this substitution in Eq. (8.4), we have

$$T^2 = \frac{4\pi^2 d^3}{G(m+M)}. \tag{8.5}$$

With d equal to the semimajor axis of the orbit for the relative motion, this relation also holds if the bodies have elliptical orbits about the center of mass. Note that Eq. (8.5) reduces to Kepler's third law if m is very small compared to M.

8.1.2 A Small Mass Orbiting a Much Larger Mass

If one of the masses is very much larger than the other, we can use Eq. (8.5) to get an estimate of the heavier mass from the orbital data of the lighter mass. For example, suppose M is the mass of the sun and m the mass of the earth. In Eq. (8.5), we neglect m compared to M. The distance $d = 150{,}000{,}000$ km is the distance from the sun to the earth and $T = 1$ year is the period of the earth's orbit. Converting these data to mks units and inserting into Eq. (8.5) we obtain $M = 2.00 \times 10^{30}$ kg for the mass of the sun. It will be convenient to use the mass of the sun as a unit for expressing the masses of other stars as well as galaxies and other large celestial objects. We denote the mass of the sun by the symbol $M_{\odot}$, which has the approximate value $M_{\odot} \approx 2 \times 10^{30}$ kg.

8.1.3 Orbits of Comparable Masses

From Eq. (8.5), we can obtain the sum of the masses of a two-body system if we know the separation of the two bodies and the period with which they orbit their center of mass. If one of the masses is negligible compared to the other, then the equation gives an estimate of the larger mass as in Section 8.1.2. If the masses are closer to the same size, it may be possible to extract from the orbital data the distance of each body from the center of mass. In this case, Eq. (8.1) gives the *ratio* of the masses. With the sum $m + M$ from Eq. (8.5) and the ratio m/M from Eq. (8.1), we can obtain the individual masses of the bodies.

For example, suppose we are able to see in a telescope that a nearby star system consists of two stars (called a *binary* system) gravitationally bound to each other and orbiting about a common center of mass. For simplicity, suppose also that the plane of the orbit happens to be oriented perpendicular to our line of sight as illustrated in Fig. 8.3. The system is shown here at two different times t_1 and t_2.

Observing this system over several years, we find that the orbit is circular with a period of about fifty years. The system is near enough that we can measure its

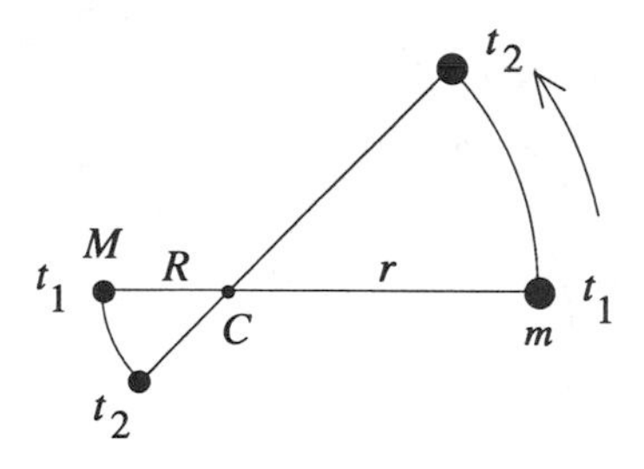

FIGURE 8.3. A binary star system.

distance from the sun geometrically* and the angular separation of the two stars directly. From these data, we are able to calculate the distance between the stars. We find this distance to be about 3.43×10^9 km. With these values $T = 50$ years and $d = 3.43 \times 10^9$ km inserted into Eq. (8.5), we obtain for the mass of system $m + M = 9.6 \times 10^{30}\text{kg} \approx 4.8\text{M}_\odot$.

From Fig. 8.3, we see that the data also show that $r = 3R$. Thus, according to Eq. (8.1), $M = 3m$, and

$$m + M = 4m = 4.8\text{M}_\odot,$$

from which we get $m = 1.2\text{M}_\odot$ and $M = 3.6\text{M}_\odot$ for the masses of the stars in this binary system. In practice, binary orbits are elliptical, the orbital planes are not necessarily oriented perpendicular to our line of sight, and we cannot always resolve both components with our telescope. However, even in these cases by careful measurements of the motions and the spectra of the stars in the system, it is possible to extract enough information to carry out these calculations to determine the masses of the binary components.

8.2 Artificial Satellites in Earth Orbit

The space age began on October 4, 1957, when the USSR launched Sputnik, the first artificial earth satellite. The orbit was nearly circular and required ninety six minutes for the satellite to complete it. Sputnik orbited the earth at an altitude of about 575 km. In this case, the mass of Sputnik is negligible compared to the mass of the earth and it is easily verified that these data are consistent with Eq. (8.5) using the values for the mass of the earth and the radius of the earth given in Appendix B.

---

*This method of determining stellar distances will be discussed in more detail in Chapter 15.

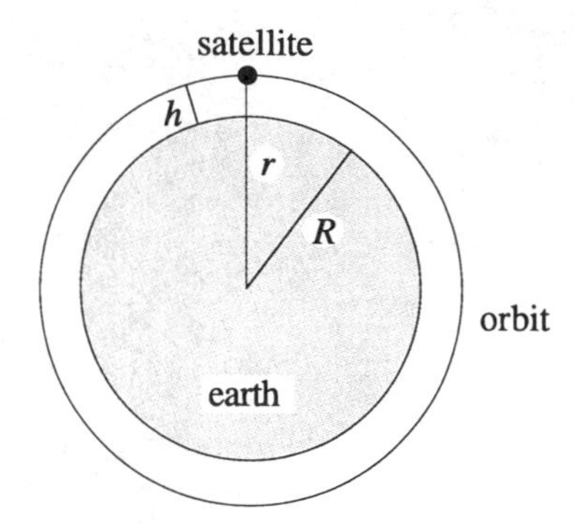

FIGURE 8.4. Spy satellite in earth orbit.

EXAMPLE. A spy satellite is placed in a circular orbit around the earth at an altitude of 389 km. How long does it take this satellite to make a complete trip around the earth? The orbit is illustrated in Fig. 8.4.

SOLUTION. The only force exerted on the satellite is the earth's gravitation which gives rise to centripetal acceleration of the satellite. For Newton's second law, we have

$$\frac{GmM}{r^2} = m\frac{v^2}{r}, \tag{8.6}$$

where m is the mass of the satellite and M is the mass of the earth. The radius of the orbit is $r = h + R$, where h is the altitude and R is the radius of the earth. With the orbital speed of the satellite given by $v = 2\pi r/T$, we obtain from Eq. (8.6) an expression for the period of the orbit T in terms of the radius of the orbit r,

$$T = \sqrt{\frac{4\pi^2 r^3}{GM}}. \tag{8.7}$$

For the parameters involved here, we have the values, $M = 5.98 \times 10^{24}$ kg, $R = 6{,}370$ km, $h = 389$ km, and $G = 6.67 \times 10^{-11}$ N·m²/kg². With these data, Eq. (8.7) yields the value $T = 5{,}528$ seconds $= 92.1$ min for the period of the satellite's orbit.

Exercises

8.1. A spy satellite is placed in a circular orbit around the earth. The satellite makes fourteen complete orbits in each twenty-four hour period. Calculate the altitude of this satellite.

Answer: (904 km)

8.2. In May 1973, Skylab was launched into a nearly circular orbit 430 km above the surface of the earth. Calculate the number of orbits Skylab made around the earth each day.

Answer: (15.5)

8.3. A spacecraft in a circular orbit around the moon makes one complete orbit in exactly 2 h 10 min. How high is the spacecraft above the moon's surface?

Answer: (224 km)

8.4. Use orbital data to determine which is larger, the acceleration of the earth as it orbits the sun or the acceleration of the moon as it orbits the earth. How many times larger?

8.5. Space engineers wish to place a satellite in a circular orbit around the planet Mercury. They want the satellite to circle Mercury at an altitude of 150 km. Calculate the period of the orbit.

Answer: (93 min)

8.6. A cannon on the moon's surface fires a cannonball horizontally. Calculate the speed with which the ball must leave the muzzle of the cannon in order to make a circular orbit about the moon.

Answer: (1.68 km/s)

8.7. A communications satellite remains above the same point on the earth's surface at all times. This means that it must make exactly one complete orbit about the earth each time the earth makes a complete rotation on its axis. Calculate the approximate height of this satellite above the earth's surface.

Answer: (35,880 km)

8.8. In 1609, with his newly constructed telescope Galileo discovered four satellites orbiting the planet Jupiter. These objects were later given the classical names, Io, Europa, Ganymede, and Callisto. Ganymede, is one of the largest satellites in the solar system—larger, in fact, than the planet Mercury. It moves in a nearly circular orbit around Jupiter once in about 7.16 days at an average distance of 1.07×10^6 km from the center of the planet. From these data, estimate the mass of Jupiter.

8.9. The minor planet Ceres orbits the sun with a period of 4.61 yr. Assume that the orbit is circular and use Newton's gravitational force law to estimate how far Ceres is from the sun.

Answer: (4.14×10^8 km)

8.10. The densities of brass and steel are 8,600 kg/m^3 and 7,380 kg/m^3, respectively. Two uniform metal spheres have equal diameters. One sphere is solid brass and the other is solid steel. Their centers of mass are separated by a distance of 67.3 cm. How far from the center of the brass sphere is the center of mass of this two-body system?

Answer: (31.1 cm)

8.11. The two stars in a binary system are found to describe circular orbits with a period of 26.7 yr. The radius of one orbit is seen to be 1.68 times as large as the other. The stars are 2.45×10^9 km apart. Calculate the masses of the components of this system.

Answer: (2.30 $M_\odot$; 3.86 $M_\odot$)

8.12. The two stars, A and B, in a binary system are 8.38 AU apart as illustrated in the scale drawing below. Their orbits are circular with a period of 9.7 yr. Calculate the masses of these stars.

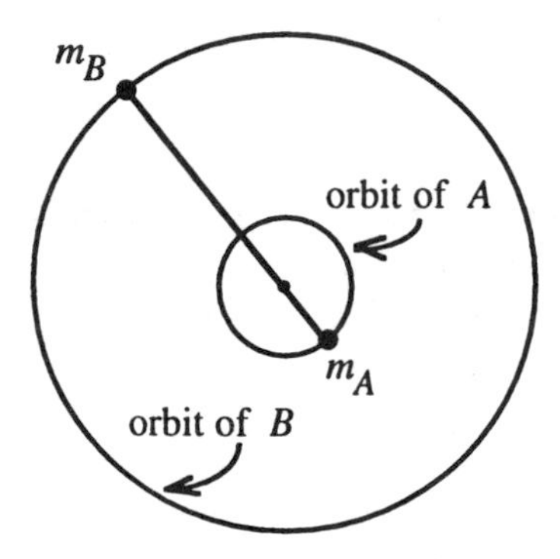

8.13. A certain nearby star is observed for several decades. Its path in the sky shows a slight wobble, suggesting that it has a dimmer companion star. From the observations, it is determined that the orbits are circular and that the radius of the orbit of one of these stars is 0.82 AU and that of the other is 2.68 AU. The orbital period of this binary system is 1.97 yr. Calculate the masses of these stars.

8.14. Two stars gravitationally bound to each other travel in circular orbits about the center of mass of this binary system with a period of 3.16 yr. The orbital radius of one of these stars is 1.05 AU and that of the other is 2.84 AU. Calculate the mass of each of these stars.

Answer: (1.60 $M_\odot$; 4.34 $M_\odot$)

8.15. Two stars separated by a distance of 7.29×10^8 km revolve in circular orbits about a common center of mass. The center of mass of the two-body system is located 2.78×10^8 km from the center of the heavier star, which has a mass of 4.18×10^{30} kg. How long does it take for one of the stars to make a complete orbit around the center of mass of the system?

Answer: (5.84 yr)

8.16. The five large satellites of Uranus were discovered prior to 1950 using earth-based telescopes. The presence of ten smaller satellites orbiting around Uranus was revealed in 1986 when the spacecraft Voyager II passed through the vicinity of the planet. Orbital data for each of these are given in the table. Use the data to plot b^3 versus T^2. From the slope of your graph, estimate the mass of Uranus.

Answer: (8.73×10^{25} kg)

Orbital Data for Satellites of Uranus[a]

satellite	semimajor axis b (km)	period T (hours)
Cordelia	49,700	8.0
Ophelia	53,800	9.0
Bianca	59,200	10.4
Cressida	61,800	11.1
Desdemona	62,700	11.4
Juliet	64,600	11,8
Portia	66,100	12.3
Rosalind	69,900	13.4
Belinda	75,300	14.9
Puck	86,000	18.3

[a] K. R. Lang, *Astrophysical Data*, Springer-Verlag, New York, 1991.

8.17. If all of the planets in the solar system were replaced by a single star with the same mass as the sun and this star were placed at the same distance from the sun as the earth is now, what would be the period of the orbit as these two stars revolve about their common center of mass?

Answer: (259 days)

8.18. How far from the center of the earth is the center of mass of the earth–moon system?

Answer: (4,700 km)

8.19. Two planets are gravitationally bound to each other. In addition to their orbital motions about their central star, they describe circular orbits around a common center of mass. The planets are assumed to have the same composition and the same average density. However, the diameter of one planet is twice as large as that of the other. The distance between the centers of mass of the two planets is 609,000 km and the period of their orbits about their center of mass is 47.3 days. Calculate the mass of each planet.

Answer: (8.89×10^{23} kg and 7.12×10^{24} kg)

8.20. A planet has a single natural satellite. Both have the same average density of 4,730 kg/m^3. The center of mass of this two-body system lies at the surface of the planet, which has a radius of 8,320 km. The distance between the centers of mass of the planet and its satellite is 427,000 km. Calculate the time it takes for the planet and its satellite each to make a complete orbit about the center of mass of the system.

Answer: (23.0 days)

8.21. The ancients discovered by geometrical means that the sun is much farther away from the earth than the moon. Similar geometrical constructions based on modern observations of the moon and the sun reveal that the sun is approximately 390 times as far from the earth as the moon. The orbital period of the earth has long been known to be about 365.25 days and that of the moon about 27.3 days. Using only these data and Eq. (8.5), obtain a value for the mass of the earth in units of the solar mass $M_\odot$. Assume $M_{\text{earth}} \ll M_{\text{sun}}$ and $M_{\text{moon}} \ll M_{\text{earth}}$.

8.22. Sirius is the star that appears brightest in our night sky. Around the middle of the last century, it was discovered that Sirius has a tiny companion, which is massive enough that the orbital motion of the primary (brighter) component can be detected. From observing Sirius over several decades, it has been determined that the orbital period of the system is about 50 yr and the semimajor axis of the relative orbit is about 3.0×10^9 km. The observations also reveal that the companion is always seen to be about 2.3 times farther from the center of mass than the primary star. From these data, estimate the mass of the bright star Sirius and the mass of its small, dim companion.

8.23. Algol is a moderately bright star in the constellation of Perseus. It varies in brightness periodically because it is a binary system with the plane of the orbit oriented so that as seen from the earth, one star passes in front of the other, briefly blocking some of its light. From this fluctuation in brightness, the period of the orbit is determined to be about 2.87 days. The semimajor axis of the relative orbit of the system is about 9,800,000 km. An analysis of the system's light spectrum reveals that the mass ratio for the two components is about 5/23. Use these data to calculate the mass of each of these stars. Express your answer in terms of $M_\odot$.

9

As the World Turns

9.1 Mechanical Equilibrium

If the motion of a body is not changing with time, we say the body is in mechanical *equilibrium*. By this definition, the glider moving with constant velocity along the horizontal air track in Fig. 2.1 is in equilibrium. Equilibrium is not concerned with whether the body is in motion, but whether the motion is changing. How can we change the state of motion of a body? What are the conditions for equilibrium? Let's see.

Consider a rectangular block of wood lying on a smooth, horizontal table. We apply two forces to the block as shown in Fig. 9.1*a*. If the magnitudes of the forces are equal ($F_1 = F_2$), they cancel and the block remains at rest. The block is in equilibrium. The sum of the forces on the body is zero and there is no acceleration. This is simply an example of Newton's first law of motion. Clearly, for a body to be in equilibrium, the sum of the forces exerted on it must be zero. In symbolic language, if there are N forces exerted on a body, equilibrium requires that

$$\sum_{i=1}^{N} \mathbf{F}_i = 0. \tag{9.1}$$

This is *one* condition for equilibrium that must be satisfied—a *necessary* condition. Is it the *only* condition that must be met? Is it a *sufficient* condition? Let's consider the block under slightly different conditions.

Again, we apply the two forces to the block with $F_1 = F_2$ as before. But this time, we apply the forces as shown in Fig. 9.1*b*. The two forces are still equal in magnitude and opposite in direction. They cancel and the condition in Eq. (9.1) holds. However, the motion of the block is *not* constant. The block starts to rotate. Its motion changes. It is *not* in equilibrium. Clearly, Eq. (9.1) by itself is not enough to guarantee equilibrium. It is a necessary condition, but not a sufficient condition for equilibrium.

If Eq. (9.1) holds, we can say that the body is in *translational* equilibrium, meaning that the translational motion is constant. We need another statement analogous to Eq. (9.1) that will hold when the body's rotational motion is constant, that is,

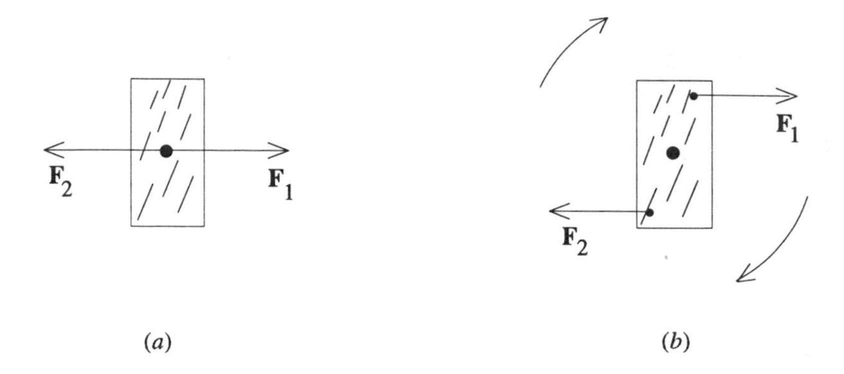

FIGURE 9.1. Equal and opposite forces applied to a rectangular block.

when the body is in *rotational* equilibrium. We seek a relation of the form

$$\sum_{i=1}^{N} (?)_i = 0, \tag{9.2}$$

where $(?)_i$ is some quantity yet to be defined that plays a role in rotational motion analogous to that the force plays in changing translational motion.

To get a better handle on the nature of the quantity (?), let us consider a very simple experimental arrangement. A uniform meter stick is suspended by means of a string attached to the stick at its midpoint. The stick hangs in the horizontal position. Light strings are used to suspend two weights of 40 g and 80 g from the stick so that the stick remains in equilibrium in the horizontal position. If we change the position of one of the weights, the equilibrium will be upset and the stick will rotate about the point of suspension (the axis in Fig. 9.2). Clearly, the equilibrium of the stick depends not only on how large the forces are, but also on where the forces are applied. With the stick in equilibrium, Eq. (9.2) must hold. We have identified three forces here that are exerted on the stick. These are listed in Table 9.1 along with the position on the meter stick at which each is applied.

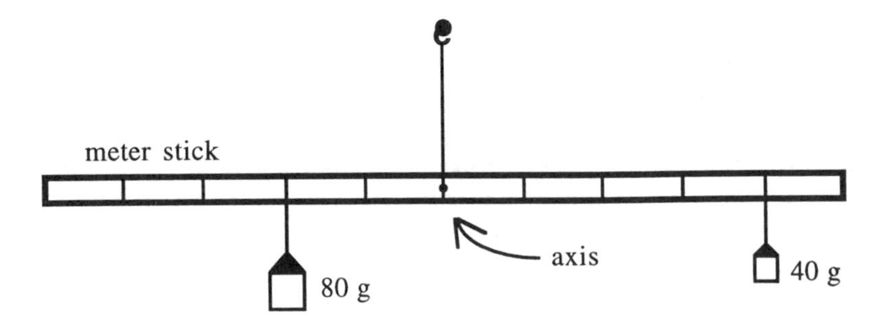

FIGURE 9.2. Meter stick.

Table 9.1. Forces on a Meter Stick

"force" (g)	position of force (cm)	distance from axis (cm)
support	50	0
80	30	20
40	90	40

Let's look for combinations of these quantities that we can use to define (?) so that Eq. (9.2) is satisfied. In Table 9.1, the positions themselves do not immediately suggest anything useful. However, if we look at the *distance* of each force from the point of suspension, we can construct the sum

$$[(0.080\,\text{kg})(9.8\,\text{m/s}^2)\cdot(0.20\,\text{m})] + [\text{force of support}\cdot 0] - [(0.040\,\text{kg})(9.8\,\text{m/s}^2)\cdot(0.40\,\text{m})] = 0. \tag{9.3}$$

Each term in rectangular brackets is a product of a force exerted on the stick and the corresponding distance of that force from the axis. Taking this product for $(?)_i$, Eqs. (9.2) and (9.3) are consistent.

9.1.1 Torque

The quantity represented by (?) is called the *torque*, denoted by **T**. The condition for rotational equilibrium, Eq. (9.2), can now be written

$$\sum_{i=1}^{N} \mathbf{T}_i = 0. \tag{9.4}$$

More generally, the torque due to a force applied to an extended body is given by

$$\text{torque} \equiv \text{force} \times \text{lever arm},$$

where the *lever arm* is *the perpendicular distance from the axis of rotation to the line along which the force acts*. The meaning of these terms can be seen from Fig. 9.3. Notice that the value of the torque for a given force depends on the choice of axis of rotation, which is completely arbitrary. However, if a body is in rotational equilibrium and the same axis is used to calculate all torques exerted on the body, Eq. (9.4) holds *regardless* of what point we choose for the axis of rotation.

The torque is a vector quantity, but a more complicated kind of vector than those we have encountered so far. In applying the vector relation, Eq. (9.4), it will be sufficient to calculate the magnitude of each torque exerted on an extended body according to the definition given above and assign a + or − sign to it according to the following convention,

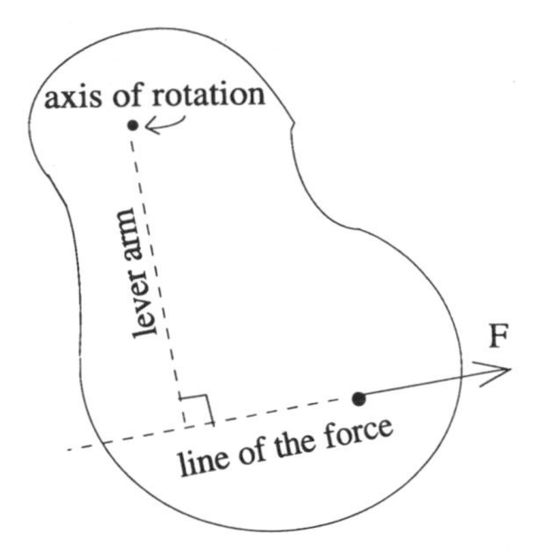

FIGURE 9.3. Extended body of arbitrary shape.

- a torque tending to produce counterclockwise rotation about the axis is +
- a torque tending to produce clockwise rotation about the axis is −

There is one thing we have left out of our discussion of the equilibrium of the meter stick. To see this, let's consider the balanced meter stick without the weights hanging on it. This is illustrated in Fig. 9.4. The stick hangs horizontally without rotating. The meter stick is in equilibrium and Eq. (9.1) holds. However, Fig. 9.4 shows only *one* force F_{support} exerted on the stick. What is the other force needed to balance this out so that Eq. (9.1) is satisfied? We have left out the force of the earth's gravity exerted on the stick—the weight of the stick. The support must supply a force equal to the weight of the stick for equilibrium to obtain. Where do we apply this force? In practice, the earth exerts a gravitational force on every part of the stick—on every molecule, billions of them all along the length of the stick. We don't want to have to draw billions of vectors on our diagram. We can represent the gravitational force on the stick by a single vector pointing downward. But where do we place it on the stick? If we apply a downward force to the left of the support point, the stick will rotate counterclockwise about the axis. If we apply the force to the right of the support, the stick will rotate clockwise about the

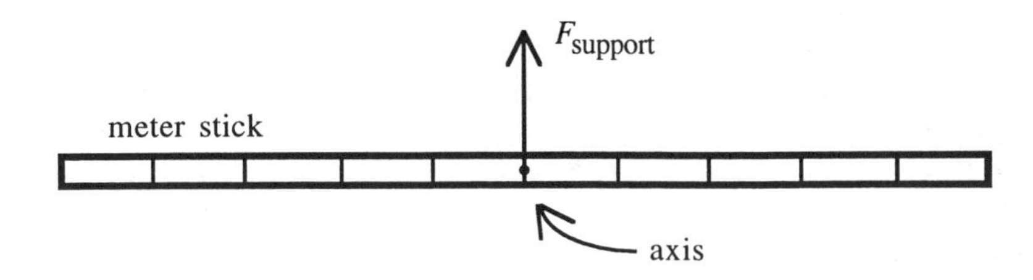

FIGURE 9.4. Unloaded meter stick.

axis. In neither case is the stick in rotational equilibrium. The only single place along the stick where we can consider the gravitational force to be acting on it is at the point of support. This is the *center of gravity* of the stick. In general, for any extended body the center of gravity of the body is *the single point at which the force of gravity may be considered to be acting on the body.*

EXAMPLE. A meter stick is supported by a thin wire as shown in Fig. 9.5. The center of gravity of the stick is located at the 50.7-cm mark. Light strings are used to suspend a 100-gram mass from the stick at the 73.4-cm mark and a 200-gram mass at the 22.1-cm mark. The mass of the stick is 103.7 g. Find the position of the support force F_{support} such that the stick hangs horizontally in equilibrium.

SOLUTION. Applying Eq. (9.1) to Fig. 9.5, we have

$$F_{\text{support}} - (0.2\,\text{kg})(9.8\,\text{m/s}^2) - (0.1037\,\text{kg})(9.8\,\text{m/s}^2) - (0.1\,\text{kg})(9.8\,\text{m/s}^2) = 0,$$

from which we find that $F_{\text{support}} = 3.956$ N. Starting at the axis, we calculate the torque for each of the forces on the stick and construct Eq. (9.4) using the sign convention above,

$$\begin{aligned} &-(0.200\,\text{kg})(9.8\,\text{m/s}^2)(0.221\,\text{m}) + (3.956\,\text{N})x \\ &\quad -(0.1037\,\text{kg})(9.8\,\text{m/s}^2)(0.507\,\text{m}) - (0.100\,\text{kg})(9.8\,\text{m/s}^2)(0.734\,\text{m}) = 0. \end{aligned}$$

On solving this equation for x, we find that the supporting wire must be at the 42.15-cm mark.

9.1.2 ANGULAR MOMENTUM

In general, the concepts associated with rotational motion are more difficult to grasp than those associated with translational motion. We can achieve a level of understanding of rotational motion that is sufficient for our purpose by considering the linear analogues. These are displayed in Table 9.2.

The first line in Table 9.2 states the conditions for translational and rotational equilibrium, respectively. On the second line, these two conditions are restated in a symbolic way. They are statements of Newton's first law of motion and its

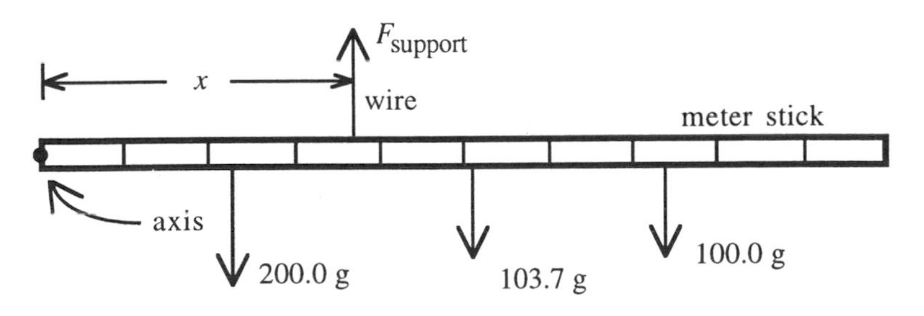

FIGURE 9.5. Loaded meter stick.

Table 9.2. Translation–Rotation Analogues

translation	rotation
sum of forces = 0	sum of torques = 0
$\sum_i \mathbf{F}_i = 0$	$\sum_i \mathbf{T}_i = 0$
$\sum_i \mathbf{F}_i = \frac{\Delta \mathbf{p}}{\Delta t}$	$\sum_i \mathbf{T}_i = \frac{\Delta \mathbf{L}}{\Delta t}$
$\mathbf{p} = m\mathbf{v}$	$\mathbf{L} = I\omega$

rotational analogue. On the third line, we have Newton's second law of motion and its rotational analogue, where we have introduced the *angular* momentum $\mathbf{L}$ in analogy with the *linear* momentum $\mathbf{p}$. The fourth line shows that the structure of the angular momentum is analogous to the linear momentum in that it depends on the (rotational) motion—angular velocity ω—and the (rotational) inertia—moment of inertia I—of the body.

The moment of inertia I is an inertial quantity that represents the body's ability to resist changes in its *rotational* motion. It plays a role in angular motion similar to that played by the mass m in linear motion. The moment of inertia depends not only on the mass of the body, but also on how that mass is distributed. For example, suppose we have a spoked wheel and a solid disk, which have equal outside radii R and equal masses M. These are illustrated in Fig. 9.6. For the wheel, most of the mass is out near the rim, whereas the mass of the disk is uniformly distributed over the body. In a given time interval, to get these bodies rotating with the same angular velocity about axes through their centers, we find that we have to apply a greater force tangent to the rim of the wheel than to the rim of the disk. The wheel has more resistance to change in its rotational motion than does the disk. That is, $I_{\text{wheel}} > I_{\text{disk}}$.

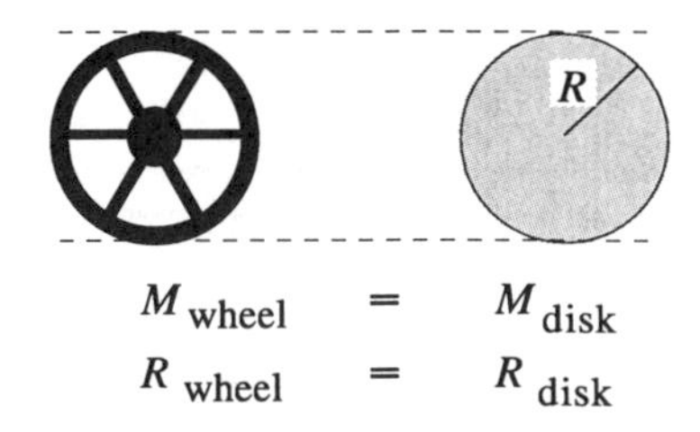

FIGURE 9.6. Different rotational intertia for bodies with equal masses.

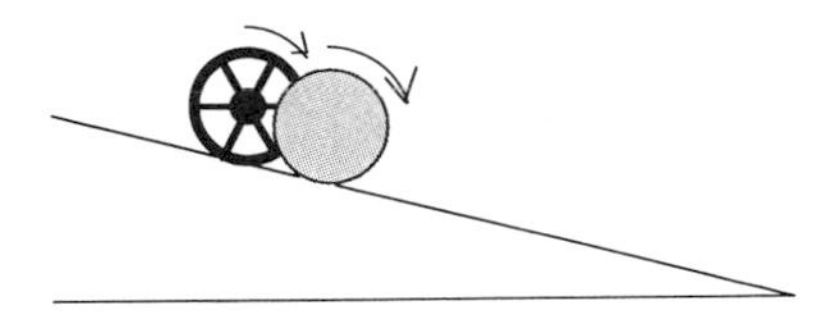

FIGURE 9.7. Demonstration of different rotational inertia.

This can also be seen by allowing the bodies to roll down an incline as in Fig. 9.7. If both are released at the same time from the top of the incline and allowed to roll without slipping, the disk will reach the bottom first.

We see from the third line in Table 9.2 that if the sum of the torques on a rotating body is zero, the angular momentum **L** is a constant. In this case, we say that the angular momentum is *conserved.*

To illustrate angular momentum conservation, consider the pirouette of a ballerina. She holds her arms extended out from the sides of her body as she whirls on her toes. On drawing her arms in toward her body, she spins faster—a manifestation of the conservation of angular momentum. In this case, the external torques are negligible and angular momentum is conserved,

$$L = I\omega = \text{constant.}$$

Part of the ballerina's mass is in her arms. As she draws them toward her body as she spins, she is redistributing her mass, pulling some closer in to the axis of rotation—making her more like the disk and less like the wheel in the example above. This reduces her moment of inertia I. To maintain constant L, ω must increase—she spins faster.

The moment of inertia of a single particle of mass m is

$$I = mr^2,$$

where r is the distance from the reference point O to the particle. The orbital angular momentum (relative to O) for the particle is then

$$L = mr^2\omega. \tag{9.5}$$

Let $v_\perp$ denote the component of the particle's velocity **v** that is perpendicular to the position vector **r** as illustrated in Fig. 9.8. Clearly, $v_\perp = r\omega$. Substituting this result into Eq. (9.5), we have

$$L = mrv_\perp. \tag{9.6}$$

For a circular orbit about the point O, $v_\perp = v$ and

$$L = mrv. \qquad \text{(circular orbit)} \tag{9.7}$$

In terms of the orbital period T, Eq. (9.7) is written

$$L = \frac{2\pi mr^2}{T}. \qquad \text{(circular orbit)}$$

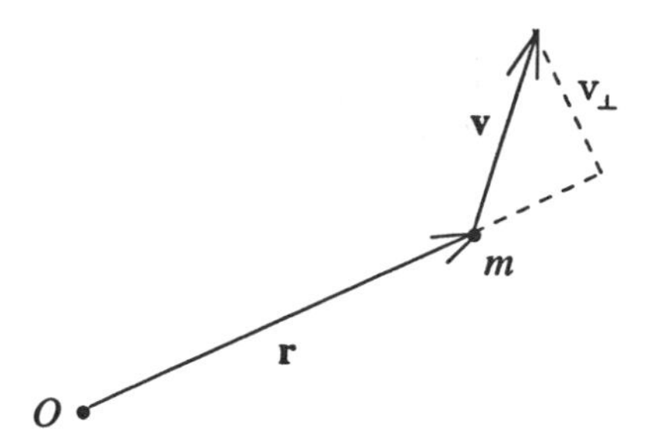

FIGURE 9.8. Definition of $v_\perp$.

9.2 The Origin of the Solar System

We are now ready to apply these ideas toward an understanding of the origin and evolution of the solar system. According to current ideas, stellar evolution begins with a huge cloud of gas—mostly hydrogen—with some dust mixed in it. Local condensations of matter occur in the cloud making the density a little higher at these places. Because the density is higher—more matter there—the gravity is stronger at any one of these places than in its immediate neighborhood. This tends to draw in more matter from the surrounding region, which further increases the density and the gravity which pulls in more matter and so on. The matter in the region of one of these condensations begins to collapse under gravity forming a protostar and eventually a star. We shall examine this process in more detail later. There is some swirling motion of the collapsing matter, therefore the protostar has an angular momentum associated with it. The external torque on a protostar is negligible, which means that the angular momentum is conserved. As the matter falls in closer to the center of the protostar, the moment of inertia of the system decreases causing the system to rotate faster. (Remember the ballerina.)

In 1755, Immanuel Kant envisioned the formation of the solar system in this way. He was able to show from Newton's theory of gravitation that a rotating cloud of gas would flatten into a disk. Later in the eighteenth century, Pierre LaPlace developed this into a more comprehensive theory that came to be called the *nebular hypothesis* (*nebula* comes from a Latin word meaning cloud) that included the formation of planets from the solar nebula. This hypothesis was criticized on several grounds. One of these was that the sun's angular momentum is too small. The criticism was so severe and so compelling that the nebular hypothesis fell out of favor among astronomers and cosmologists until relatively recently.

What specifically is the criticism regarding angular momentum? The sun accounts for most of the mass of the solar system. An estimated distribution of solar system mass among the constituents of the solar system is given in Table 9.3. Because the sun has most of the mass it should also have most of the angular momentum. However, from Table 9.3 we see that most of the angular momentum

Table 9.3. Mass and Angular Momentum in the Solar System

object	mass (%)	angular momentum (%)
sun	99.87	0.6
outer planets	0.13	99.0
terrestrial planets	< 0.01	0.2
asteroids	< 0.01	0.1
comets	< 0.01	0.1

of the solar system is carried by the large planets—due largely to the fact that they are far from the center of rotation of the system. The fact that the sun has over ninety-nine percent of the mass and less than one percent of the angular momentum of the solar system represents a serious discrepancy.

What happened to the sun's angular momentum? For a long time, there was no acceptable answer to this question. This changed earlier in this century with the discovery of the solar wind. The sun is very hot so the gas particles that make up the atmosphere of the sun are moving rapidly, but they are bound by gravity to the sun. Farther out from the center of the sun—at distances of several solar radii—the gravity is weaker but the particles are still hot enough to be ionized (electrically charged) and to have velocities that exceed the escape velocity at that distance from the sun. As a result, these particles escape at high speed from the gravitational field of the sun thereby producing a high velocity wind of charged particles streaming away from the sun—the solar wind. The sun has a magnetic field that extends into the space around it. As the sun rotates, it carries its magnetic field with it. The magnetic field sweeps through the solar wind exerting a force on the charged particles that deflects them from their radial path away from the sun thereby dragging them around toward the direction of the sun's rotation and imparting an angular momentum to them.* By Newton's third law, through the magnetic field the solar wind particles exert an equal and opposite force on the sun which tends to slow its rotation and decrease its angular momentum. By this mechanism, angular momentum is transferred from the sun itself to the particles streaming toward the remote reaches of the solar system where it is not directly observable. It is believed that in the early history of the solar system, the solar wind was many times more intense than it is today. At that time, this mechanism would have been even more effective in removing angular momentum from the sun.

*This sidewise magnetic force on a moving electric charge is discussed more fully in Chapter 10.

Exercises

9.1. A straight, uniform rod of length 83 cm has a mass of 486 grams. The rod is supported in the horizontal position by a string attached to one end and a finger at the other. When the supporting finger is removed, what is the torque on the rod relative to an axis through the end of the rod with the string support?

Answer: (1.98 N·m)

9.2. The handle of a mechanic's wrench is 30 cm long. To tighten a bolt with this wrench, the mechanic applies a force of 100 N. What is the maximum torque the mechanic can apply to the bolt using only this wrench and this force? Describe clearly where and how the force must be applied to achieve the maximum torque.

9.3. A wooden pin is glued into a 1.83-m wooden rod at a point 27 cm from one end of the rod. The pin is fastened to a vertical wooden post so that the rod is held fixed at an angle with respect to the horizontal as shown in the scale drawing. A 250-g ball attached to a string hangs vertically from the upper end of the rod. Calculate the torque exerted on the rod by the string about an axis through the pin.

Answer: (3.61 N·m)

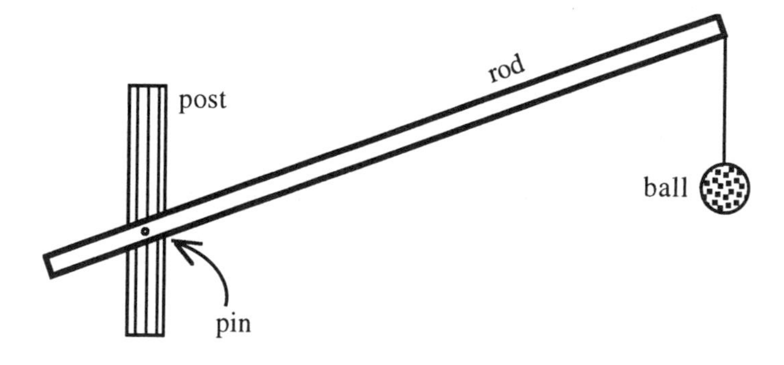

9.4. A tapered rod has a mass of 780 grams and a length of 2.85 meters. When suspended at its midpoint, the rod hangs horizontally with a 260-gram weight hanging from its narrow end. Find the position of the center of gravity of the rod.

Answer: (95 cm from the thicker end)

9.5. A nonuniform, rigid rod of mass 566 g is made to hang horizontally by means of a string attached to a point 33.6 cm from the left-hand end of the rod. A 443-g weight hangs from a point on the rod 57.6 cm from this same end. How far from this end of the rod is the center of gravity of the rod located?

Answer: (14.8 cm)

9.6. A 52-lb uniform walkboard 12 ft long is supported at its ends by two vertical walls. Find the force exerted on each wall when a 160-lb carpenter stands on the walkboard 3.5 ft from one end.

Answer: (72.7 lb; 139.3 lb)

9.7. A 55-kg window washer is 1.45 m from one end of a uniform, 35-kg platform that is 7 m long. Her water bucket weighing 220 N is 2.75 m from the same end. The platform is supported by ropes at the two ends. Calculate the tension in each rope.

Answer: (370 N; 732 N)

9.8. After some holes are drilled in a meter stick, the stick has a mass of 107 g with its center of gravity at the 527-mm mark. By means of a light string, a 185-g weight is attached to the stick at the 831-mm mark. At what single point on the stick must the system be supported so that the stick is horizontal and in equilibrium?

Answer: (720-mm mark)

9.9. A 25.4-kg, uniform beam 3.97 m long is fastened to a vertical wall at one end by a pin. The beam is held horizontal by means of a cable attached to the other end of the beam as shown in the scale drawing. Use the scale drawing to find the tension in the cable.

Answer: (280 N)

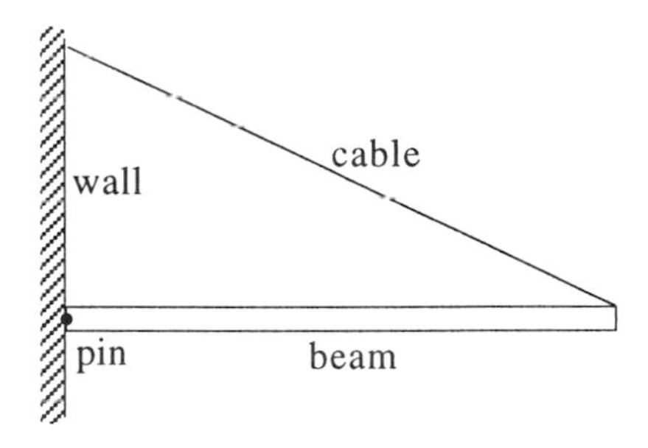

9.10. A 44.0-kg traffic light is suspended from one end of a rigid, uniform metal strut of mass 6.4 kg. The other end of the strut is fastened to a vertical pole. The system is held in equilibrium by means of a cable as illustrated in the scale drawing. Use the scale drawing to calculate the tension in the cable.

Answer: (420 N)

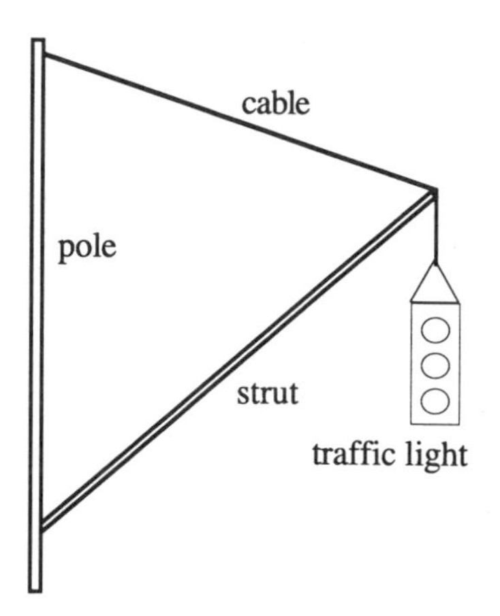

9.11. A rigid, uniform metal beam is fastened to a vertical wall by means of a hinge at one end of the beam. The length of the beam is 3.30 m and it has a mass of 10.45 kg. A 13.92-kg light fixture is suspended from the beam. One end of a thin, horizontal rod is attached to the beam. The other end of this rod is fastened to the wall. This holds the beam at an angle with respect to the wall as shown in the scale drawing. Calculate the tension in the rod and the force (magnitude and direction) that the hinge exerts on the beam.

Answer: (250 N; 350 N at 44° above the horizontal to the right)

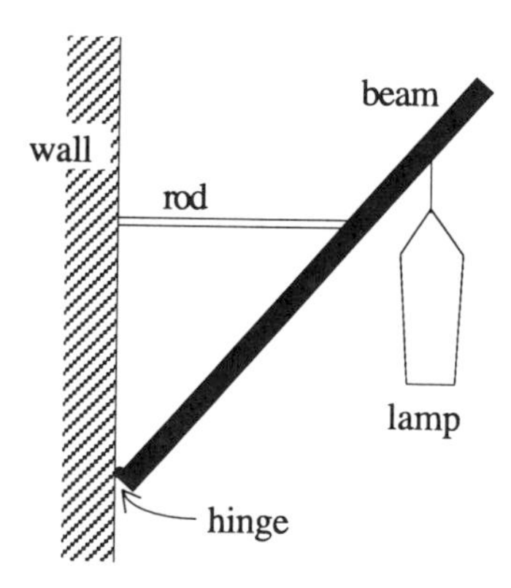

9.12. A man stands at one end of a uniform plank that is 3.25 meters long. The plank is supported by a vertical post at the other end and by a heavy duty spring scale connected to the plank at a point 1.08 m from the man. The plank weighs 53 N and the scale reads 1,140 N. How much does the man weigh?

Answer: (735 N)

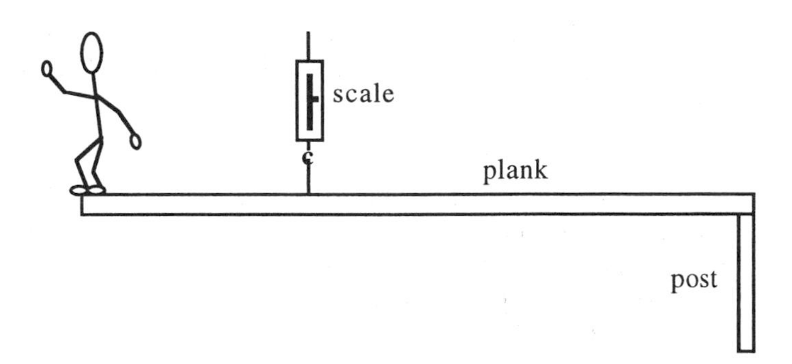

9.13. Weights of 8, 5, 3, and 10 N are located, respectively, at the 25-, 50-, 75-, and 100-cm marks on a meter stick whose weight is negligible. What are the magnitude and location of the single upward force that will balance the system?

Answer: (26 N at 64.4-cm mark)

9.14. A rigid metal plank of uniform density is 19.6 ft long and weighs 63.2 lb. It is supported by a slim vertical post, which is located 6.75 ft from one end of the plank. A cable is fastened to the bottom of the plank as shown in the *scale* drawing. The other end of the cable is fastened to the floor so that the cable makes an angle with the plank as shown. A 182-lb man stands 1.58 ft from the opposite end of the plank. With the plank horizontal, calculate the tension in the cable.

Answer: (430 N)

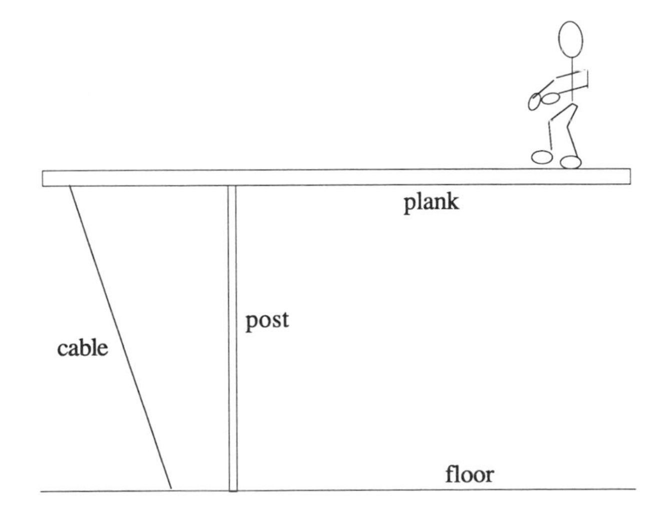

9.15. Assume circular orbits and calculate the orbital angular momentum of the earth and Jupiter relative to the sun. How many times larger is Jupiter's orbital angular momentum than the earth's?

Answer: (720 times larger)

9.16. The mass of Pluto is only about two thousandths of the earth's mass, but Pluto is very far from the sun compared to the earth. Which has the larger orbital angular momentum, the earth or Pluto? How many times as large? (Assume that both planets have circular orbits. This is not a very good approximation for Pluto.)

9.17. In 1957, the Soviet Union launched the world's first artificial satellite—the 84-kg Sputnik. Sputnik circled the earth every ninety six minutes at an altitude of 575 km. Calculate the orbital angular momentum of Sputnik.

Answer: (4.42×10^{12} J·s)

10

Let There Be Light!

What is light? Clearly, it comes from matter, but how? To answer these questions, we must come again to a consideration of the attributes of matter—how matter makes its existence known to us.

We have seen that all matter has an intrinsic ability to resist efforts to change its motion. Mass is the quantity we use to measure this characteristic, which we call *inertia*. Another fundamental attribute of matter is *electric charge*. Electric charge is responsible for two phenomena observed in nature—electricity and magnetism. Both were known to the ancients, although they did not realize they derived from the same property of matter. This recognition did not come until early in the nineteenth century.

10.1 Electricity

By the fifth century B.C., the Greeks knew that a piece of amber that had been rubbed with fur would attract and pick up small bits of straw or fiber. A glass rod similarly rubbed also exhibits this property. In the early eighteenth century it was established experimentally that there are two distinct kinds of electrification. We now call the quantity that is exchanged when two bodies are rubbed together *electric charge*. Although he had a view different from the one we hold today, Benjamin Franklin, the American statesman, is responsible for naming the distinction between the two kinds of electric charge. He arbitrarily called the type of charge on a glass rod that has been rubbed with silk cloth *positive* (+). The charge on a hard rubber rod after rubbing with wool cloth he called *negative* (−).

Electrically charged bodies exert forces on each other. If the charges are alike (both + or both −), the bodies repel each other. If the charges are opposite (one + and the other −), they will attract each other.

10.1.1 Coulomb's Law

In the late eighteenth century, the French scientist Charles Coulomb carried out a series of measurements of the force that two electrically charged spheres exert on each other. His apparatus included a torsion balance similar to the one Cavendish used to measure the force between two spherical masses. The Cavendish experiment involved the gravitational force, whereas Coulomb was measuring the electric force.

Coulomb's experiments led him to the following law for the force that an electric charge q_1 exerts on a second charge q_2,

$$\mathbf{F}_{\text{electric}} = k\,\frac{q_1 q_2}{r^2}\,\hat{\mathbf{r}}, \tag{10.1}$$

where $\hat{\mathbf{r}}$ is a unit vector (i.e., a dimensionless vector of unit length) in the direction of the vector $\mathbf{r}$, which gives the position of charge q_2 relative to charge q_1 as shown in Fig. 10.1.

Clearly, Eq. (10.1) shows that if q_1 and q_2 are alike, $\mathbf{F}_{\text{electric}}$ is in the same direction as $\mathbf{r}$ and the force is repulsive. If the two charges have opposite polarity (one $+$ and the other $-$), $\mathbf{F}_{\text{electric}}$ is in the direction opposite to $\mathbf{r}$ and the force is attractive. The unit we use for electric charge is called the *coulomb* (C). The quantity k in Eq. (10.1) is the Coulomb constant and has the value $k = 9 \times 10^9$ N$\cdot$m^2/C^2.

10.1.2 Electric Fields

It is convenient to think of the electric interaction in terms of a *field*. For example, we envision the charge q_1 in Fig. 10.1 having associated with it an electric field $\mathbf{E}$, which extends into the space surrounding q_1. It is this field (due to q_1) that exists at the position of q_2 and interacts with q_2 to exert the force on it.

This idea is expressed more clearly by removing the charge q_2 and considering the electric field $\mathbf{E}$ at an arbitrary point P in the neighborhood of q_1. This is illustrated in Fig. 10.2 where we have removed the charge q_2 from the point P and dropped the subscript on q_1 to represent the electric field produced at P by a point electric charge q.

Note that there is *no electric charge at P*, only the electric field $\mathbf{E}$ due to the charge q. We have assumed that q is a positive charge. If q were negative, $\mathbf{E}$ would have the opposite direction.

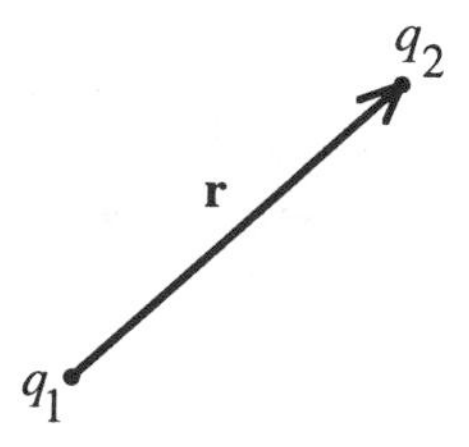

FIGURE 10.1. Two electric charges separated by a distance r.

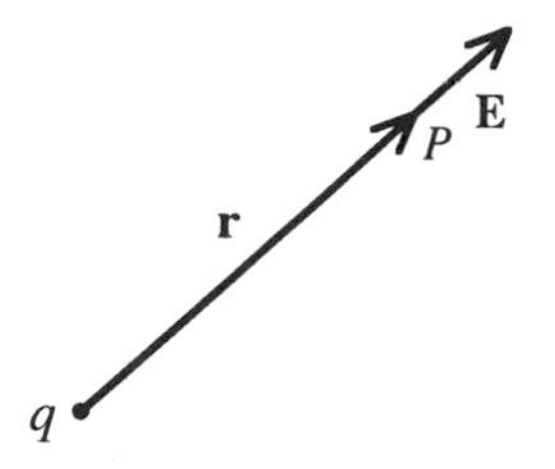

FIGURE 10.2. Electric field due to a single point charge q.

The electric field is a vector quantity. It has a magnitude and a direction. The direction of the field $\mathbf{E}$ at a given point P in space is the direction in which a small, positive charge placed at P would experience a force. The magnitude of the field is proportional to the force that a unit of electric charge would experience at the field point.

To make this notion general and precise, we consider the force $\mathbf{F}$ that a small, positive point charge q' experiences in an electric field as illustrated in Fig. 10.3. By definition, the electric field $\mathbf{E}$ at the point in space where the test charge q' is located is

$$\mathbf{E} = \lim_{q' \to 0} \frac{\mathbf{F}}{q'}. \tag{10.2}$$

In this limit, the charge q' used to test the presence of the field becomes vanishingly small, hence makes no contribution of its own to the field $\mathbf{E}$.

Now let us apply the definition in Eq. (10.2) to calculate the electric field due to a single point charge q. We imagine a small, positive charge q' placed in the neighborhood of q as shown in Fig. 10.4a. From Eqs. (10.1) and (10.2), we obtain for the electric field at P

$$\mathbf{E} = \lim_{q' \to 0} \frac{1}{q'} \left(k \frac{qq'}{r^2} \hat{\mathbf{r}} \right) = \frac{kq}{r^2} \hat{\mathbf{r}}. \tag{10.3}$$

If we have a distribution of electric charges, say N of them, then the electric field due to the distribution is the vector sum of the contributions from the individual charges in the distributions. This is illustrated in Fig. 10.4b.

At the point P, the electric field due to the distribution of N electric charges is

$$\begin{aligned} \mathbf{E} &= \mathbf{E}_1 + \mathbf{E}_2 + \cdots + \mathbf{E}_N \\ &= \sum_{i=1}^{N} \mathbf{E}_i = \sum_{i=1}^{N} \frac{kq_i}{r_i^2} \hat{\mathbf{r}}_i. \end{aligned} \tag{10.4}$$

FIGURE 10.3. Charge q' experiencing a force in an electric field $\mathbf{E}$.

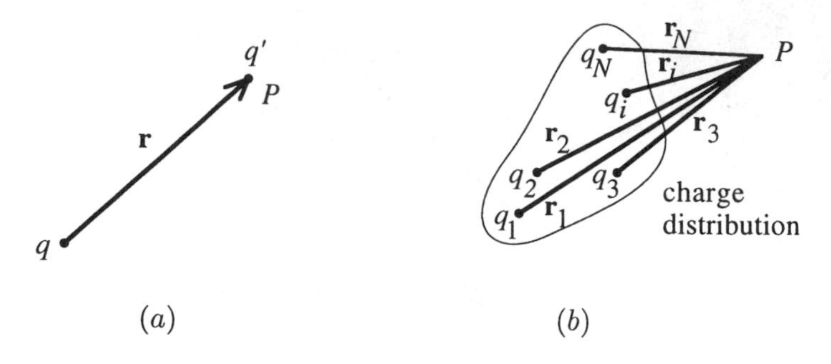

FIGURE 10.4. Fields due to electric charge distributions.

EXAMPLE. Three electric charges of $+5\ \mu$C, $-7\ \mu$C, and $+3\ \mu$C are placed in that order on the same straight line as shown in Fig 10.5. Each is 15 cm from its neighbor. Calculate the electric field at the point P, which is 20 cm to the right of the $+3\ \mu$C charge.

SOLUTION. The electric field given by Eq. (10.4) is a *vector* sum. In carrying out this sum, we consider separately the magnitude and the direction of the field $\mathbf{E}_i$ contributed by each charge q_i. To determine the direction of the field that the ith charge contributes, we imagine placing a small, *positive* charge at P. The direction in which this small test charge would experience a force due to q_i alone is the direction of $\mathbf{E}_i$. For the example here, the individual field contributions will be directed toward the right or the left. We (arbitrarily) choose those to the right to be positive and those to the left negative. In calculating the magnitude of each field contribution, we use only the *magnitude* of each charge q_i in Eq. (10.4), which gives the field of a distribution of point charges. For the field at P, we have

$$\begin{aligned}\mathbf{E} &= E_1 - E_2 + E_3 \\ &= \frac{(9\times10^9\ \mathrm{N\cdot m^2/C^2})(5\times10^{-6}\ \mathrm{C})}{(0.50\ \mathrm{m})^2} \\ &\quad - \frac{(9\times10^9\ \mathrm{N\cdot m^2/C^2})(7\times10^{-6}\ \mathrm{C})}{(0.35\ \mathrm{m})^2} \\ &\quad + \frac{(9\times10^9\ \mathrm{N\cdot m^2/C^2})(3\times10^{-6}\ \mathrm{C})}{(0.20\ \mathrm{m})^2} = +3.41\times10^5\ \mathrm{N/C}.\end{aligned}$$

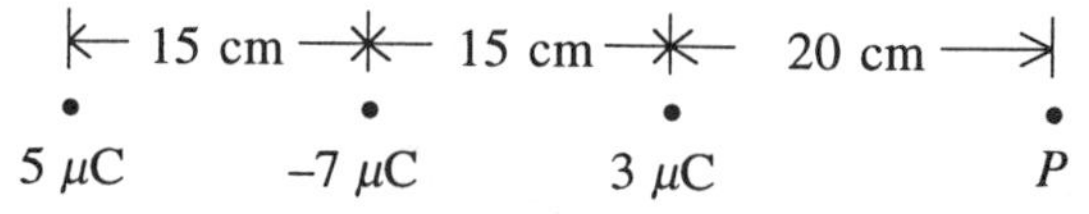

|← 15 cm →|← 15 cm →|← 20 cm →|

5 μC −7 μC 3 μC P

FIGURE 10.5. Field due to a linear charge distribution.

The plus sign tells us that the direction of the field is toward the right. The resultant electric field at P due to the three charges is therefore 3.41×10^5 N/C to the right.

10.1.3 Electric Current

Ultimately, it is the so-called elementary particles of matter that carry the electric charge. For example, an electron has a negative electric charge of magnitude $e = 1.60 \times 10^{-19}$ C. In certain materials, metals for example, some of the charge-carrying particles are free to move. An electric field applied to the material will cause these charged particles to move along the direction of the field giving rise to an *electric current*.

A useful analogue is a water current. Water flows along river beds from the mountains to the sea. In this case, gravity causes the mass (water molecules) to fall from a region of high gravitational potential energy (the mountains) to a region of lower potential energy (the sea). To maintain the current, work must be done on the water molecules to carry them from the sea (low potential energy) back to the mountains (high potential energy). The source of this energy is the sun. Solar radiation evaporates water from the ocean surface and drives the earth's weather systems, which carry the evaporated molecules back to the mountains where they fall as precipitation.

A similar energy source is required to maintain a current in an electrical circuit. One such device is the familiar dry cell, which, through chemical reactions, converts chemical energy into electric potential energy. The dry cell, "pumps the electric charge back up hill" so to speak. A simple electric circuit is illustrated in Fig. 10.6. The dry cell maintains an electric current through the filament of the light bulb.

We can make this notion of electric current quantitative by defining the electric current I as the time rate of flow of electric charge,

$$I \equiv \lim_{\Delta t \to 0} \frac{\Delta q}{\Delta t},$$

where Δq is the electric charge flowing past a given point in the time interval Δt. The unit of electric current is the *ampere* (A). One ampere is equal to one coulomb/second.

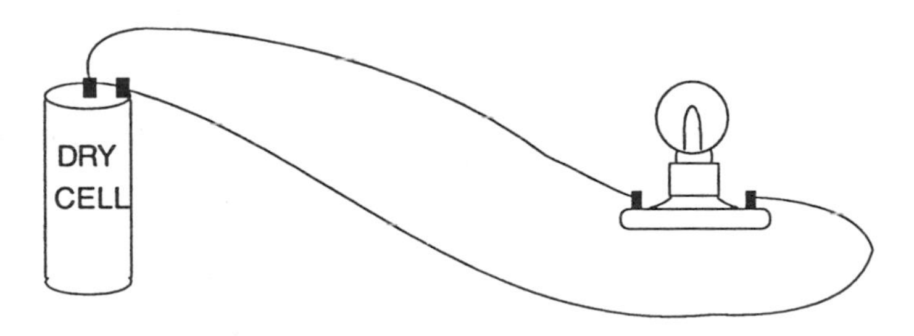

FIGURE 10.6. A simple electric circuit.

EXAMPLE. Suppose that during a 2.5 minute time interval 99 coulombs of electric charge pass through the filament of the light bulb in Fig. 10.6. Calculate the electric current in the circuit.

SOLUTION.

$$I = \frac{\Delta q}{\Delta t} = \frac{99\ \mathrm{C}}{150\ \mathrm{s}} = 0.66\ \mathrm{A}.$$

The current in the circuit is 0.66 A.

10.2 Magnetism

Already in the sixth or seventh century B.C., the Greeks knew that pieces of a certain mineral attracted other pieces of the same mineral. The mineral is an oxide of iron (Fe_3O_4). A sample of it is called a *lodestone*. Deposits of this mineral were known to exist at the ancient city of Magnesia in Asia Minor near the Aegean Sea. For this reason, materials that exhibit this property came to be called *magnets*, an abbreviation for *Mágnēs líthos*—stone of Magnesia. By stroking a steel rod or needle with a lodestone, the steel can acquire these properties, that is, it can become a magnet itself.

A suspended needle-shaped lodestone constrained to rotate in a horizontal plane will align itself with an approximate north-south orientation. This is the effect on which the operation of a magnetic compass is based. By the twelfth century A.D., mariners had learned to rely on it as an aid to navigation. A compass needle has two magnetic poles—one associated with each end. These are distinguished by calling the one on the end generally toward the geographic north the *N-pole* and the opposite end, the *S-pole*. In fact, as far as we know all magnetic poles occur in pairs—an N-pole with an S-pole. For example, a bar magnet can be constructed with opposite poles at its ends. If we bring such a bar magnet near a compass, we find that one of the poles repels the N-pole of the compass and the other attracts it. A similar bar magnet will exhibit this same effect. If we place the two bar magnets near each other, we find that the poles of the magnets that repelled the N-pole of the compass repel each other. Similarly, the two poles of the magnets that attracted the N-pole of the compass also repel each other. Clearly, like poles (those that affect the compass needle in the same way) repel each other. On the other hand, opposite poles (those on the bar magnets that affect the compass needle in the opposite manner) attract each other. Thus, in principle we can determine the polarity of any magnetic pole by observing its effect on a compass needle—an N-pole will repel the N-pole of a compass needle and an S-pole will attract it.

10.2.1 Coulomb's Law for Magnetic Poles

Coulomb also investigated the force that two magnetic poles exerted on each other and arrived at a force law similar to Eq. (10.1) for electric poles. For an arrangement

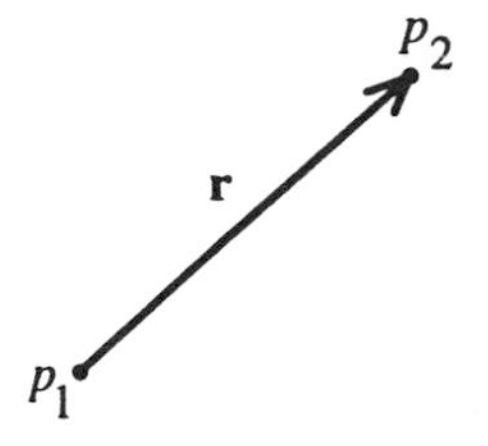

FIGURE 10.7. Two magnetic poles separated by a distance r.

like that illustrated in Fig. 10.7, he found that a magnetic pole of strength p_1 exerts a force $\mathbf{F}_{\text{magnetic}}$ on a second pole p_2 given by

$$\mathbf{F}_{\text{magnetic}} = k' \frac{p_1 p_2}{r^2} \hat{\mathbf{r}}. \tag{10.5}$$

10.2.2 Magnetic Fields

In metric units, the constant k' has the value $k' = 10^{-7}\ \text{N}\cdot\text{s}^2/\text{C}^2$. If we arbitrarily assign to an N-pole a (+) sign and to an S pole a (−) sign, then Eq. (10.5) shows that the force p_1 exerts on p_2 is directed along $\mathbf{r}$ *away from* p_1 (repulsive) if p_1 and p_2 are alike and toward p_1 (attractive) if p_1 and p_2 have opposite magnetic polarity.

A *magnetic field* exists in a region in which a magnetic pole experiences a force. We define the direction of the magnetic field $\mathbf{B}$ at a point P to be the direction in which a small N-pole p' experiences a force $\mathbf{F}$,

$$\mathbf{B} = \lim_{p' \to 0} \frac{\mathbf{F}}{p'}.$$

An electric current produces a magnetic field. Consider the arrangement shown in Fig. 10.8. The electric field from the battery can cause an electric current in the

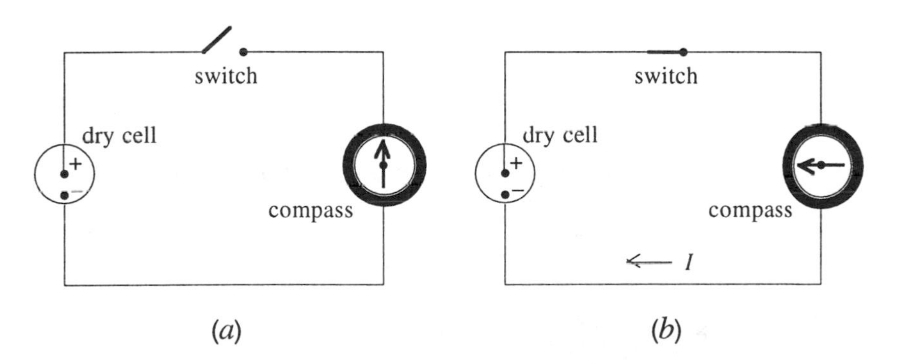

FIGURE 10.8. Deflection of a compass needle by an electric current.

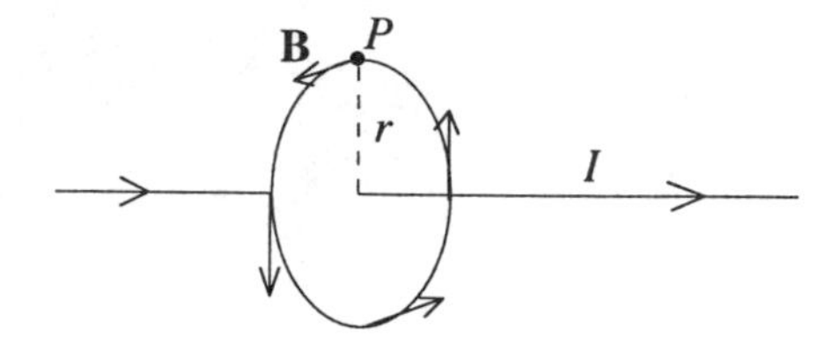

FIGURE 10.9. Magnetic field around a long, current-carrying wire.

metal wire of the circuit when the switch is closed. In Fig. 10.8*a* with the switch open (no current) the compass needle aligns with the earth's magnetic field. Closing the switch results in an electric current in the circuit and the compass needle is deflected as shown in Fig. 10.8*b*.

The electric current produces a magnetic field in the vicinity of the wire. At any point P, the direction of the magneitc field is tangent to a circle centered on the current and passing through P as shown in Fig. 10.9. The direction of **B** along the tangent line can be determined by a rule using the right-hand.

Rule: Curl the fingers of the right-hand around the wire with the thumb in the direction of the current (positive charge flow). At the field point, the tips of the fingers point in the direction of the magnetic field.

The N-pole of a compass needle placed at P points in the direction of **B**.

If it is a very long, straight wire that carries the current and the field point P is not near the ends of the wire, the magnitude of the magnetic field at a perpendicular distance r from the wire is

$$B = \frac{2k'I}{r}. \tag{10.6}$$

EXAMPLE. The long, straight wire in Fig. 10.10 carries an electric current of 1.89 A. Find the magnitude and direction of the magnetic field at a point P located 4 mm below the wire and in the plane of the page.

SOLUTION.

$$B = \frac{2k'I}{r} = \frac{2(10^{-7}\ \mathrm{N/A^2})(1.89\ \mathrm{A})}{0.004\ \mathrm{m}} = 9.45 \times 10^{-5}\ \mathrm{N/A \cdot m}$$

The magnetic field at P is 9.45×10^{-5} T down into the plane of the page. One tesla (abbreviated T) is equal to one N/A·m.

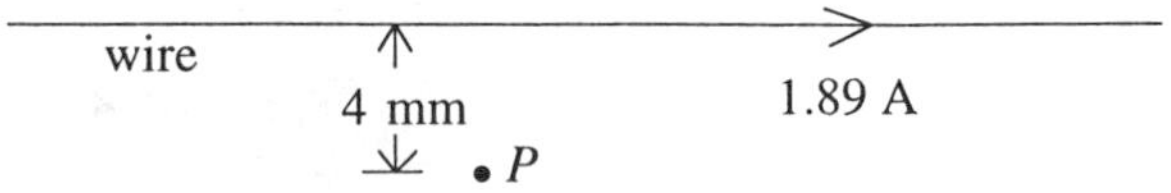

wire

4 mm

1.89 A

P

FIGURE 10.10. Magnetic field at a point near a long current.

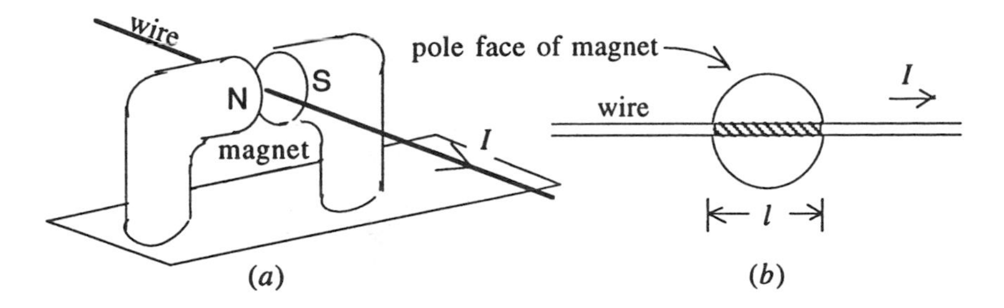

FIGURE 10.11. Force on an electric current in a magnetic field

10.2.3 Magnetic Force on an Electric Current

An electric current experiences a force in a magnetic field. In Fig. 10.11a, a metal wire carrying an electric current I is shown between the pole faces of a strong magnet.

The current (hence, the wire) experiences a force in a direction perpendicular to the plane defined by the magnetic field and the electric current. For example, if the pole on the left is an N-pole and the pole on the right is an S-pole (so that the field is directed from left to right), then the force on the current will be directed toward the top of the magnet. That is, the wire will be ejected *up* from between the pole faces of the magnet. The magnitude of the force on the current is given by

$$F = IlB_{\perp}, \tag{10.7}$$

where l is the length of the current element within the magnetic field (shaded region in Fig. 10.11b) and $B_{\perp}$ is the component of the magnetic field $\mathbf{B}$ perpendicular to the current.

From a more fundamental point of view, we recognize that the electric current arises from the motions of individual charge carriers (electrons, and so on). These carriers drift along the length of the wire with some average velocity $\mathbf{v}$. An enlarged section of the wire is represented in Fig. 10.12 where we show a typical charge q moving with the drift velocity $\mathbf{v}$ in the direction of the current.

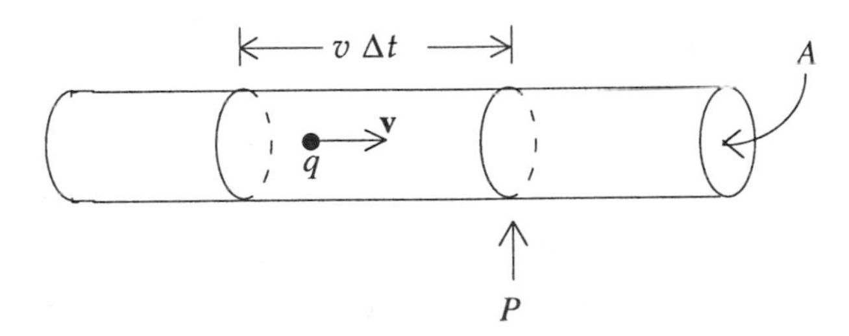

FIGURE 10.12. Flow of individual electric charges.

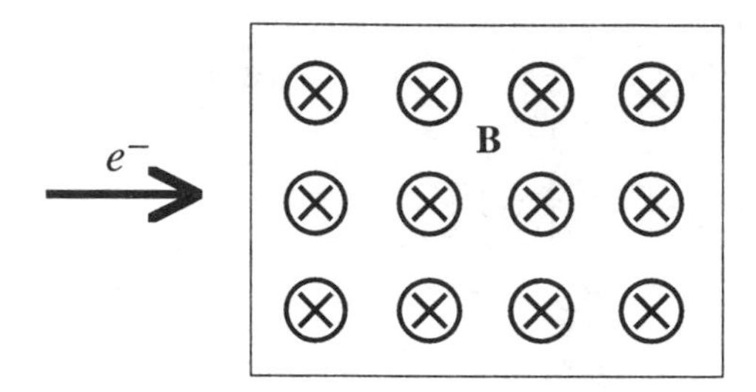

FIGURE 10.13. Electron entering a magnetic field.

The wire has a cross-sectional area A. In a time Δt, all of the charges in the cylinder of length $v\Delta t$ pass the point P in Fig. 10.12. Thus, for the current I we have,

$$I = \frac{\Delta q}{\Delta t} = \frac{qn(Av\Delta t)}{\Delta t} = qnvA,$$

where q is the electric charge on each carrier and n is the number of carriers per unit volume. If we insert this last expression for I in Eq. (10.7), we get for the force on an element of current of length l,

$$F = (qnvA)lB_{\perp} = q(nAl)vB_{\perp}.$$

Now nAl is the total number of charges q in a length l of the wire. Thus, the force F_q on *each charge* q in this volume is

$$F_q \equiv \frac{\text{force on element of length } l}{\text{number of charges in length } l} = \frac{q(nAl)vB_{\perp}}{nAl} = qvB_{\perp}. \qquad (10.8)$$

Note that there are three mutually perpendicular vectors (**v**, $\mathbf{B}_{\perp}$, and **F**) involved in Eq. (10.8). We can relate them by means of another rule using the right-hand (for $q > 0$).

Rule: Hold the right-hand flat with palm open, thumb in the direction of **v**, and fingers in the direction of $\mathbf{B}_{\perp}$. The open palm faces in the direction of **F**.

EXAMPLE. An electron traveling to the right with a speed 6.0×10^5 m/s enters a uniform magnetic field of 1.75 T that is directed down into the plane of the page as shown in Fig. 10.13. Calculate the force on the electron as it enters the field.

SOLUTION.

$$\begin{aligned} F_e = evB_{\perp} &= (1.60 \times 10^{-19}\,\text{C})(6.0 \times 10^5\,\text{m/s})(1.75\,\text{N/A}\cdot\text{m}) \\ &= 1.68 \times 10^{-13}\,\text{N}. \end{aligned}$$

As it enters the field, the electron is deflected down toward the bottom of the page with a force of 1.68×10^{-13} N. (A *positive* charge would be deflected toward the *top* of the page.)

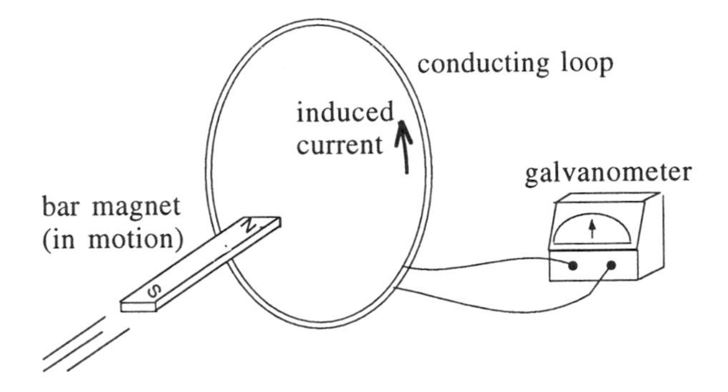

FIGURE 10.14. Current induced in a conducting loop.

10.3 Induced Electric Currents

By the middle of the nineteenth century, these and other connections between electricity and magnetism were well established. Michael Faraday demonstrated that changing the magnetic field passing through a conducting loop of wire induces an electric current in the wire. For example, thrusting a bar magnet through a loop of wire as illustrated in Fig. 10.14 causes a deflection of the meter, indicating a flow of electric charge in the circuit.

It is the *changing* magnetic field that is responsible for the induced current in the experiment described above. Consider two circular coils of wire placed side by side as shown in Fig. 10.15. Coil A is connected through a switch to a dry cell. Coil B is connected to a galvanometer. With the switch open, no current exists in either coil. Closing the switch completes the circuit and the dry cell causes a current in coil A. This current starts at zero and rises to a steady value, which is maintained in coil A as long as the switch remains closed. Opening the switch, causes the current in coil A to drop to zero. The behavior of the coil-A current as

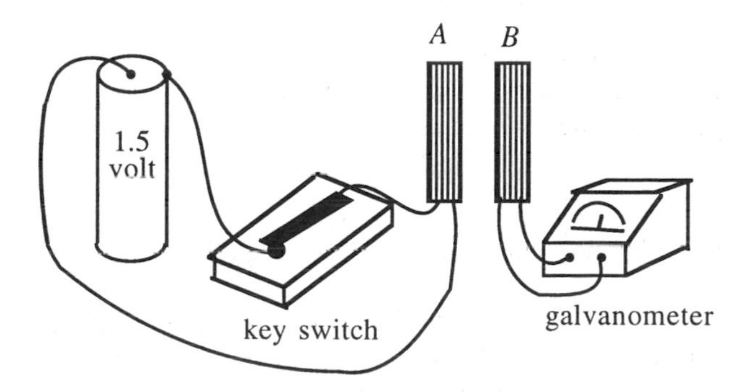

FIGURE 10.15. Magnetic coupling between two conducting coils.

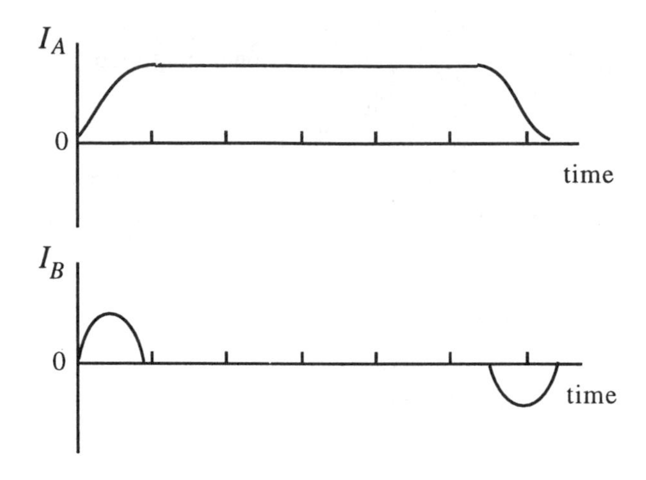

FIGURE 10.16. Currents in the coils as functions of time.

a function of time is shown graphically in the upper plot in Fig. 10.16. The lower plot in Fig. 10.16 shows the current in coil B during the same time interval.

Note that coil B is not directly connected to the dry cell or to coil A. So, how is the current in coil B produced? During the brief time interval from the instant the switch is closed at $time = 0$ until the current in coil A reaches its steady value, the magnetic field produced by this current is *changing*, hence coil B experiences a *changing* magnetic field passing through it. This changing magnetic field induces a current in coil B causing the galvanometer to deflect briefly as indicated by the current pulse on the left in the lower plot of Fig. 10.16. While the current is steady in coil A, there is a magnetic field around coil A extending into coil B. But this field is not changing, hence no current is induced into coil B during this time, as the lower plot shows.

When the switch is opened, the current in coil A drops to zero, again producing a *changing* magnetic field that reaches into coil B and induces a current. This is represented by the brief current pulse on the right in the lower plot.

Summarizing the electric and magnetic connections considered so far, we have

- An electric current produces a magnetic field.
- An electric current experiences a force in a magnetic field.
- A changing magnetic field can produce an electric current.

10.4 Electromagnetic Radiation and Light

In 1864, James C. Maxwell combined the laws of electricity and magnetism in a unified electromagnetic theory. Maxwell saw that his equations could be cast in the form of a well-known wave equation—a mathematical equation that was known to describe the behavior of waves. This implied that the energy of the electromagnetic

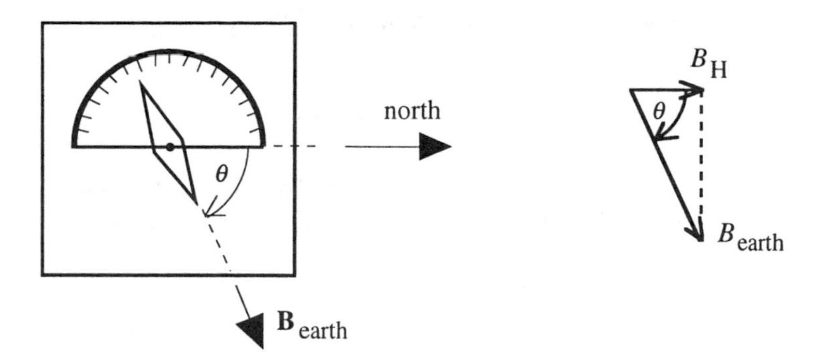

FIGURE 10.17. Measurement of the dip angle.

field propagates as a wave. The wave propagation velocity v appeared in Maxwell's wave equation in terms of Coulomb's electric and magnetic force constants k and k'. That is,

$$v = \sqrt{\frac{k}{k'}} = \sqrt{\frac{9 \times 10^9\ \text{Nm}^2/\text{C}^2}{10^{-7}\ \text{Ns}^2/\text{C}^2}} = 3 \times 10^8\ \text{m/s}.$$

This is a remarkable result! Maxwell realized that, within experimental error, this value was the same as the experimentally known speed with which light waves propagate. Maxwell's conclusion: *Light* is electromagnetic radiation! All of a sudden, quite unexpectedly, one has a complete mathematical theory of light Maxwell's equations.

The Maxwell theory predicts that an accelerated electric charge radiates electromagnetic energy. That is how light is produced. It originates with that fundamental property of matter we call electric charge.

10.5 The Earth's Magnetic Field

Although the way in which the earth's magnetic field is produced is not completely understood, it is thought to be related to the fact that a part of the earth's core is in the form of molten metal (princpally iron and nickel). The rotational motion of the earth gives rise to convection currents in this liquid, which contains electrically charged matter, resulting in electric currents that generate the earth's magnetic field.

At a given point on the earth's surface in the northern hemisphere, the direction of the earth's magnetic field is roughly toward the geographic north, but directed downward at an angle θ below the horizontal. This angle is called the "dip angle" and is easily measured with a device that includes a compass needle free to move in a vertical plane as illustrated in Fig. 10.17. The transparent frame supporting the compass needle is marked as a protractor so that the dip angle θ can be read directly.

It is clear from Fig. 10.17 that the earth's magnetic field is related to its horizontal component B_{H} and the dip angle by

$$B_{\text{earth}} = \frac{B_{\mathrm{H}}}{\cos\theta}. \tag{10.9}$$

A simple apparatus for measuring the strength of the horizontal component of the earth's magnetic field is illustrated in Fig. 10.18. A small magnetic compass is placed on a platform at the center of a circular coil of wire. A power source supplies an electric current that produces a magnetic field at the center of the coil where the compass is located. This field is directed perpendicular to the plane of the coil. Its magnitude is given by

$$B_{\text{coil}} = \frac{2\pi k' N I}{r}, \tag{10.10}$$

where I is the current, N is the number of turns in the coil, and r is the radius of the coil.

By orienting the coil in a vertical plane that is aligned with the direction of the earth's magnetic field, the field of the coil $\mathbf{B}_{\text{coil}}$ will be perpendicular to the direction of B_{H}. The earth's field B_{H} tries to align the compass needle toward the north, while $\mathbf{B}_{\text{coil}}$ tries to align it along the east-west line. The compass needle settles somewhere between these directions depending on the relative strengths of the two fields. The angle α through which the needle is deflected away from the direction of B_{H} gives a measure of the strength of the field B_{coil} relative to the strength of the horizontal component of the earth's magnetic field B_{H}. In fact,

$$B_{\mathrm{H}} = \frac{B_{\text{coil}}}{\tan\alpha}. \tag{10.11}$$

The deflection angle α is shown in Fig. 10.18*b*.

Thus, by calculating B_{coil} from the data and reading the deflection angle directly off the compass as illustrated in Fig. 10.18*b*, we obtain a value for B_{H}. Combining

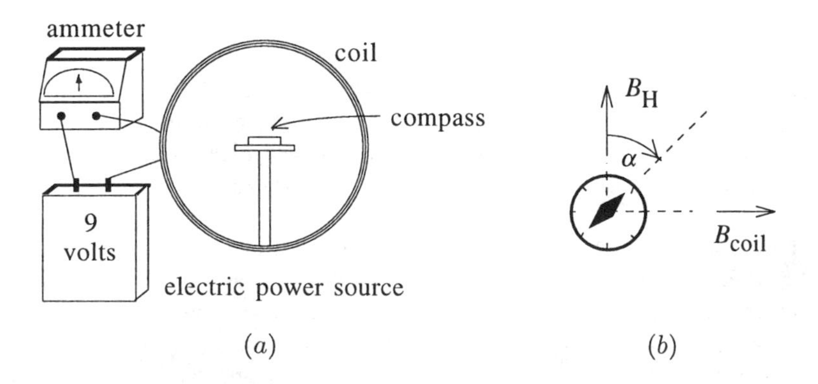

FIGURE 10.18. Measurement of the earth's magnetic field.

this result with the dip angle according to Eq. (10.9) gives the measured strength of the earth's magnetic field at a given location.

10.6 Electric Potential Energy

An electric charge q in a uniform electric field $\mathbf{E}$ experiences a constant force $\mathbf{F}$ given by

$$\mathbf{F} = q\mathbf{E}.$$

Assume that q is small and positive. The work ΔW done by an external force in displacing q by an amount $\Delta\mathbf{s}$ in the field $\mathbf{E}$ as illustrated in Fig. 10.19 is

$$\Delta W = -qE\Delta s \cos\theta, \tag{10.12}$$

where θ is the angle between the vectors $\mathbf{E}$ and $\Delta\mathbf{s}$.

This work is recoverable in the sense that if q is released following the displacement, the field $\mathbf{E}$ can do the same amount of work *on* the charge in displacing it by $-\Delta\mathbf{s}$ back to its original position. As the charge is accelerated by the field, it acquires a kinetic energy equal to the work given in Eq. (10.12). Clearly, with an electrically charged particle in an electric field we can associate an electric potential energy PE_{electric} analogous to the gravitational potential energy of a particle of mass m in a gravitational field. The change in electric potential energy in moving a charged particle from one point to another in an electric field is just equal to the work done *on the charge* in moving it. That is,

$$\Delta(PE_{\text{electric}}) = \Delta W. \tag{10.13}$$

We have seen that an electric charge Q produces in the space around it an electric field $\mathbf{E}_Q$ given by Eq. (10.3) as

$$\mathbf{E}_Q = \frac{kQ}{r^2}\hat{\mathbf{r}}.$$

A small charge q located at $\mathbf{r}$ relative to Q experiences a force in this field. Therefore, in general, work is done in moving q from one point to another in the field of Q. That is, q gains or loses electric potential energy. By analogy with the expression for gravitational potential energy in Eq. (7.11), the electric potential energy

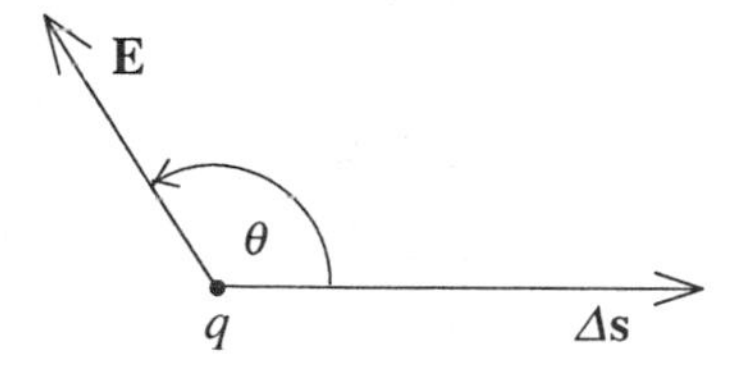

FIGURE 10.19. Work done in displacing a charge in an electric field.

of a small electric charge q at a distance r from a charge Q is

$$PE_{\text{electric}} = \frac{kqQ}{r}, \tag{10.14}$$

where we have arbitrarily chosen the zero of potential energy to be at infinity. Notice that if Q and q have the same sign, the charges repel each other and $PE_{\text{electric}} \geq 0$. If Q and q have opposite polarities, the force is attractive and $PE_{\text{electric}} \leq 0$.

A related quantity that depends only on the source of the field and not on the charge q experiencing the field is the *electric potential*, denoted by V and defined by

$$\text{electric potential} = V = \lim_{q \to 0} \frac{PE_{\text{electric}}}{q}. \tag{10.15}$$

For the point charge Q in Eq. (10.14), we have

$$V = \lim_{q \to 0} \frac{1}{q} \left(\frac{kQq}{r} \right) = \frac{kQ}{r}. \tag{10.16}$$

It is clear that the electric potential at any point in space represents the electric potential energy per unit charge. Thus, the electric potential energy of an electric charge q at a point where the electric potential is V is

$$PE_{\text{electric}} = qV. \qquad (10.17).$$

The electric potential *difference* between two points is equal to the work done in moving one unit of electric charge from one point to the other. From Eq. (10.15), we see that the unit of electric potential is that of an energy unit divided by a electric charge unit. In the mks system, we have

$$\text{unit of electric potential} = \frac{\text{one joule}}{\text{one coulomb}} \equiv \text{one volt.}$$

The unit of electric potential in this system is the *volt.*

Exercises

10.1. What is the force that an electric charge of one coulomb exerts on a second charge of one coulomb when the two charges are separated by a distance of one meter? Express your result in pounds. Based on your result, what statement would you make about the unit of electric charge, the coulomb?

10.2. Three point charges are arranged along a line as shown in the drawing.

|← 10 cm →|← 15 cm →|← 22 cm →|

•	•	•	•
9 μC	–4 μC	7 μC	P

Find the magnitude and direction of the force which the 9-μC charge exerts on the -4-μC charge. Do the same for the force which the 7-μC charge exerts on the -4-μC charge. If the -4-μC charge were free to move under the influence of these forces, in which direction would it move? Explain clearly and completely your reasoning. Calculate the resultant electric field at the point P due to all three charges.

Answer: (32.4 N to the left; 11.2 N to the right; to the left; 1.40×10^6 N/C)

10.3. Two point charges separated by a distance of 25 cm repel each other with a force of 0.140 N. Find the force if the separation is increased to 50 cm. If one of the charges is $+1.20$ μC, what is the value of the other charge? Explain clearly and completely your reasoning.

Answer: (0.035 N; 0.810 μC)

10.4. An electric charge $+9.80$ μC is placed to the left of a -3.20-μC charge. How far apart are these charges if the electric field at a point 20 cm to the right of the -3.20-μC charge is exactly zero? Calculate the force that each charge exerts on the other when they are separated by this distance.

Answer: (15 cm; 12.5 N)

10.5. Two charged particles exert a force of 4.72 N on each other. What will be the force each will exert on the other if they are moved so that they are only one-seventh as far apart?

10.6. Assume that the innermost electron in a copper atom is 1.4×10^{-12} meter from the nucleus of the atom. Calculate the magnitude of the electric force between the nucleus (charge $= 29e$) and this electron.

10.7. Two electric charges separated by a distance of 1.47 m exert a force of 2.72 N on each other. What will be the force each exerts on the other if one of the charges is moved so that they are only 42 cm apart?

10.8. Calculate the work done in moving a 125-nC charge a distance of 30 cm in a uniform electric field of 2,300 N/C. Consider the displacement to be in the direction opposite to the field.

Answer: (8.625×10^{-5} J)

10.9. Three electric charges are arranged in a straight line as shown in the drawing. Find the electric field (magnitude and direction) at the point P, which lies on the same line as the charges. What would be the electric force (magnitude and direction) on a -4-μC charge placed at the point P?

15 cm | 20 cm | 25 cm

9 μC | –5 μC | *P* | –7 μC

10.10. A certain point charge exerts a force of 0.054 N on a small, 3-nC point charge situated 50 mm away from it. Calculate the electric field at the position of the 3-nC charge due to the first charge. What is the magnitude of the first charge?

Answer: (1.80×10^7 N/C; 5 μC)

10.11. A 9-nC electric charge stands 0.138 mm to the left of a −4-nC charge. Calculate the total electric field at the point 0.217 mm to the right of the −4-nC charge.

Answer: (1.22×10^8 N/C toward the left)

10.12. A point charge of 3.21 μC moves in a circular orbit of radius 62.7 cm. The charge makes five thousand complete orbits each second. Calculate the electric current represented by this moving electric charge.

Answer: (16 mA)

10.13. Three point charges +8 nC, +5 nC, and −7 nC are arranged along a line with the +5-nC charge located at a point exactly halfway between the other two. The charges on the ends are 4.24 mm apart. Calculate the electric field (magnitude and direction) at a point exactly halfway between the +5-nC charge and the −7-nC charge.

Answer: (1.032×10^8 N/C toward the right)

10.14. For the system of charges in Exercise 13, calculate the total electric force exerted on the 8-nC charge by the other two charges.

Answer: (0.0521 N toward the left)

10.15. Two very long parallel wires 17.3 mm apart carry the same current in opposite directions as shown. Find the magnetic field (magnitude and direction) at the two points P_1 (midway between the wires) and P_2 15.2 mm below the bottom wire).

Answer: (1.51×10^{-4} T into the page; 2.29×10^{-5} T out of the page)

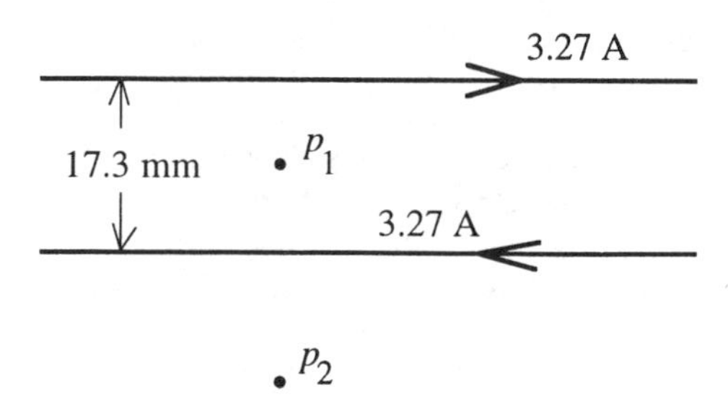

10.16. A proton carries a positive electric charge of magnitude e. A proton traveling to the right with a speed 2.33×10^6 m/s enters a uniform magnetic field 2.38 T directed toward the top of the page as shown. What is the magnitude and direction of the force on the proton as it enters the field?

Answer: (8.87×10^{-13} N out of the plane of the page)

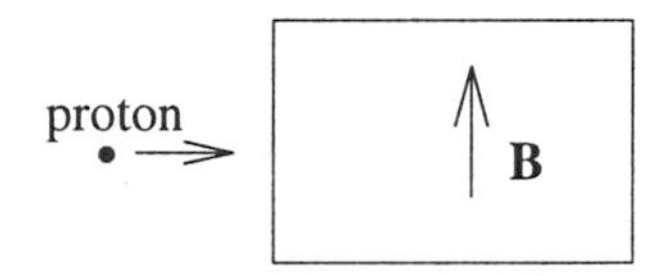

10.17. In the example on page 136, the magnetic force on the electron is *perpendicular* to the velocity. Given this fact, describe the path of the electron. Explain clearly your reasoning. Calculate the acceleration of the electron. What is the radius of curvature of its path?

Answer: (1.84×10^{17} m/s^2; 1.95 μm)

10.18. An electrically charged particle with linear momentum 8.32×10^{-21} kg·m/s enters a uniform magnetic field of 0.0635 T. The direction of motion of the particle is perpendicular to the magnetic field. The path of the particle is a circular arc of radius 27.3 cm. Find the electric charge on this particle.

10.19. A proton enters a uniform magnetic field of strength 0.040 T and moves along a circular arc of radius 14.0 cm. Calculate the speed of this proton.

Answer: (5.4×10^5 m/s)

10.20. A rectangular loop of wire lying in the plane of the page carries a steady current 1.73 A as shown. A long wire carrying a current 2.39 A lies in the same plane. Calculate the force on the current loop due to the current in the long wire.

Answer: (8.93×10^{-7} N down toward the bottom of the page)

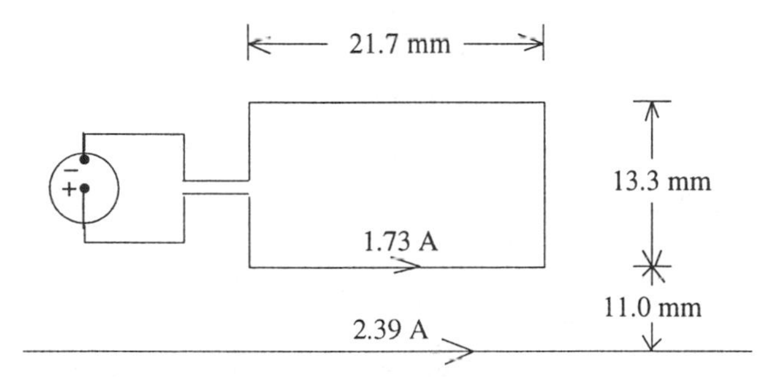

10.21. A uniform magnetic field **B** is directed into the plane of the page. A beam of small particles enters this region as shown. In which direction is the beam deflected by the field if the particles (a) have a positive electric charge, (*b*) have a negative electric charge, (*c*) are electrically neutral? Be very clear.

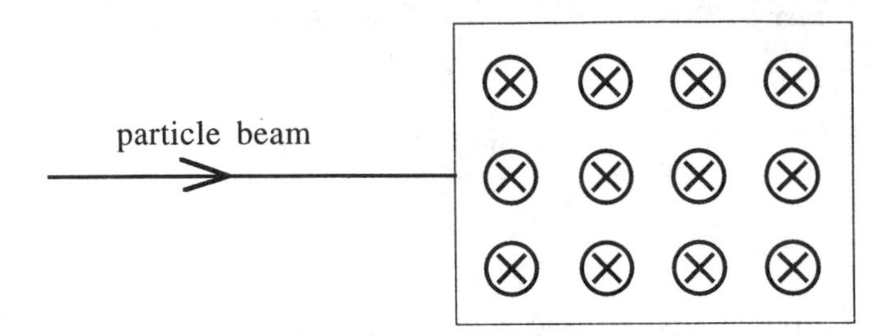

10.22. A nearly uniform electric field can be created in the region between two closely-spaced parallel metal plates that carry equal, but opposite electric charges. This is illustrated in the drawing with the upper plate positively charged and the lower one negatively charged. A small particle of mass 0.60 μg carries ten thousand extra electrons. This particle is placed in this uniform field as shown in the drawing. Calculate the electric field required for the gravitational force on the particle to be exactly balanced by the electric force.

Answer: (3.68×10^5 N/C)

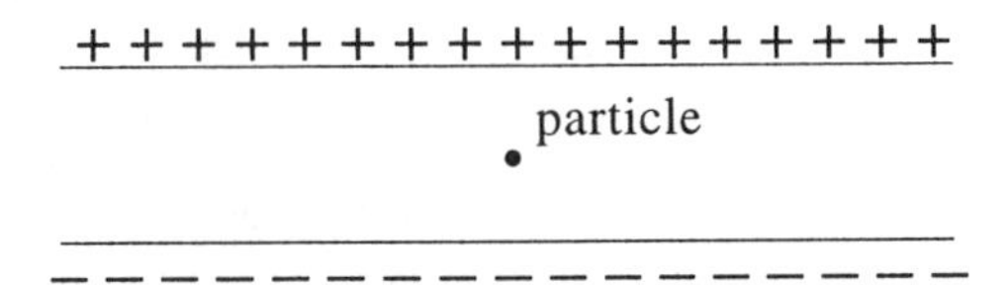

10.23. An electron is placed in a uniform electric field that is directed vertically downward. Suppose the electron is in equilibrium in this field with the gravitational force exactly balanced by the electric force. Calculate the magnitude of the electric field. What would be the magnitude of the electric field required to hold a proton in equilibrium in this same position? What other change would be necessary for the proton to be in equilibrium?

Answer: (5.58×10^{-11} N/C; 1.02×10^{-7} N/C)

10.24. In a simple model, a hydrogen atom consists of an electron and a proton separated by a distance of about 5.28×10^{-11} m. Calculate the electric potential energy of the electron in the field of the proton.

Answer: (-4.36×10^{-18} J)

10.25. The physical law associated with the currents introduced in the conducting loops of Section 10.3 is known as *Faraday's law*. It is conveniently expressed in terms of a scalar quantity called *magnetic flux*. The magnetic field vector **B** is related to this quantity and is often called the magnetic flux *density*. Consult an introductory physics text to find the precise connection between the magnetic flux and the flux density **B**. In terms of magnetic flux, write a general statement regarding currents induced in conducting loops by magnetic fields. Based on your general statement, can you conceive of a way in which a *constant* magnetic field could induce a current into a closed conducting loop of *constant* size and shape? Explain.

10.26. In the experiment with the two coils described in Section 10.3, the lower plot in Fig. 10.16 shows that when the switch is opened the current induced in coil B is in the direction opposite to the direction of the current induced by closing the switch. This behavior is dictated by the principle of conservation of energy and is expressed in the form of a general rule known as *Lenz's law*. Look up Lenz's law in an introductory physics text and use it to explain clearly why opening the closing the switch induces currents in opposite directions in coil B.

11

What's the Matter?

What is matter anyway? Really. In the fifth century B.C., a Greek—Democritus, by name—argued on philosophical grounds that one could not indefinitely subdivide matter into smaller and smaller bits. He expected that on cutting matter into finer and finer pieces, eventually one would come to a bit of matter that could not be cut. In his view, there existed tiny fundamental building blocks or particles, which could not be divided any finer and from which all other matter is constructed. These fundamental entities came to be called *atoms*, from a Greek word meaning *uncuttable*. Of course, Democritus had no means of testing his conjecture experimentally. Such evidence did not come until the early nineteenth century with the work of Lavoisier, Proust, and Dalton.*

Dalton was led to postulate that chemical reactions could be explained on the basis of a model in which all matter is composed of small indestructible, indivisible particles that he called atoms. For a given chemical element, he considered these atoms to be identical. However, they differed from element to element in mass as well as other properties. Dalton was premature, if not presumtuous, in applying Democritus' term "atom"—meaning uncuttable—to his particles. Indeed, by the end of the century it was clear that Dalton's atoms themselves have structure and can be subdivided or "cut." Nevertheless, the name atom has stuck and still applies to the basic particles of the chemical elements.

Our purpose in this chapter is to examine developments that occurred around the turn of the century that led to a deeper penetration of the atom. There are two important discoveries made about that time that were extremely useful in opening up the atom for closer examination,

- J. J. Thomson's discovery of the electron (1897),
- Henri Becquerel's discovery of radioactivity (1896).

*Antoine Lavoisier and Joseph Proust were French chemists. John Dalton was an English school teacher and amateur scientist.

11.1 Thomson's Model of the Atom

During the nineteenth century, much effort was directed toward the study of cathode rays—rays that were emitted by hot, metal filaments inside a partially evacuated glass envelope with a high electric potential between the filament and a positively charged collector plate. Evidence for the existence of the rays was provided by the glow coming from the low pressure gas that filled the glass tube. These rays were found to be deflected by electric and magnetic fields, showing that they carried electric charge. In 1897, J. J. Thomson established that the cathode rays are beams of negatively charged particles now called *electrons*. He was also able to demonstrate that these electrons are roughly two thousand times lighter than a hydrogen atom, which is the lightest of all the atoms. Presumably, the electrons were constituents of the atoms in the metal filament. The high voltage ripped them out of the metal and accelerated them toward the collector. In transit, they collided with the gas molecules inside the glass tube, causing the glow.

From this, what can we say about the structure of the atom? The light weight electron can only account for a tiny fraction of the total mass of the atom. Normally, matter—hence, an atom—is electrically neutral, so there must be some positively charged component to balance the negative charge of the electrons. At the time, the approximate sizes of atoms were also known with diameters typically of the order of 10^{-10} m.

These observations led Thomson to the picture of the atom illustrated in Fig. 11.1. In his view, the atom consisted of a spherical distribution of positively charged matter with the required diameter. Embedded in this matter and distributed uniformly, were the electrons—all identical and carrying negative electric charge. The positively charged stuff accounts for most of the mass of the atom and has exactly the right amount of positive charge to cancel the combined electric charge of all of the electrons.

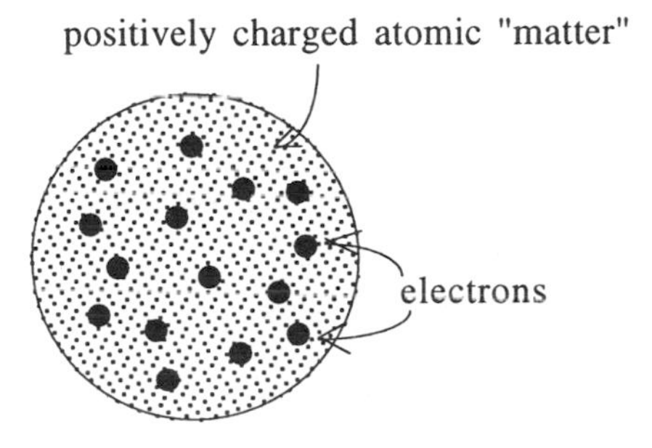

FIGURE 11.1. Thomson's model of the atom.

11.2 The Discovery of Radioactivity

Henri Becquerel, a French chemist, discovered in 1896 that certain materials containing the heavy element, uranium, emitted radiations that could expose photographic plates. The exact origin and nature of the emissions were not immediately clear. A research student of Becquerel, Marie Curie, soon discovered that other heavy elements, thorium, polonium, and radium, also emitted these mysterious radiations.

Experiments were carried out to determine whether the radiations were electrically charged. A laboratory arrangement that illustrates a typical test for electric charge is shown in Fig. 11.2. A hole is drilled into a solid block of lead. A small sample of radioactive material is placed at the bottom of the hole. The lead itself is not radioactive and is a good absorber of radiation. This arrangement permits a narrow beam of radiation to be emitted along the axis of the hole and fall on the photographic plate above. Any radiation reaching the plate will interact with the emulsion in the plate and leave an image. To test the radiation for electric charge, a uniform magnetic field is applied perpendicular to the beam. Any electrically charged particles in the beam will be deflected by the magnetic field—positive charges to the left and negative charges to the right. Electrically neutral radiations will be unaffected by the magnetic field. By using different radioactive materials, it turns out that all three types of radiations are observed. Three kinds of radioactive emissions were recognized, but their identities were still unknown. To distinguish them, they were assigned the first three letters of the Greek alphabet, α, β, and γ. The alpha rays are positively charged, the beta rays negatively charged, and the gamma rays electrically neutral. Within a few years, all of these were identified.

- α Alpha particles are the common type of helium atoms with both electrons removed, making a positively charged ion.
- β Beta particles are electrons.
- γ Gamma rays are high energy (short wavelength) photons—electromagnetic radiation.

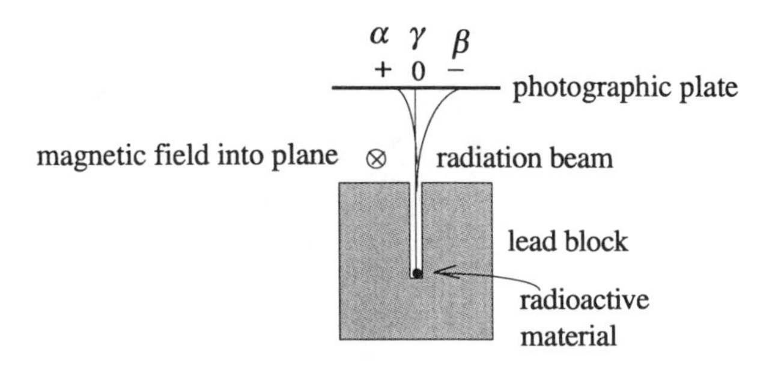

FIGURE 11.2. Detection of radiations from a radioactive source.

11.3 The Geiger–Marsden Experiment

In 1908, one of Thomson's former students, Ernest Rutherford, received the Nobel Prize in Chemistry for his experimental investigations using alpha particles to disintegrate atoms. By then, Rutherford had become professor of physics at the University of Manchester and was continuing his study of the structure of matter using alpha particles as probes. His experimental setup is illustrated in Fig. 11.3. It consisted of a beam of alpha particles produced by an arrangement similar to that shown in Fig. 11.2. The beam was directed onto a thin foil of heavy metal. Presumably, the heavy metal atoms would scatter the lighter alpha particles, which were detected by means of a scintillator placed at various positions on the circular arc around the foil. The scintillator was constructed of material that emits a flash of light when struck by an alpha particle or other radiation.

How would one expect an alpha particle to interact with an atom? In those days, the only two fundamental forces known were gravity and the electromagnetic force. For objects the size of atoms, gravity is too weak to produce any noticeable effect. Now, an alpha particle (helium ion) carries an electric charge and an atom, although electrically neutral, has some electric charge distributed around inside it.

We saw in Chapter 10 that electric charge, like mass, is a fundamental property of matter. The electric force between two electrically charged particles with charges q_1 and q_2 and separated by a distance r is given by Coulomb's law,

$$F_{\text{electric}} = k\,\frac{q_1 q_2}{r^2}\,. \tag{11.1}$$

Whereas the gravitational force between two masses can only be attractive, the electric force between two charges can be either attractive or repulsive, depending on the kinds of charge. If the two charges are alike (both positive or both negative), the charges will repel each other. If they are unlike charges (one positive and the other negative), they will attract each other. An alpha particle (positive charge) would be affected by a much heavier, positively-charged particle as illustrated in

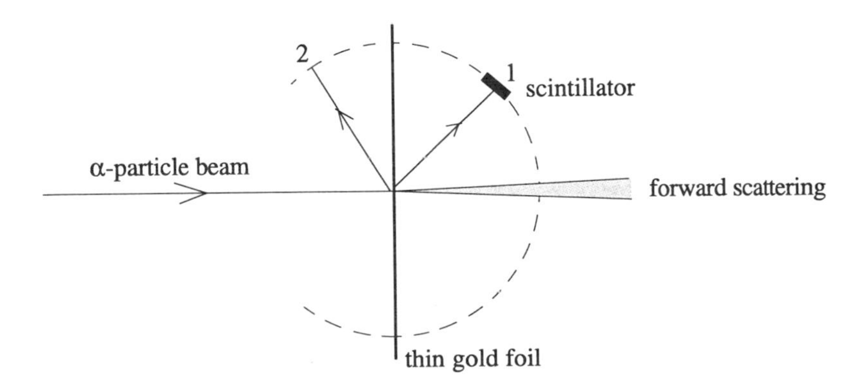

FIGURE 11.3. The scattering of alpha particles incident on a thin gold foil.

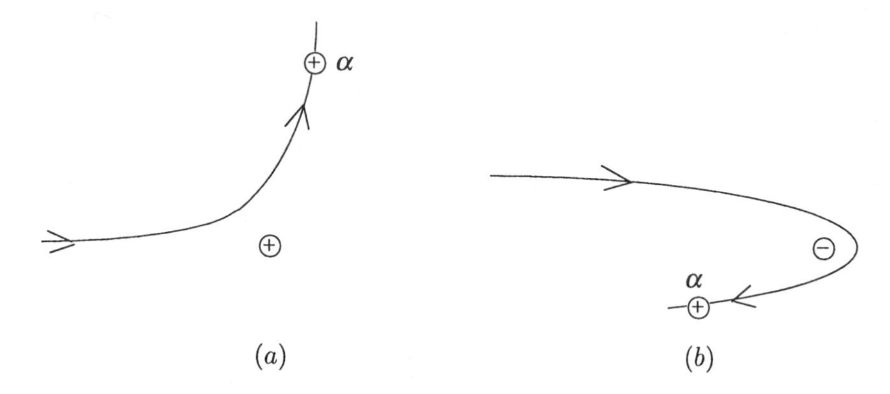

FIGURE 11.4. Coulomb scattering by a heavy particle.

Fig. 10.4*a*. The influence of a heavy, negatively-charged particle on the trajectory of an alpha particle is shown in Fig. 11.4*b*.

Rutherford assigned the task of making the measurements to two assistants in his laboratory, H. Geiger and E. Marsden. Geiger and Marsden used gold to make a foil thin enough for the alpha particles to pass through. Geiger used Thomson's model to calculate the deflection of an alpha particle by a gold atom. An alpha particle is about 8,000 times heavier than an electron. Therefore, the scattering of an alpha particle by an electron is predicted to be very tiny indeed. The lightweight electron is swept aside by the massive projectile barreling through the atom as illustrated in Fig. 11.5.

Geiger also calculated the scattering expected as the alpha particle passes through the continuous distribution of positively charged matter making up the rest of Thomson's atom. Again, he found the expected deflection to be very small. Multiple scattering resulting from encounters with many atoms as an alpha particle travels through the foil is entirely random. Geiger's calculations led him to the conclusion that the alpha particles in his experiment should be scattered into a small cone in the forward direction within an angle of one or two degrees of the incident beam.

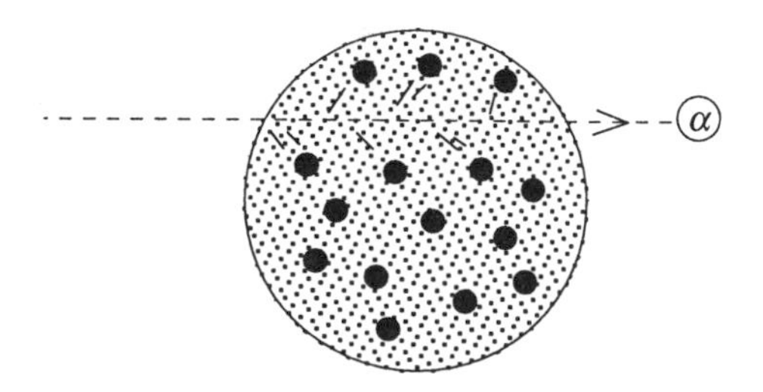

FIGURE 11.5. An alpha particle penetrating Thomson's atom.

For the most part, this is just what Geiger and Marsden saw. However, on very rare occasions they did see alpha particles scattered at large angles (e.g., at 1 in Fig. 11.3, sometimes even at back angles as at 2 in Fig. 11.3. How could this be? There are two important characteristics here. First, there were large-angle scattering events. Second, they didn't occur very often. What are the implications?

11.4 Rutherford's Model of the Atom

Presumably, a single gold atom is responsible for the large angle scattering of the alpha particle in the Geiger-Marsden experiment, since the random scatterings from many atoms would tend to cancel out. It cannot be the electrons that are causing this effect. They simply do not have enough mass to produce that kind of change in the momentum of an alpha particle. Therefore, it must be the heavy part of the atom that is doing it—the part associated with the positive electric charge. That is certainly massive enough to deflect an alpha particle. A gold atom is about fifty times as heavy as an alpha particle. This brings us to the other attribute of the observations—the large angle scattering is a rare event. This led Rutherford to the conclusion that the heavy, positively charged part of the gold atom must be very difficult for an alpha particle to encounter—perhaps because that part of the atom is very small. From these observations, Rutherford arrived at the following picture of an atom. The positive electric charge and most of the mass of the atom are confined to a very tiny volume at the center. The rest of the atomic volume is occupied by the cloud of electrons. The heavy, central part of the atom is called the *nucleus*. Rutherford's model is pictured in Fig. 11.6. This figure is not to scale.

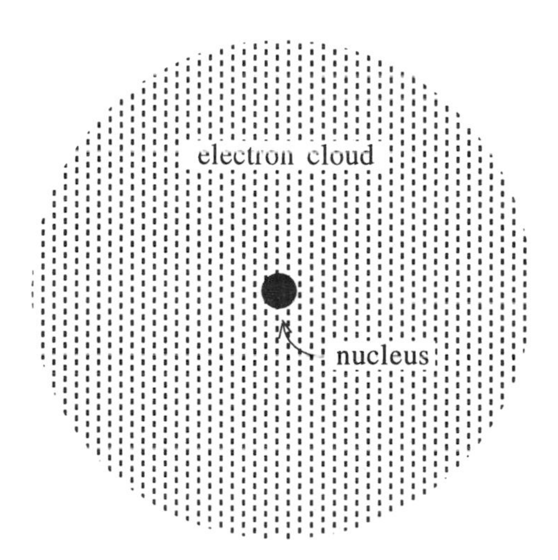

FIGURE 11.6. Rutherford's nuclear atom.

For a typical atom, the diameter of the electron cloud is about 100,000 times bigger than the diameter of the nucleus.

Exercises

11.1. Two pieces of information obtained from the scattering of helium ions on thin gold foils were:
- some ions were scattered at large angles,
- these large angle events were very rare.

How did these observations lead Lord Rutherford to his view of an atom? Be clear, logical, and complete.

11.2. Explain clearly why the results of the Geiger–Marsden experiments were incompatible with Thomson's model of the atom.

11.3. Calculate the electric force between an electron and an alpha particle separated by a distance 2×10^{-10} m (approximately the radius of an atom). Is this force attractive or repulsive? Explain.

Answer: (1.15×10^{-8} N)

11.4. How many times larger is the electric repulsion than the gravitational attraction between two electrons?

11.5. One coulomb is a substantial electric charge. Two electric charges of one coulomb each are separated by a distance of one meter. Calculate the electric force that each of these charges exerts on the other. Suppose an object at rest on the earth's surface experiences a gravitational force (due to the earth) equal to the force that these electric charges exert on each other. What would be the mass of this object? Suppose this object is a ball of solid ice. What is the diameter of this ice ball? The density of ice is about 920 kg/m^3.

Answer: (9×10^9 N; 9.16×10^8 kg; 124 m)

11.6. Explain clearly why Lord Rutherford required the atomic nucleus to have a positive electric charge.

11.7. According to Rutherford's interpretation of the Geiger–Marsden experiments, the nucleus of the gold atom was responsible for the large deflection of the alpha particle. Use words and a drawing to explain clearly and completely how the gold nucleus was able to do this. Do not omit any relevant fact.

11.8. Explain clearly and completely why Lord Rutherford required the atomic nucleus to be very small.

11.9. When the kinetic energy of the α particle in the Geiger–Marsden experiment exceeds about 5×10^6 eV, the distribution of scattered α particles begins to deviate from classical Coulomb scattering. Rutherford interpreted this "anomalous" scattering as evidence that at this energy the α particle begins to penetrate the nucleus. In a head-on collision at this energy, how close does the α particle (charge = $2e$) come to the center of the gold nucleus (charge = $79e$)? (Assume that the gold nucleus remains stationary during the collision.)

12

Hot Stuff

For decades, classical physics was eminently successful in describing the world we know—on earth and in the heavens; a world in which two things were known to exist—matter and radiation. The behavior of matter was described by the system of the world synthesized by Isaac Newton in the seventeenth century. The unified electromagnetic theory of James C. Maxwell, which appeared in the late nineteenth century, provided a complete description of the wave characteristics of electromagnetic radiation, including light. These two—Newton's mechanics (1686) and Maxwell's electrodynamics (1864)—are the pillars of what we now call *classical physics*.

Defects in classical physics began to appear just around 1900. The trouble first became evident in phenomena involving the interaction of these two ingredients of the universe—matter and radiation. We shall look at three prominent early instances of the failure of classical physics to account for experimental observations, namely

- The spectral distribution of radiation from an ideal radiator,
- The photoelectric effect (ejection of electrons from illuminated metal surfaces),
- The discrete spectrum of light emitted from hot hydrogen gas.

The first of these will be taken up in this chapter and the other two in subsequent chapters.

12.1 Radiation from Hot Matter

In the last century, it was already known that all matter radiates and absorbs energy. For a given body, two quantities are of interest here

- *emissive power* (total energy emitted per unit time by the body)
- *absorptivity* (fraction of radiant energy incident that the body absorbs)

Both of these are properties of the radiating surface. In 1859, G. R. Kirchhoff proved that the ratio of these two—emissive power to absorptivity—is exactly the same for all bodies at the same temperature. This tells us that the effectiveness of a body as an emitter of radiation is determined by its effectiveness as an absorber (and vice versa). For our purpose, this is quite useful, because it isn't so obvious, *a priori*, what the characteristics of a "perfect" radiator would be. But from Kirchhoff's theorem, we see that the best radiator is also the best absorber. What is the best absorber? That's easy. It's the one that absorbs *all* of the radiation that falls on it. You can't do any better than that. Such a body that absorbs all radiation incident on it is perfectly black and therefore is technically known as a *black body*.

All black bodies have exactly the same radiation characteristics, regardless of their structure and composition. An ideal radiator can be approximated in a variety of ways. For example, one could coat a metal surface with a layer of lamp black scraped from the chimney of a kerosene lamp. A surface like this can absorb perhaps 99 percent of the radiation incident on it. A more useful arrangement for analytical purposes is the following. Consider a body that has a closed cavity inside it. The cavity can be connected to the outside by drilling a small hole in the object as shown in Fig. 12.1. The *hole* behaves like a black body, since it absorbs virtually all of the radiation incident on it.

We know from everyday experience that hot objects not only get brighter, but change color as well when they get hotter. For example, the burner on an electric kitchen range glows a dull red when turned on at low heat. Turn it up higher, and it gets brighter, but the color also changes to a cherry red. In principle, all bodies emit radiation at all wavelengths, but the relative intensity at each wavelength depends on the temperature.

If we measure the intensity (power per unit area) of light emitted at various wavelengths λ, from a hot object at temperature T_1, we get a distribution that looks qualitatively as shown by the solid curve in Fig. 12.2. Clearly, most of the radiation falls in a region around λ_1. This is the dominant wavelength and determines the *color*.

At a higher temperature T_2, the distribution of radiated energy is represented by the dashed curve in Fig. 12.2. On comparing the two curves, we see that the higher temperature one has more area under it (more energy emitted—accounting for the increase in brightness) and the peak of the curve shifts toward shorter wavelength at higher temperature (accounting for the change in color).

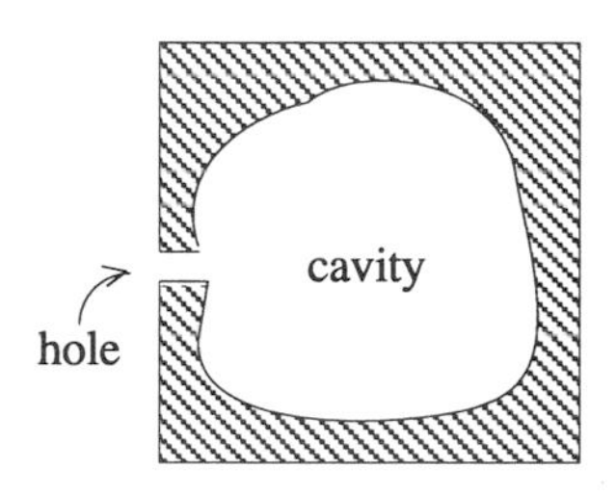

FIGURE 12.1. A practical black body.

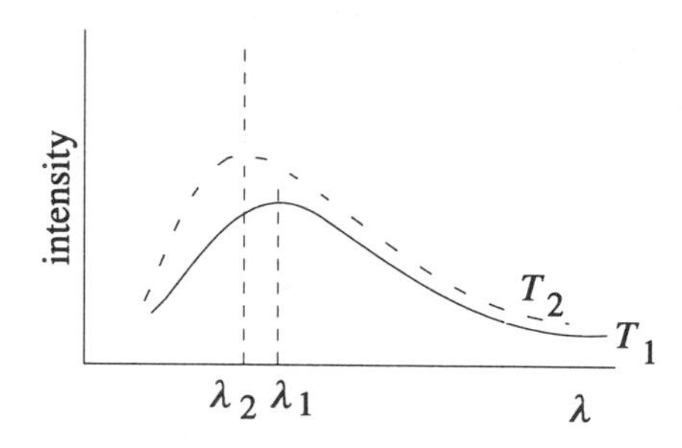

FIGURE 12.2. Spectral distribution of radiation from a black body.

12.2 The Failure of Classical Physics

From a theoretical point of view, it is simpler to calculate the radiation distribution from an ideal radiator—a black body. To confront theory with experiment, it is convenient to adopt the arrangement shown in Fig. 12.3. Since the hole connecting the cavity to the outside behaves like a black body, the radiation emitted from the hole is just that of a black body. For example, if we disperse the emitted light (by means of the prism) into its component colors, we get a distribution characteristic of a black body at the given temperature. A black body is much simpler to treat theoretically because it is an ideal (perfect) radiator.

The radiation emitted from the hole has the same distribution as the radiation inside the cavity. Because of this, black body radiation is often called *cavity radiation*—both have the same spectral distribution. Since the radiation characteristics are the same for all black bodies regardless of their composition or structure, this holds for radiation inside closed cavities as well.

If we heat the cavity walls, the spectral distribution changes qualitatively as described in Fig. 12.2. The advantage here is that it is quite straightforward to calculate from the Maxwell theory the spectral distribution of electromagnetic waves inside a closed box (cavity) at a given temperature. This was first carried out around 1900. The result is compared with experiment in Fig. 12.4. The

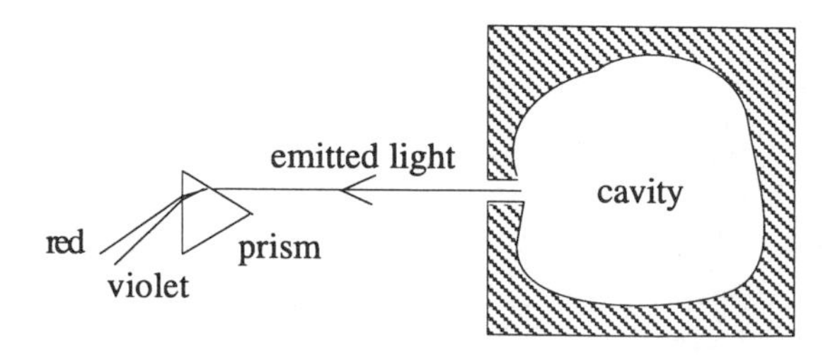

FIGURE 12.3. Spectrum of light from an ideal radiator.

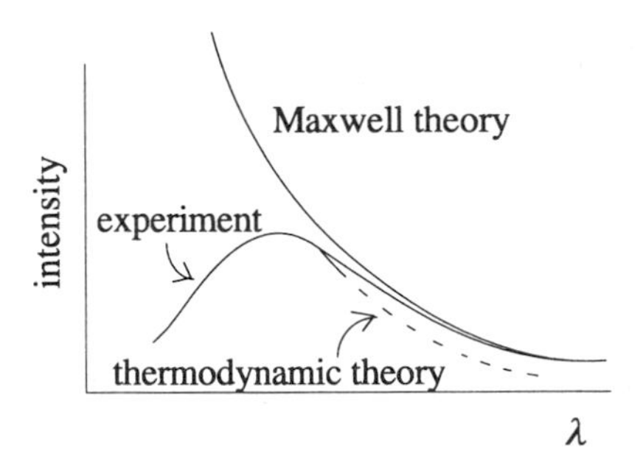

FIGURE 12.4. Black body radiation and classical theories

agreement is quite good for long wavelengths, but disastrously bad for short wavelengths.

A different approach based on thermodynamic arguments was also introduced about this same time. Acclerating electric charges emit radiation. The cavity walls contain bound electric charges (e.g., electrons) that can oscillate about their equilibrium positions. Light—electromagnetic radiation—has an oscillating electric field component. This field incident on one of the charged oscillators in the wall accelerates the charge, causing it to radiate energy back into the cavity. There is a continual exchange of energy between the field inside the cavity and the oscillators in the walls. From statistical mechanics, one can calculate the average energy of the charged oscillators in terms of the temperature of the wall and from this, the spectral distribution of the radiation inside the cavity follows. The result is shown in Fig. 12.4 as the broken line. The agreement with experiment is good for short wavelengths and poor for long ones—just the reverse from the result obtained from the Maxwell theory.

12.3 Planck's Quantum Hypothesis

Clearly, we need a bridge between the two theories. This was provided by a German physicist, Max Planck, just about this same time—1900. He followed the idea of the oscillators in the cavity walls with one completely unjustified, but very important, condition—an oscillator is not allowed to have just any energy, but only a discrete amount that is a whole-number multiple of a fundamental unit proportional to the frequency ν of the oscillator. That is, the energy E_{osc} of an oscillator could only be

$$E_{\text{osc}} = nh\nu \qquad n = 0, 1, 2, 3, \cdots,$$

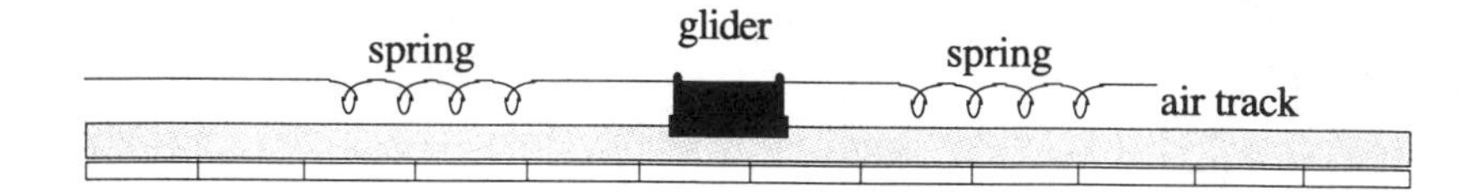

FIGURE 12.5. A classical oscillator.

where h is a fundamental constant of nature known as *Planck's constant.**

What was so radical about this condition that Planck imposed? First of all, it was purely *ad hoc*. There was no ground whatever in classical physics for imposing it. The only thing it had to commend it was that it worked! Exactly how is it in conflict with classical physics? Let's examine the behavior of a classical harmonic oscillator. For this purpose, we return to the glider on the air track with identical elastic springs connected to the ends of the glider as shown in Fig. 12.5.

We push the glider to one side along the track. The work we do in stretching (compressing) the spring is transferred to the glider as elastic potential energy. On releasing the glider, it oscillates back and forth about its equilibrium position. Neglecting friction, the energy of the oscillator is just equal to the work we did in displacing it from equilibrium. This work depends on two things—the stiffness of the springs and the distance we displaced the glider from equilibrium. We can stretch the spring by *any* amount up to the elastic limit of the spring (beyond which the spring is permanently distorted). Therefore, we can give the glider *any amount* of energy up to this limit. So it is with any classical oscillator—it can oscillate with any energy we choose to give it. Planck ignored this freedom allowed by classical physics, and arbitrarily restricted the oscillators in the cavity walls to whole-number units of energy $h\nu$. Thus, was the beginning of a revolution in physical thought, culminating in 1925 with the invention of *quantum mechanics*.

Exercises

12.1. What innovation did Planck introduce to account for the spectral distribution of light from a hot black body? In what way was this a contradiction of classical physics?

12.2. What two noticeable changes occur in a hot black body when its temperature is increased?

---

*Planck's constant is a very tiny quantity. In mks units, it has the approximate value $h = 6.63 \times 10^{-34}$ J·s.

13

Einstein's Bundles

13.1 The Photoelectric Effect

In the late nineteenth century, it was discovered that a negatively charged, clean metal plate lost its electric charge when illuminated by ultraviolet light. Later, after Thomson's discovery of the electron, it was recognized that the negative charge on the metal plate was due to an excess of electrons. These electrons absorbed energy from the electromagnetic field (ultraviolet radiation), increasing their kinetic energies sufficiently to break the bonds that held them and escape from the metal surface. This phenomenon in which electrons are ejected from metal as a result of light shined on the surface is known as the *photoelectric effect*. All metals exhibit this property. For some, it can be demonstrated with visible light.

In an application of classical physics, the electromagnetic field is described by the Maxwell theory and the mechanical behavior of the electron by Newton's laws. First, let's look at the experimental observations and then the predictions of classical physics. The laboratory apparatus includes several monochromatic (single-color) light sources, some samples of various metals, and some instruments for determining the intensity of light incident on the metal surface, the rate at which electrons are ejected from the surface, and the kinetic energies of these electrons. Among the notable observations were

- The number of electrons ejected per unit time varied directly with the intensity of the incident light.
- The kinetic energy ($KE_{\max}$) of the *fastest* electrons ejected from the surface did *not* depend on the intensity of the incident light.
- There was no noticeable time delay (less than a billionth of a second) between the time the light first shined on the surface and the observation of ejected electrons. Electrons were ejected immediately as soon as the light source was turned on. This observation was also independent of the intensity of the incident light.

13.1.1 Predictions of Classical Physics

From the point of view of classical physics, an idealized experimental arrangement is illustrated in Fig. 13.1. Light from the (point) source spreads out as a spherical wave with the intensity diminishing inversely as the square of the distance from the source.*

The light emitted into a narrow cone falls on a part of the metal surface where an electron sits. After some time, depending on the intensity of the light, the electron can absorb enough energy from the electromagnetic field of the incident light wave to break loose from the surface and be observed as a photoelectron.

What should we expect from this picture? If we turn up the intensity of the light source, we supply more energy to the metal surface and are thereby able to give more electrons the energy they need to escape. This is in accord with the first of the experimental observations—increasing the light intensity increases the number of electrons ejected. However, by turning up the intensity, we dump more energy per unit area on the surface. That is what we mean by intensity—energy incident per unit area per unit time. Since each point on the surface receives more energy, then we expect that an electron located at one of these points should also receive more energy and escape with a higher kinetic energy than it would with lower intensity light. That is, the classical theory predicts that the kinetic energy of the fastest electrons ejected should *increase* with increasing intensity of incident light. That is in flat contradiction to the second experimental observation above.

Since the energy from the light source spreads out over the wave front with the intensity diminishing with distance from the source, an electron sitting at a given point on the illuminated surface will require some time to absorb enough energy from the light to escape. In fact, the classical theory predicts that we should be able to lower the intensity of the incident light sufficiently to be able to see time delays of several seconds—perhaps, even up to one minute. Even under these conditions, the measured time delay is still of the order of one billionth of a second.

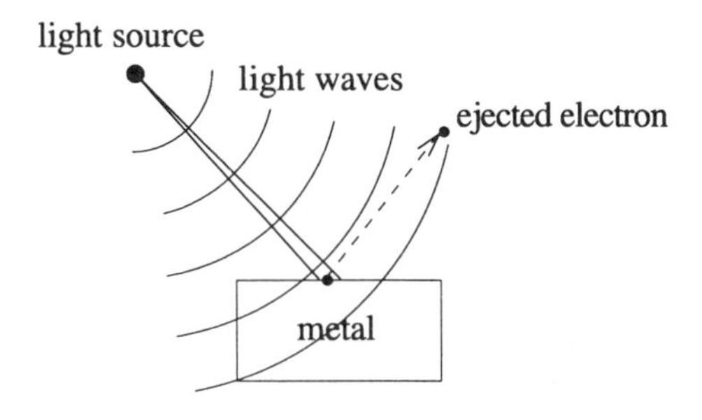

FIGURE 13.1. Classical wave theory account of the photoelectric effect.

*This effect is discussed in more detail in Chapter 16.

So even though the classical theory is able to account for one of the observations, it fails dismally to explain the other two.

13.1.2 Enter Einstein

In 1905, Albert Einstein published a paper in which he proposed a different picture for explaining the photoelectric effect. He imagined that the energy of the electromagnetic field did *not* spread out from the source as a wave, but rather was emitted in little bundles that remained localized in space. In this, Einstein has ascribed the notion of discreteness or quantization—introduced into physics by Planck—to the electromagnetic field itself. Since these little bundles of energy behave like particles—they are localized in space and follow particle-like trajectories—they eventually acquired a particle-like name—the *photon*. Einstein's idea is represented pictorially in Fig. 13.2.

According to Einstein's model, each little bundle or photon carries an amount of energy given by

$$E_{\text{photon}} = h\nu = \frac{hc}{\lambda}, \qquad (13.1)$$

where ν is the frequency of the light and h is Planck's constant. We have also used the relationship, $\lambda\nu = c$, connecting frequency and wavelength with the speed of light c. An electron in the metal can only absorb the energy of the photon as an entire bundle—the *whole* bundle or *none* of it.

Now let's apply this model to the photoelectric effect. The photons fly out of the source, strike the metal, and give up their energy to electrons which leave the surface. If we turn up the intensity of the source, more photons fly out and knock out more electrons, consistent with the first experimental observation. So far, so good.

What about the kinetic energy of the ejected electrons? According to the model, an electron absorbs all of the energy of the photon. It has to use up some of this energy to overcome the force binding it to the metal. Whatever energy that is left over after separating itself from the metal surface is just the kinetic energy with which it escapes. In a naive view, we expect that the electrons that would have

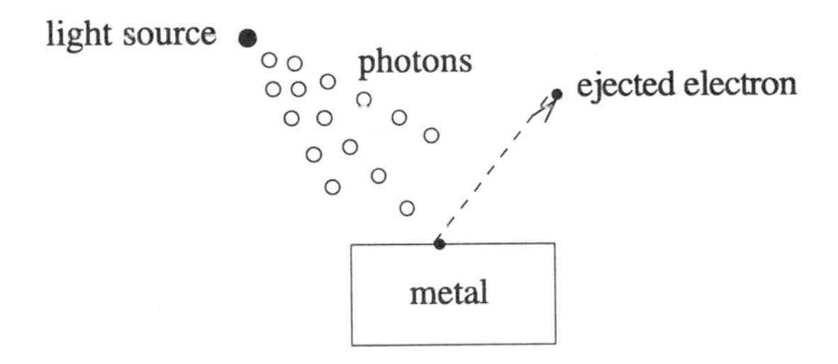

FIGURE 13.2. Photon account of the photoelectric effect.

the least difficulty getting loose—and therefore have the most kinetic energy left over—would be those sitting on the surface of the metal. In any case, the binding energy of the *least* tightly bound electrons depends on the atomic arrangement of the particular metal and therefore, varies from metal to metal. This energy is called the *work function* of the metal and is denoted by W.

A convenient unit for expressing the very small amounts of energy associated with atomic and subatomic particles is the *electron volt*, abbreviated eV. One electron volt is the amount of energy acquired by an electron when it is accelerated through an electric potential difference of one volt. That is, from Eq. (10.17) we have,

$$\text{electric energy} = qV = (1.60 \times 10^{-19}\,\text{C})(1\,\text{volt}) = 1.60 \times 10^{-19}\,\text{J} \equiv 1\,\text{eV}.$$

Work functions for various metals have been measured extensively and are tabulated in handbooks listing the properties of materials. Values for some selected metals are given in Table 13.1.

It is clear then, that the kinetic energy of the fastest electrons ejected from the metal is given by,

$$KE_{\text{max}} = \tfrac{1}{2} m v_{\text{max}}^2 = \frac{hc}{\lambda} - W. \tag{13.2}$$

According to this equation, the maximum kinetic energy of the ejected electrons depends on some property of the metal and on the *color* of the incident light, but not at all on the *intensity* of the incident light. This is entirely consistent with the second experimental observation.

Finally, we look at the time delay predicted by Einstein's model. The photon travels with the speed of light. It *is* light! A quick calculation shows that at this speed, it takes about a billionth of a second for the photon to travel a few tens of centimeters—typical laboratory distances. As soon as the light source is turned on, a single photon can fly at this speed to an electron in the metal and immediately knock it out. There is no appreciable time delay and, since a single photon ejects a single electron, the incident intensity plays no role in considering a time delay.

Table 13.1. Photoelectric Work Functions

metal	work function (eV)	metal	work function (eV)
aluminum	4.08	lithium	2.36
cadmium	2.70	lead	4.14
calcium	2.71	potassium	2.25
chromium	4.38	silver	4.40
copper	4.81	sodium	2.28
gold	5.26	uranium	3.63

Here is another instance where a crack appeared in the structure of classical physics. It failed to explain correctly the photoelectric effect—a phenomenon involving the interaction of electromagnetic radiation with matter. Einstein invoked Planck's notion of discreteness in nature by quantizing the electromagnetic field itself and was able to account for the photoelectric effect admirably.*

Kinetic energy never takes on negative values. Notice that if the wavelength λ of the light is too large, the right-hand side of Eq. (13.2) is negative. This means that *no* electrons are ejected from the metal by such light. This explains why the photoelectric effect can be observed for some metals only with ultraviolet light while for others it can be seen with visible light.

EXAMPLE. When light of wavelength 411 nm is shined on the surface of a certain metal, electrons are ejected from the metal. The electrons that leave the metal with the greatest speed have kinetic energy 0.57 eV. Find the kinetic energy of the fastest electrons ejected from this metal if the light source is replaced by one emitting light of wavelength 439 nm.

SOLUTION. First, we calculate the work function for this metal using the original wavelength and Eq. (13.2)

$$KE_{\text{max}} = \frac{hc}{\lambda} - W$$

$$0.57\ \text{eV} = \frac{1{,}242\ \text{eV} \cdot \text{nm}}{411\ \text{nm}} - W,$$

from which we find that $W = 2.45$ eV.

With this value of W, we apply Eq. (13.2) again

$$\begin{aligned} KE_{\text{max}} &= \frac{hc}{\lambda} - W \\ &= \frac{1{,}242\ \text{eV}}{439\ \text{nm}} - 2.45\ \text{eV} - 0.38\ \text{eV}. \end{aligned}$$

Thus, with incident light of wavelength 439 nm, the fastest photoelectrons leave the metal with kinetic energy 0.38 eV.

13.2 Momentum of Light

Electromagnetic radiation carries momentum. An example of this is seen in the formation of the dust tail of a comet.† The sunlight incident on the dust in the head of a comet exerts a pressure on the dust and pushes it away from the comet head to form the dust tail.

---

*Alternative explanations that do not require the existence of photons have been proposed. For a discussion of this point, see P. W. Milonni, *Am. J. Phys.* **65** (1997), p. 11.

†Comets are discussed in more detail in Chapter 14.

Maxwell's theory provides the relationship between the energy E of an electromagnetic wave and its momentum p. It is simply $E = pc$, where c is the speed of light. Expressing this result now in the language of Einstein's photon model of light, we see that the energy E_{photon} and momentum p_{photon} of a photon are related by

$$E_{\text{photon}} = p_{\text{photon}}c. \tag{13.3}$$

13.3 Equivalence of Mass and Energy

In 1906,* Einstein proposed a *Gedankenexperiment* (thought experiment) to establish his famous relationship between mass and energy,

$$E = mc^2. \tag{13.4}$$

He imagined a box of length L at rest with no external forces exerted on it. A photon is emitted (say, from an atom in the left wall of the box) at one end of the box at time $t = 0$ and is absorbed at the other end of the box at a later time t as illustrated in Fig. 13.3.

To conserve momentum (no *external* forces involved), the box must recoil in the direction opposite to the flight of the emitted photon. During the time t that the photon is in flight, the center of mass of the box will move a distance d as seen in Fig. 13.3b. However, the center of mass of the entire photon-box system cannot have moved during this time, because it was initially at rest and no external forces are exerted on the system. Since the center of mass of the box moves to the left,

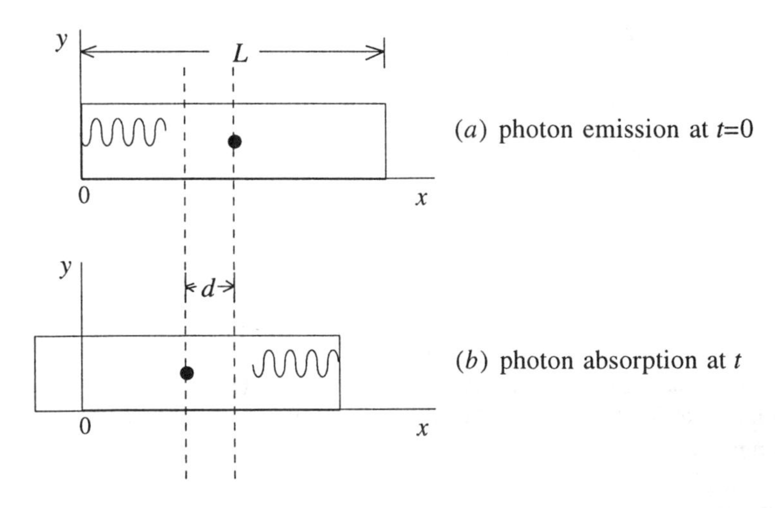

FIGURE 13.3. Einstein's *Gedankenexperiment* demonstrating $E = mc^2$.

*A. Einstein, *Annalen der Physik* **20** (1906), p. 627.

the photon must carry a mass m' to the right such that the position of the center of mass of the entire system remains fixed. Let m be the mass of the box alone (that is, its mass while the photon is in flight). For the two times, we have:

time $= 0$: The photon is at $x = 0$ and the center of mass of the box is at $x = L/2$ as shown in Fig. 13.3*a*. Thus, the center of mass of the system is at

$$x_{\text{CM}} = \frac{m'(0) + m(\frac{L}{2})}{m + m'} \quad (\text{time} = 0).$$

time $= t$: The photon is at $x = L - d$ and the center of mass of the box is at $x = \frac{L}{2} - d$ as seen in Fig. 13.3*b*. Hence, the center of mass of the system is at

$$x_{\text{CM}} = \frac{m'(L - d) + m(\frac{L}{2} - d)}{m + m'} \quad (\text{time} = t).$$

Because the center of mass has not moved, these expressions must yield the same value for x_{CM}. On setting the two expressions equal, we find that

$$m'(L - d) = md. \tag{13.5}$$

From momentum conservation, the momentum mv of the recoiling box must be equal to the momentum p' of the emitted photon. Invoking Eq. (13.3), this leads to

$$v = \frac{E'}{mc}$$

for the speed v of the recoiling box. The photon travels a distance $L - d$ with speed c. Thus, the time of flight of the photon is $t = (L - d)/c$. From these results, we obtain the distance d traveled by the box,

$$d = vt = \frac{E'}{mc^2}(L - d). \tag{13.6}$$

Eliminating $L - d$ from Eqs. (13.5) and (13.6), we get

$$m' = \frac{E'}{c^2}. \tag{13.7}$$

The interpretation of this result is that on emitting a photon the box loses mass in the amount m', which has an energy equivalent $E' = m'c^2$. On reabsorption of the photon, the mass of the box increases by the same amount m'. Putting it another way, some of the mass of the box is transformed into energy in the form of the emitted photon. This energy is then converted back into mass when the photon is absorbed by the box. Thus, we have established the equivalence of mass and energy expressed in Eq. (13.4).

EXAMPLE. The mass of a proton is 1.67×10^{-27} kg. Calculate the energy equivalent of this mass.

SOLUTION. From Eq. (13.4), we have

$$E = mc^2 = (1.668 \times 10^{-27}\ \text{kg})(3 \times 10^8\ \text{m/s})^2 = 1.50 \times 10^{-10}\ \text{J} = 938\ \text{MeV}.$$

The energy equivalent of the mass of a proton is about 1.50×10^{-10} J or 938 MeV.

Exercises

13.1. A photon has an energy of 2.50 eV. Find its frequency and its wavelength in free space.

Answer: (6.04×10^{14} Hz; 497 nm)

13.2. The work function of mercury is 4.53 eV. Find the longest wavelength light that can produce photoelectric emission from this element and the maximum kinetic energy of photoelectrons emitted when the mercury is irradiated with ultraviolet photons of wavelength 193 nm.

Answer: (274 nm; 1.90 eV)

13.3. When green light falls on the surface of a piece of calcium metal, electrons are found to be ejected from the metal. When the same green light shines on a piece of tin, no photoelectrons are observed. Use Einstein's theory of the photoelectric effect to explain the difference in observations.

13.4. Blue light of wavelength 476 nm is shined on the surface of a piece of metal in a vacuum. The most energetic photoelectrons ejected from this metal are observed to have kinetic energy 0.22 eV. Give a complete quantitative description of what is observed if the light source is replaced by another one of wavelength 514 nm. Describe quantitatively what is observed if this second light source is replaced by a third one of wavelength 537 nm.

13.5. Electrons are ejected from a certain metal when light is shined on the surface of the metal. The kinetic energy of the fastest electrons ejected is observed for several different colors of incident light. The data are shown in the graph. What is the value of the photoelectric work function for this metal? To what physical quantity is the slope of this graph related?

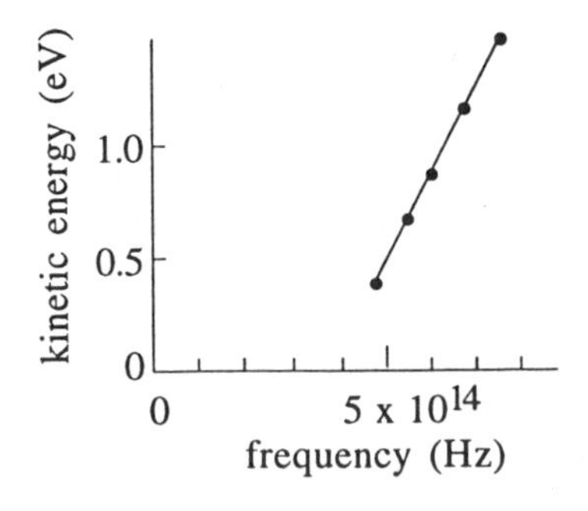

13.6. A human eye can detect light at very low intensity. The eye is most sensitive to yellow light of wavelength around 550 nm. Suppose electromagnetic energy (light) of this wavelength from a low-intensity source enters the eye at a rate of 10^{-16} watt (about the threshold of detection for the eye). Calculate the number of photons entering the eye each second.

Answer: (276)

13.7. Ultraviolet light of wavelength 212 nm is shined on the surface of one of the metals listed in Table 13.1. The most energetic photoelectrons are emitted with kinetic energy 1.05 eV. Which metal is it?

13.8. Calculate the kinetic energy of the fastest electrons ejected when violet light of wavelength 410 nm is shined on a piece of cadmium.

Answer: (0.33 eV)

13.9. When light of a certain wavelength is shined on the clean surface of a piece of lithium metal, the fastest electrons leave the metal with speed 3.08×10^5 m/s. Calculate the wavelength of the incident light.

Answer: (472 nm)

13.10. From the graph, estimate the value of Planck's constant h and the work function W for each of the three metals, sodium, calcium, and lead.

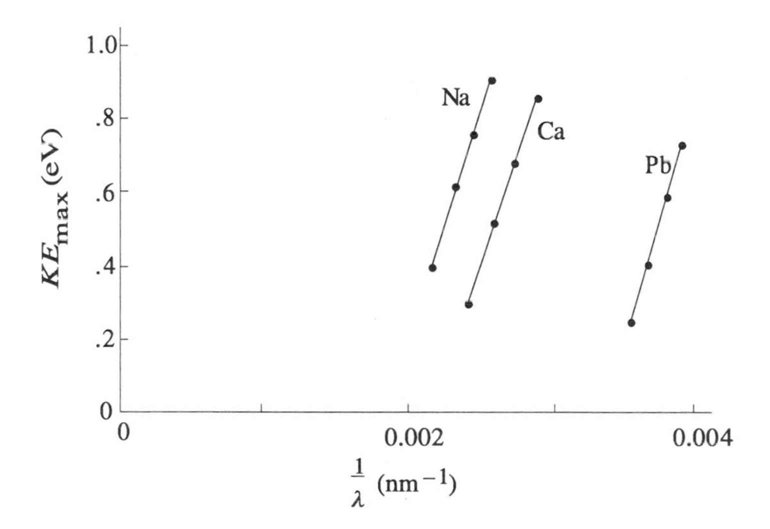

13.11. The fastest photoelectrons ejected from the surface of copper when irradiated by a certain light source have kinetic energy 2.38 eV. Calculate the wavelength of the radiation. Describe the radiation. Explain.

13.12. Photoelectrons are ejected from a certain piece of metal when ultraviolet light of wavelength 350 nm is shined on its surface. The speed of the fastest electrons emitted is measured at 4.39×10^5 m/s. Calculate the work function (in eV) for this metal.

Answer: (3.00 eV)

13.13. The surface of a sample of potassium is irradiated with monochromatic light. Photoelectrons are observed with energies up to 0.234 eV. What color is this light? Explain clearly your reasoning.

13.14. Photoelectrons are observed when ultraviolet light of wavelength 291.5 nm is shined on the clean surface of a metal bar. The kinetic energy of the electrons emitted with the greatest speed is 0.18 eV. Calculate the kinetic energy of the fastest electrons ejected when ultraviolet light of wavelength 279.0 nm is shined on this metal surface. How fast are these electrons moving as they leave the surface of the metal bar?

Answer: (0.37 eV; 3.6×10^5 m/s)

13.15. Calculate the energy available from one gram of matter if it could be converted completely into energy. Approximately how long could a 1,200-watt electric heater be operated continuously with this energy source?

Answer: (9×10^{13} J; 2,400 years)

13.16. Estimate the mass equivalent (in kg) of a photon of yellow light (λ = 550 nm). Approximately what fraction of the mass of a hydrogen atom (1.67×10^{-27} kg) does this represent?

Answer: (4.0×10^{-36} kg; 2.4×10^{-9})

13.17. The speed of a five-gram bullet fired from a rifle is 300 m/s. Calculate the kinetic energy of the bullet. What is the mass equivalent (in kilograms) of this energy? Approximately what fraction of the mass of the bullet does this represent?

Answer: (2.5×10^{-15} kg; 5×10^{-13})

13.18. Under certain conditions, electrons are ejected from a metal surface when light is shined on it. This is the photoelectric effect. These electrons are ejected with a range of speeds from zero up to some maximum value. Contrary to the prediction of the classical electromagnetic theory of light, measurements show that the energy of the fastest of these photoelectrons does not depend on the intensity of the incident light. Explain clearly and completely *in words* how Einstein's theory accounts for this observation.

13.19. The self-energy of an electron is given by Eq. (10.14) with $q = Q = e = 1.60 \times 10^{-19}$ C. The so-called classical radius of the electron ($r = r_c$ in Eq. 10.14) is obtained by setting the electron self-energy equal to the rest energy of the electron. Calculate r_c.

Answer: (2.82×10^{-15} m)

13.20. A bare 500-W light source radiates energy uniformly in all directions. Calculate the intensity of the radiation at a distance of 50 cm from the source. (The intensity of light at a distance r from the source is equal to the total power radiated divided by the surface area of a sphere of radius r.) Suppose the radiation falls on the clean surface of a sodium block placed 50 cm from the source. According to the classical wave theory of the photoelectric effect, how long would it take for an electron sitting on the surface to absorb enough energy to escape from the sodium block? See Exercise 13.19 for the classical radius of the electron.

Answer: (159 W/m^2; 2.9 years)

13.21. When green light of wavelength 493.1 nm is shined on a certain metal surface, photoelectrons are ejected from the metal. Blue light of wavelength 440.0 nm shined on this surface yields photoelectrons with 3.2 times the kinetic energy of the fastest electrons ejected with the green light. Calculate the photoelectric work function for this metal.

Answer: (2.37 eV)

13.22. When light of wavelength 472.2 nm is shined on the clean surface of a certain metal, the fastest electrons leave the metal with speed 3.08×10^5 m/s. When the light source is changed to a different wavelength, electrons with speeds up to 3.97×10^5 m/s are observed. Calculate the wavelength of this second light source.

Answer: (442.2 nm)

13.23. Photoelectrons with a maximum kinetic energy of 0.146 eV are ejected from a metal surface when light of wavelength λ is shined on the surface. Shining light of wavelength $\lambda/3$ on the same metal surface produces photoelectrons with a maximum kinetic energy of 4.527 eV. Calculate λ.

Answer: (567.0 nm)

13.24. The energy released in burning a ton of coal is about 7,300 kW·h. How many joules is this? What is the mass equivalent of this energy? What is the energy equivalent of the mass of a ton of coal? What fraction of the mass of the coal is equivalent to the energy released in burning it? (The mass of one ton of a substance is about 907 kg.)

Answer: (2.6×10^{10} J; 2.9×10^{-7} kg; 8.2×10^{19} J; 3.2×10^{-10})

13.25. A 3.5-g ballpoint pen is lifted from the top of a desk to a height of 10 cm above the top of the desk. What is the change in gravitational potential energy of the pen? If this amount of energy is converted into yellow light of wavelength 583 nm, how many photons would be produced? What is the mass equivalent (in kilograms) of this amount of energy?

14

The Great Dane

We cannot go to the stars to sample them directly. So how do we get information about the stars? We have to extract it from something that *comes* to us *from* the stars. Most commonly this "something" is light—electromagnetic radiation. To decode the information contained in the light, we have to know some things about how it is produced in the first place. What does the light tell us about the thing that is emitting the light? Ultimately, where does the light come from anyway? In this chapter, we seek answers to these questions.

14.1 The Structure of the Atom

Matter emits light. Matter is made of atoms. Therefore, atoms emit light. Reasonable. And true. But how do they do that? What does light coming from individual atoms look like? We can answer this second question by examining the light emitted by a hot, dilute gas or vapor of a single element. In a dilute (low density) gas, the atoms are on the average far enough apart that we can be sure the light we see comes directly from isolated atoms. We analyze this light by first passing it through a narrow slit and then dispersing it (e.g., by means of a prism). The dispersion fans the light into a series of slit images or lines of different colors at different positions. This pattern of different colored images is called a *line spectrum*. There are two remarkable observations:

- The line spectrum for each element is discrete—not continuous. Not all colors are present.
- The line spectrum for each element is different from that of every other element. The spectrum is unique.

14.1.1 The Hydrogen Spectrum

In the next chapter, we shall endeavor to understand both of these features in terms of atomic structure. For the present, we concentrate on the line spectrum of the lightest element, hydrogen. Production of this spectrum is illustrated in Fig. 14.1.

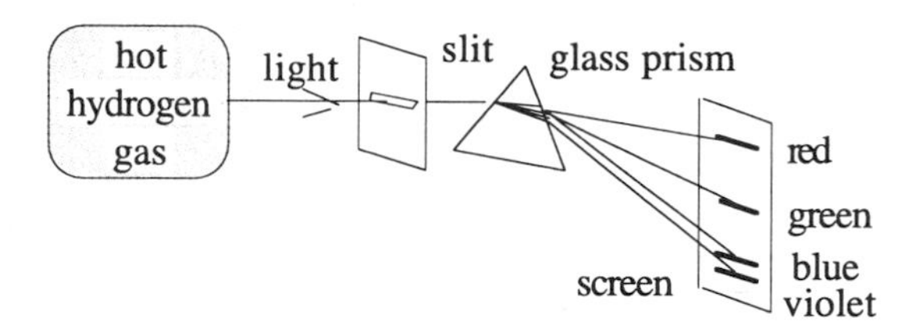

FIGURE 14.1. Dispersion of light from hot hydrogen gas.

The hydrogen spectrum is characterized by four prominent lines—red, green, blue, and violet as shown qualitatively in Fig. 14.2 with the red line at 656 nm.

A high resolution spectrometer reveals that the violet line (410 nm) actually consists of a number of separate lines very close together. Several of these lines had been observed and their wavelengths measured quite precisely already in the nineteenth century.

14.1.2 The Balmer Formula

In 1885, J. J. Balmer published a paper in which he noted that the wavelengths of all of the observed lines in the hydrogen spectrum could be given by a single formula, which we write in the form,

$$\frac{1}{\lambda} = R\left(\frac{1}{4} - \frac{1}{n^2}\right), \qquad n = 3, 4, 5, 6, \ldots \tag{14.1}$$

where R is a constant ($R = 0.01097\,\text{nm}^{-1}$). This is a remarkable formula that suggests some deeper physical significance concerning the structure of the hydrogen atom itself. However, the formula is purely empirical and provides us with no insight whatever toward any deeper understanding of the nature of the atom emitting the light. It is just number juggling. The formula works, but that is as far as it goes. On the other hand, any acceptable theoretical model we construct for the atom, must yield this formula. Thus, the Balmer formula, Eq. (14.1), provides a test for models of the hydrogen atom.

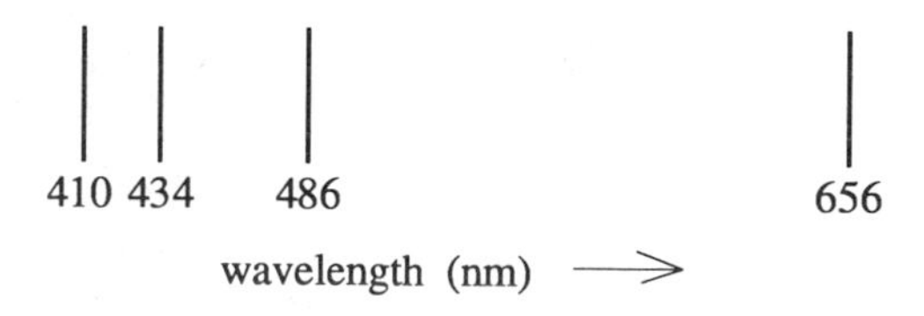

FIGURE 14.2. The hydrogen spectrum.

14.2 The Man from Copenhagen

From Denmark, in 1912, a brilliant and energetic young physicist came to work in Manchester. He was already familiar with Rutherford's interpretation of the Geiger–Marsden experiments—the nuclear model of the atom. Rutherford's picture is sometimes referred to as the planetary model because of its similarity to the arrangement of the solar system with the electrons describing closed orbits about the nucleus as planets orbit the sun. It is the sun's gravity that binds the planets in their orbits. In Rutherford's atom, the electrons are bound to the nucleus by the electric force.

This gravitational analogue is useful in providing us with an intuitive sense of what goes on inside an atom. We shall return to it often. However, there is a very important flaw in the argument—a flaw that Rutherford had recognized, but could not repair within the framework of classical physics. The task of finding a remedy awaited the young man from Copenhagen—Niels Bohr.

Bohr addressed his attention to the simplest of all the atoms—hydrogen. In Rutherford's view, hydrogen has a single electron electrically attracted to a nucleus with a single unit of electric charge. This is illustrated in Fig. 14.3*a* where the electron is shown in a circular orbit around the nucleus.*

As we have already noted, classical electrodynamics—the Maxwell theory—predicts that an accelerated electric charge emits electromagnetic radiation. That is how a radio station works. High frequency electric currents are run up and down the antenna and the energy given to the electric charge in accelerating it in this way is radiated away as electromagnetic waves to be detected by a distant radio receiver.

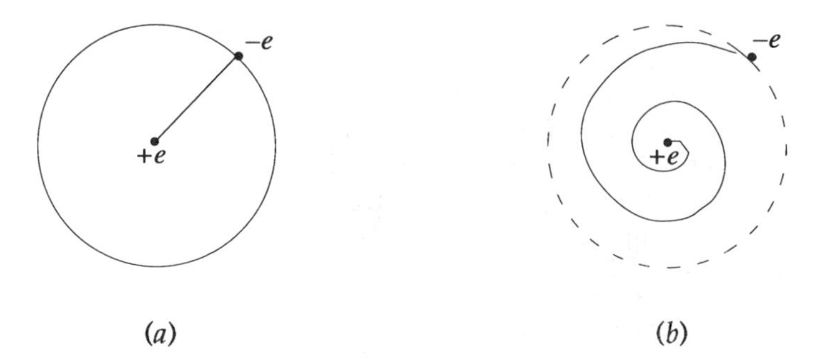

FIGURE 14.3. Classical model of the hydrogen atom.

*More correctly, both electron and nucleus revolve about a point directly between them (the center of mass). Because of the large mass of the nucleus compared to that of the electron, this point is very close to the center of the nucleus. For our purpose, we can consider the nucleus to be at the center of the electron's orbit.

14.2.1 Classical Physics Fails Again

The orbiting electron is continuously accelerated by the electric force attracting it to the nucleus. Therefore, according to the Maxwell theory, it must lose energy in the form of electromagnetic radiation. With this continuous loss of energy, its orbit becomes smaller and smaller until it spirals into the nucleus as depicted in Fig. 14.3*b*. Within a very short time, the atom shrinks to the size of the nucleus. Rutherford knew that atoms don't do that. Atomic sizes and nuclear sizes could be determined. Typically, the diameter of an atom is about 100,000 times larger than that of the nucleus. An atom does not collapse to the size of its nucleus. Classical physics fails to support the picture of the atom suggested by the experiment of Geiger and Marsden.

14.2.2 Bohr's Quantum Hypothesis

So what is the remedy? Bohr dealt with the problem by declaring that there are certain discrete orbits that the electron can occupy without this continuous spiral toward the nucleus. His criterion for a so called "allowed" orbit is that the orbital angular momentum L for the electron in that orbit is a whole number multiple of a fundamental unit of angular momentum given by Planck's constant. That is,

$$\text{allowed Bohr orbit: } L = n\,\frac{h}{2\pi}, \qquad n = 1, 2, 3, \ldots \tag{14.2}$$

where h is Planck's constant. This is a purely *ad hoc* assumption with no justification whatever in classical physics. Classically, the angular momentum of a particle of mass m with speed v at a distance r from the point about which it is orbiting is

$$L = mvr. \tag{14.3}$$

The electric force that the nucleus exerts on the orbiting electron is given by Eq. (10.1),

$$F_{\text{electric}} = k\,\frac{e^2}{r^2}, \tag{14.4}$$

where k is the electric constant.* This force is always directed toward the nucleus causing centripetal acceleration of the electron. Therefore, from Newton's second law we find that

$$mv^2 = k\,\frac{e^2}{r}. \tag{14.5}$$

The kinetic energy of the electron is

$$KE = \tfrac{1}{2}mv^2 \tag{14.6}$$

---

*In mks units, $k = 9 \times 10^9$ N·m²/C².

and from Eq. (10.14), the electric potential energy is

$$PE = -k\frac{e^2}{r}. \tag{14.7}$$

The minus sign shows that the electric force is attractive. On comparing these expressions with Eq. (14.5), we see that $PE = -2KE$. This result coupled with Eq. (14.7) yields

$$\frac{1}{r} = \frac{2KE}{ke^2}. \tag{14.8}$$

Moreover, we can get the total energy E from the kinetic energy alone,

$$E = KE + PE = -KE. \tag{14.9}$$

According to Eqs. (14.3) and (14.6),

$$KE = \frac{1}{2}mv^2 = \frac{L^2}{2mr^2} = \frac{L^2}{2m}\frac{4(KE)^2}{(ke^2)^2}. \tag{14.10}$$

The last equality in Eq. (14.10) follows from Eq. (14.8). Finally, we replace L in Eq. (14.10) with Bohr's value given in Eq. (14.2) and obtain from Eq. (14.9) the energy of the nth Bohr orbit,

$$E_n = -\frac{2\pi^2 mk^2e^4}{h^2n^2}. \tag{14.11}$$

We calculate the combination of constants here from known values of the electron mass, Planck's constant, and so on, to obtain,

$$E_n = -\frac{13.625 \text{ eV}}{n^2}, \qquad n = 1, 2, 3, \ldots. \qquad (14.12).$$

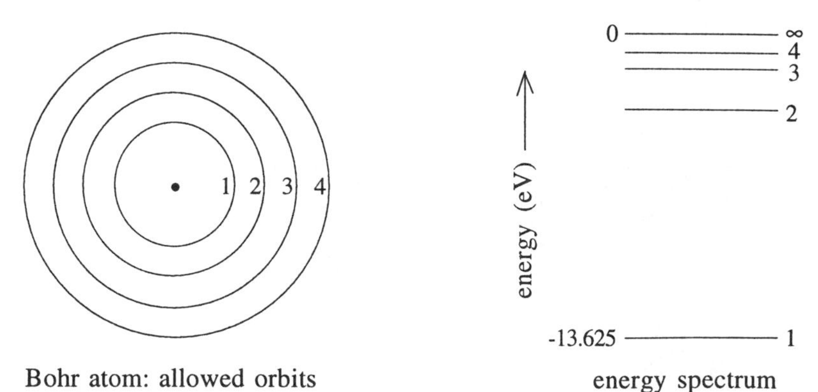

FIGURE 14.4. Allowed electron orbits in the Bohr model of hydrogen.

The lower (more negative) the energy, the closer the electron is to the nucleus. Therefore, the orbit closest to the nucleus has $n = 1$ with energy $E_1 = -13.625$ eV. With the electron in this orbit, the atom is said to be in the *ground state*. The first few low-energy orbits are shown schematically in Fig. 14.4*a*. Note that the *highest* energy given by Eq. (14.12) is zero corresponding to $n = \infty$. Thus, we have an infinite number of orbits ($n = 1, 2, 3, \ldots, \infty$) and all of them have energy between -13.625 eV and 0 eV. This energy range and part of the energy spectrum are represented in Fig. 14.4*b*.

According to this model of the atom, how is the spectrum of light from hot hydrogen gas illustrated in Fig. 14.1 produced? The electron in an atom tends to get as close to the nucleus as it can, that is, in the orbit with $n = 1$. When the gas is cool, the atoms (molecules) move relatively slowly and do not collide with sufficient energy to excite each other. In a hot gas, the atoms move around much faster and often collide violently, knocking each others' electrons into excited states (with $n \geq 2$). An excited electron tends to fall back toward the nucleus, perhaps in a cascade to lower-energy orbits, eventually stopping in the ground state.

In falling from a higher-energy orbit to a lower-energy orbit, the electron loses energy. Because it is the electric force that causes the electron to make this transition, it is electromagnetic energy that is released in the form of a photon. Invoking the energy conservation principle, Eq. (5.3), we have

$$E_{\text{initial}} = E_{\text{final}} + E_{\text{photon}}. \tag{14.13}$$

Calculating the initial and final energies of the electron with Eq. (14.12) and the energy of the emitted photon with Eq. (13.1) and rearranging, Eq. (14.13) can be written

$$\frac{1}{\lambda} = \frac{13.625\ \text{eV}}{hc}\left(\frac{1}{n_f^2} - \frac{1}{n_i^2}\right). \tag{14.14}$$

In this equation, λ is the wavelength of the light emitted when an electron jumps from an initial state with quantum number n_i to a lower-energy state with quantum number n_f. On substituting values for the constants, Eq. (14.14) becomes

$$\frac{1}{\lambda} = (0.01097\ \text{nm}^{-1})\left(\frac{1}{n_f^2} - \frac{1}{n_i^2}\right). \qquad \text{(emission)} \tag{14.15}$$

We see that this equation agrees with Balmer's formula, Eq. (14.1), if we set $n_f = 2$.

Now, with Bohr's model we can make a physical connection between Balmer's empirical formula and what happens inside the atom to produce the light. An excited electron falls from an orbit with $n_i \geq 3$ down to the orbit with $n_f = 2$ to produce the visible light. Since the red light has the longest wavelength in the Balmer series, it must correspond to the *smallest* energy change in falling down to the $n = 2$ orbit, that is, from $n_i = 3$ to $n_f = 2$. The green line is the next-to-longest visible wavelength and is emitted in the electron transition from $n_i = 4$ to $n_f = 2$. And so on.

Notice that Eq. (14.15) includes Balmer's formula, Eq. (14.1), but extends further to describe electron transitions to final states other than $n_f = 2$. For example, when the electron drops from the $n = 2$ state directly to the ground state ($n = 1$), Eq. (14.15) yields $\lambda = 121.5$ nm for the wavelength of the emitted photon. The shortest wavelength light that the human eye can detect is violet light with wavelength around 360 nm. As we have just seen, even the longest wavelength transition ($\lambda = 121.5$ nm) in hydrogen with $n_f = 1$ is much shorter than that and lies well in the ultraviolet region of the spectrum. So, with $n_f = 1$, Eq. (14.15) defines a series of ultraviolet transitions in addition to the Balmer series of visible emission lines corresponding to $n_f = 2$.

Similarly, the *shortest* wavelength light emitted from hydrogen with $n_f = 3$ has $\lambda = 820$ nm, which is also not visible, but lies in the infrared portion of the spectrum. Other infrared series are characterized by $n_f \geq 4$.

Clearly, Bohr's model accounts for the visible light from hydrogen studied by Balmer and represented by his formula, Eq. (14.1). It provides a clear picture of the production of the radiation in terms of the behavior of the particles that make up the atom. But more than that, Bohr's model predicts the existence of invisible radiations from hydrogen that Balmer's work did not include.

EXAMPLE. Ultraviolet radiation of wavelength 93.76 nm is detected from hot hydrogen gas. According to Bohr's model of the hydrogen atom, what was the initial state from which the electron fell in this transition?

SOLUTION. Since we are told that the radiation is ultraviolet, we know that $n_f = 1$. Using this information in Eq. (14.15), we obtain

$$\frac{1}{93.76 \text{ nm}} = (0.01097 \text{ nm}^{-1})\left(\frac{1}{1^2} - \frac{1}{n_i^2}\right),$$

from which we find that $n_i = 6$. Remember that n *must* be an integer.

Next we apply the model to the absorption of light by hydrogen. Suppose we allow a beam of white light emerging from a narrow slit to pass through a transparent container of hydrogen gas and then disperse the light by means of a prism to obtain a spectrum. The arrangement is illustrated in Fig. 14.5.

What do we see? The white light (for example from a hot filament in an incandescent light bulb) contains all of the colors from red to violet. Since all of the

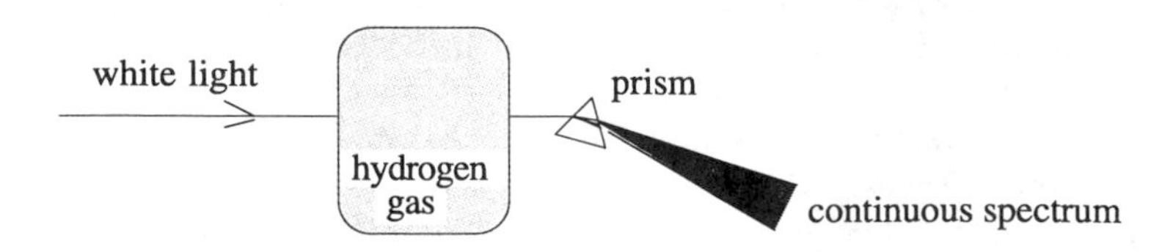

FIGURE 14.5. Absorption spectrum.

colors are present in the beam, there are red photons, for example, of *exactly* the right wavelength to have energies just equal to the energy difference between the $n = 2$ state and the $n = 3$ state in Bohr's model of the hydrogen atom. In some of the hydrogen atoms in the gas, the electron will be in the $n = 2$ state. Such an electron can *absorb* one of these red photons and move up to the $n = 3$ state. Similarly, the electrons in other atoms can absorb green photons from the white light and jump up to the $n = 4$ state. Similar transitions occur for absorption of other photons of exactly the right wavelengths to correspond to discrete energy differences between the $n = 2$ and some higher-energy state in the atom.

Because of the absorption of light of these wavelengths by atoms in the gas, the intensity of the light emerging from the container will be diminished at exactly these wavelengths while the intensity of the light at all other wavelengths will be unaffected by the gas. The light fanned out by the prism will show a continuous spectrum from red to violet with dark (low intensity) lines appearing at the wavelengths corresponding to the bright lines in the *emission* spectrum (Fig. 14.2). Such a spectrum is called an *absorption* spectrum. Applying the energy conservation principle, we have from Eq. (5.3)

$$E_{\text{initial}} + E_{\text{photon}} = E_{\text{final}}$$

leading to an equation for the wavelength of the *absorbed* photon analogous to Eq. (14.15) for emission,

$$\frac{1}{\lambda} = (0.01097\,\mathrm{nm}^{-1})\left(\frac{1}{n_i^2} - \frac{1}{n_f^2}\right). \qquad \text{(absorption)} \qquad (14.16)$$

On combining Eqs. (14.8), (14.9), and (14.11), we obtain an expression for the radius of the nth Bohr orbit,

$$r_n = \frac{h^2}{4\pi^2 mke^2} n^2, \qquad n = 1, 2, 3, \ldots \qquad (14.17)$$

For the first Bohr orbit ($n = 1$), we find that $r_1 = 5.28 \times 10^{-11}$ m.

14.3 Comets

Comets are members of the solar system. When they are near the sun, comets can be very spectacular objects in the night sky. For many comets, it is difficult to determine if their orbits are closed (elliptical) or open (parabolic or hyperbolic) although for none of them are the data inconsistent with a closed orbit. Comets with well established orbits return at regular intervals to pass near the sun. These periodic comets obey Kepler's laws. Their paths around the sun are elongated ellipses as illustrated in Fig. 14.6.

The most famous of the periodic comets is Comet Halley, which last appeared in 1985-86. Edmund Halley, Astronomer Royal of Britain and contemporary of Isaac Newton, supplied much of the coaxing and most of the money to get the *Principia* published. Halley applied Newton's theory to the calculation of the orbit

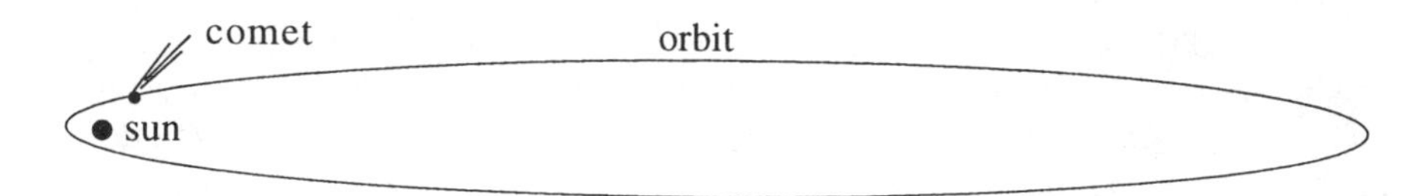

FIGURE 14.6. The orbit of a comet.

of a comet that appeared in 1682. He found that the orbital characteristics of this comet were very similar to comets that had been seen in 1607 and 1531. After some further investigation, he discovered still earlier records of the appearance of comets with these characteristics. These observations led him to conclude that these were all different instances of the same comet passing near the sun. Halley did not discover the comet that bears his name, but he was the first to recognize that the comet of 1682 was the same object that had been seen at approximately 76-year intervals for centuries. Observations of Halley's comet have been recorded for every passage it has made around the sun since the year 87 B.C.

14.3.1 The Structure and Composition of Comets

Astronomers now have a fairly good idea of what a comet really is. When it is far from the sun, it is a very small object only a few kilometers across. It consists mainly of ices (water, methane, ammonia) with bits of dust embedded in it—a kind of dirty ice ball. As it approaches the sun, radiation from the sun vaporizes the icy matter and releases some of the dust. This forms a gigantic halo around the ice ball. This halo—called the *coma*—extends out tens of thousands of kilometers from the icy core, which is the *nucleus* of the comet. Sunlight reflected off the dust paritcles makes the coma visible to observers on the earth.

Ultraviolet radiation from the sun breaks down the vapor molecules into their constituents. These components can be excited by absorbing radiation from the sun. In returning to lower-energy states, the excited atoms and ions emit light, contributing to the luminosity of the coma.

When the comet gets even closer to the sun, one of its most spectacular parts begins to form—the tail. Actually, there are two kinds of tails—the *dust tail* and the *ion tail*. The dust tail is produced by the light from the sun reflecting off the dust particles in the coma. A photon carries momentum. In bouncing off a dust particle, it imparts a tiny, but perceptible, momentum change to the dust particle driving it away from the coma. As the comet sweeps along its orbit, it leaves a curving trail of dust behind in its path. This visible dust tail can extend for tens or hundreds of millions of kilometers out from the nucleus. The dust tail is characterized by its gently curving shape and its yellowish color.

A different mechanism is responsible for the ion tail. Near the sun, ultraviolet radiation from the sun ionizes and excites the atoms in the coma. As the solar wind sweeps through the coma, the high-velocity charged particles of the solar wind interact with the electrically charged excited ions in the coma driving them away from the head of the comet. In returning to lower-energy states, these excited ions emit photons and form a luminous, bluish colored tail extending out from the comet directly away from the sun.

Since both kinds of tails are produced by radiations streaming out from the sun, they extend out from the coma in the general direction away from the sun. A comet may exhibit several tails of each kind.

The principal parts of a typical comet are shown in Fig. 14.7. The nucleus is of the order of a few kilometers in size. The diameter of the coma may be tens or hundreds of thousands of kilometers. The tails typically extend out tens or hundreds of millions of kilometers away from the coma.

A comet leaves a trail of matter behind it as it moves through the inner solar system. Some of this debris may get strewn across the earth's orbit around the sun. When the earth passes through this part of its annual path, it sweeps through the dust trail. The particles enter the earth's atmosphere at high velocity. The air friction can cause one of these bits of matter to produce a brief streak of light as it burns up in the atmosphere. This phenomenon is called a *meteor* or sometimes, more picturesquely, a "falling star" or "shooting star."

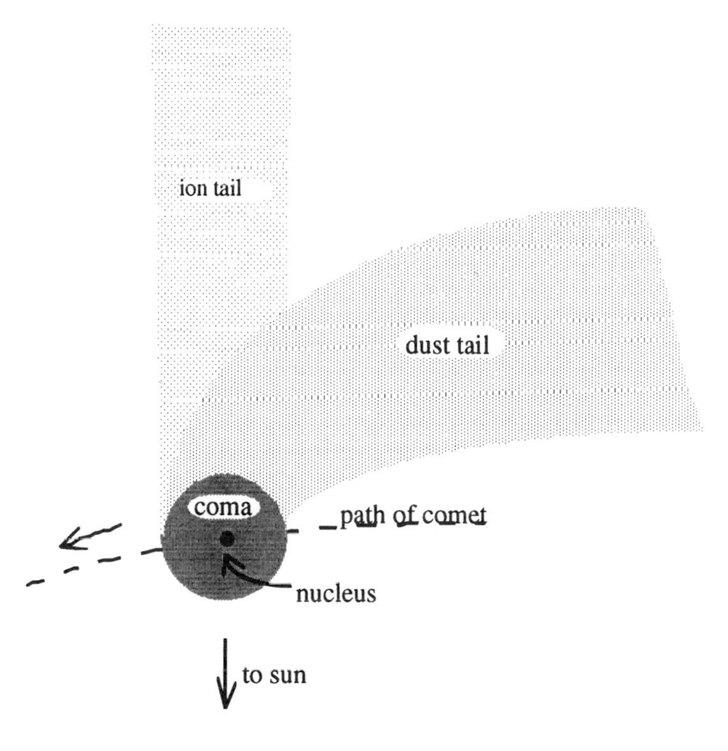

FIGURE 14.7. Model of a comet.

14.3.2 Where Do Comets Come From?

Since a comet loses matter on each pass by the sun, eventually it will be depleted to the point where it is no longer visible. Comets that approach the sun have finite lifetimes. Given the typical sizes of comets and the typical rates at which they lose matter astronomers have concluded that the lifetimes of comets with orbits that bring them near enough to the sun to be seen from the earth are very much shorter than the age of the solar system (about 4.5×10^9 years). Where do the new comets come from to replace the old ones that dissipate and vanish from view?

The Dutch astronomer Jan Oort noticed that many long-period comets have orbits that extend out about 50,000 AU from the sun,* which is about one fourth of the average distance between neighboring stars in the vicinity of the sun. This is near the limit at which the sun's gravity is strong enough to bind objects in orbit around it. At this distance, an orbiting body could be perturbed by the gravity of a neighboring star passing by the sun. Oort proposed that a giant cloud of matter left over from the formation of the solar system surrounds the sun, extending out to about 50,000 AU. This cloud, called the *Oort cloud*, contains large chunks of matter like the nuclei of comets. The gravitational influence of a passing star can be sufficient to perturb the orbit of one of these chunks to send it toward the inner solar system and bring it near the sun. The larger planets, especially Jupiter and Saturn, can further perturb its orbit, sometimes resulting in a relatively short period (less than a few hundred years) for the comet. Since comets are observed to have no preferred direction from which they approach the sun, Oort assumed the cloud to be spherical. This is the idea generally accepted by astronomers today to account for the origin of the comets.

Exercises

14.1. Use the Bohr model to calculate the wavelength of the blue light emitted by hot hydrogen gas. Explain clearly and completely your reasoning.

Answer: (434 nm)

14.2. Light from hot hydrogen gas is passed through a prism spectrometer. The wavelength of one of the violet lines in the spectrum is measured at 373.5 nm. What is the initial state quantum number (n-value) of the electron making this transition in the atom?

Answer: (13)

---

*One astronomical unit, abbreviated AU, is the mean distance between the sun and the earth. 1 AU $\approx 1.50 \times 10^8$ km.

14.3. In the Bohr model of the hydrogen atom, compare the gravitational force that the proton exerts on the electron with the electric force that the proton exerts on the electron.

Answer: (The electric force is about 2×10^{39} times larger.)

14.4. In Bohr's model of the hydrogen atom, assume that the electron travels in a circular orbit around the nucleus (proton). Obtain an expression for the time T required for the electron to circle the nucleus exactly once when it is in the nth Bohr orbit. Calculate T for the first orbit (n=1). How fast is the electron traveling in this orbit?

Answer: (1.5×10^{-16} s; 2.2×10^{6} m/s)

14.5. The boundary separating the visible and the ultraviolet parts of the electromagnetic spectrum is not sharp. Radiation that is visible as violet light to one observer may be completely invisible (ultraviolet) to another observer. Suppose we (arbitrarily) *define* the wavelength dividing the visible and ultraviolet portions of the spectrum to be 360 nm. Calculate the shortest wavelength of the Balmer series in hydrogen. Is it in the visible or ultraviolet region according to our definition? What general statement can you make about ultraviolet radiation from hydrogen? Explain.

14.6. By means of words and drawings, explain clearly and in detail how green light is produced by hot hydrogen gas according to Bohr's model. Do not omit any relevant fact.

14.7. In a cool gas, an atom of hydrogen is in its ground state. This atom absorbs a photon of a certain wavelength causing the electron to make a jump to the $n = 5$ state. Calculate the wavelength of the radiation absorbed by the atom. In a complete sentence, describe this light.

Answer: (95.0 nm)

14.8. The *ionization energy* of a hydrogen atom is the minimum energy required to remove the electron completely from the atom. Calculate the ionization energy of a hydrogen atom in the third excited state. Explain clearly your reasoning.

Answer: (0.85 eV)

14.9. Radiation from hydrogen is observed with a wavelength of 93.8 nm. Describe this radiation. According to Bohr's model, what is the initial state quantum number (n value) of the electron in this transition?

14.10. A hydrogen atom absorbs violet light of wavelength 389 nm. What are the initial and final states of the electron in this transition? Explain clearly your reasoning.

Answer: (2; 8)

14.11. An electron in a hydrogen atom makes a transition from the ninth orbit down to the third orbit. Calculate the wavelength of the radiation emitted. Describe this radiation. Explain clearly your reasoning.

Answer: (923 nm)

14.12. An electron excited to the $n = 4$ orbit in Bohr's model of the hydrogen atom can return to the ground state ($n = 1$) in a variety of ways emitting electromagnetic radiation in the process. List all of these and calculate the wavelength of the photon emitted in each case. Identify the radiation in each case as visible, ultraviolet, or infrared.

14.13. Calculate the longest wavelength of ultraviolet light emitted from hot hydrogen gas. Explain clearly in terms of Bohr's model exactly how this radiation is produced.

Answer: (121.5 nm)

14.14. In Chapter 10, we defined the electric current as the time rate of flow of electric charge. Hence, an electric current is associated with the orbiting electron in Bohr's model of hydrogen. Calculate this current for the first and second Bohr orbits.

Answer: (1,050 μA; 262 μA)

14.15. The continuous spectrum of light from an incandescent source is first passed through hot hydrogen gas and then through an infrared spectrometer. Three dark lines with wavelengths 1,005 nm, 1,094 nm, and 1,282 nm are observed as part of an infrared series. Use Bohr's model of the hydrogen atom to give a clear and complete explanation of how this dark line spectrum is produced. Find the final state quantum numbers (n-values) for each of these transitions. Explain clearly your reasoning.

Answer: (7, 6, 5)

14.16. A wavelength of 4,170 nm is measured for a line in the infrared emission spectrum of hydrogen. First, by calculating the longest and shortest wavelengths in a series, find all series to which this radiation could possibly belong. (A series of emission lines is characterized by the final state n-value for the electron transition.) Then, identify the series by finding the initial state n-value that gives the best fit to the measured line. Identify clearly the initial and final state n-values for this transition. Explain clearly your reasoning.

14.17. In Bohr's model, the centripetal force on the electron is due to the electric attraction of the nucleus. Use Newton's second law to calculate the orbital speed of the electron in the first ($n = 1$) Bohr orbit.

Answer: (2.18×10^6 m/s)

14.18. In the ultraviolet spectrum of light from hot hydrogen gas, a line of wavelength 94.96 nm is observed. Use Bohr's model of the hydrogen atom to explain completely and in quantitative detail how this radiation is produced. (A calculation is required.) Explain clearly your reasoning. Suppose the electron started from the same *initial* state, but made a transition corresponding to a line in the Balmer series. What would be the wavelength and color of this Balmer line? Explain clearly your reasoning.

14.19. A hydrogen atom in an excited state is slightly heavier than it is in the ground state. In making a transition from a higher-energy state to a lower-energy one, the atom loses some of this excess mass in the form of energy. From Eq. (13.4), the mass loss Δm is found to be

$$\Delta m = \frac{E_{\text{photon}}}{c^2}.$$

By approximately what fraction is the total mass of a hydrogen atom reduced in emitting a photon of red light? (The recoil kinetic energy of the atom is negligible.)

Answer: (2×10^{-9})

14.20. Assume that the binding of the electron and the proton in the hydrogen atom is due to the gravitational attraction rather than the electric attraction of the two particles. With this assumption, start with Eq. (14.4) and derive the expression analogous to Eq. (14.11). Evaluate the constants to obtain the expression corresponding to Eq. (14.12). What would be the energy of the atom in the $n = 1$ orbit in this case?

Answer: (2.6×10^{-78} eV)

14.21. A comet, first seen in September of 1987, is found to have an elliptical orbit about the sun as shown in the drawing. The comet is no longer visible. Approximately when (month and year) would you expect this comet to become visible again?

Answer: (May, 2054)

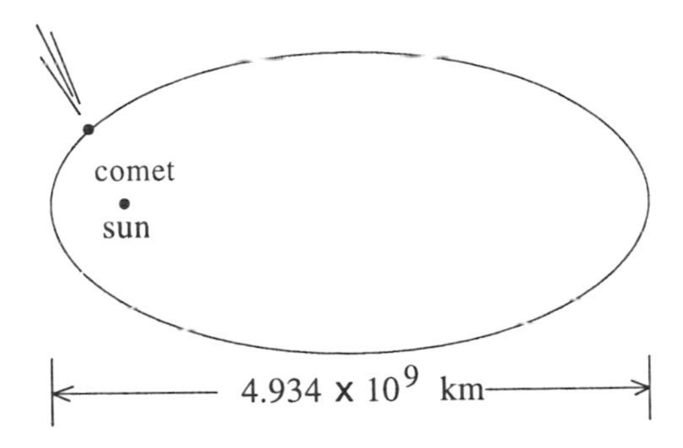

14.22. The head of a comet is surrounded by a huge envelope of relatively cool hydrogen gas. This envelope was discovered using instruments on board an earth-orbiting satellite. Use Bohr's model of hydrogen to explain clearly and completely why this envelope could not be discovered with similar instruments on the ground.

14.23. A certain comet discovered in 1953 approaches the sun as close as 0.338 AU. Its elliptical orbit carries it out as far as 4.095 AU from the sun. What is the period of the orbit of this comet?

Answer: (3.3 years)

14.24. A comet seen in 1938 has a period of 156 years. The nearest this comet came to the sun was 1.122×10^8 km. What is the greatest distance from the sun that this comet reaches? How far from the sun does it get compared to Pluto? Write your answer in a complete sentence.

15

Sugar and Spice and Everything Nice...

And what are little atoms made of? First, let's see why it is important that we know. Our motive for looking into the structure of many-electron atoms is twofold. The first is concerned with the light emitted by the atoms. If we pass the light from a hot gas or vapor of any element through a narrow slit and disperse the light (e.g., by means of a prism as in Fig. 14.1), we find that the spectrum is discrete—not continuous. Only certain colors appear. For each element, the spectrum is unique—different from all other elements. This feature provides a means of identifying the element by the spectrum of light it emits. We want to answer two questions. Why are the spectra discrete and why, for each element, is the spectrum unique?

The second motive for our interest in the structure of complex atoms has to do with the chemical behavior of the atoms. In Fig. 15.1, the chemical symbols representing the elements are arranged in numerical order according to the number of electrons in the neutral atom of the element. This number is denoted by Z and is called the *atomic number* of the element. It is the number that appears at the top of each block* in Fig. 15.1.

Clearly, the chemistry of each element is determined by the number of electrons in the atom. For example, let us start with lithium, element number 3, a very active metal. We add another electron (and another unit of positive electric charge to the nucleus to keep the atom neutral) to get beryllium, number 4, a less active metal and chemically different from lithium. Another electron (and another positive unit of charge on the nucleus) gives us boron, number 5, with still different chemical behavior and not a metal at all. And so on. As we move *across* the chart in Fig. 15.1 adding electrons to obtain new elements, the chemistry changes with each addition until we reach neon, number 10. Neon is an inert gas. The atoms of neon do not readily react chemically—not even with each other. Adding another electron brings us to sodium, number 11. But the chemical behavior of sodium is quite similar to that of lithium. It too, is a very active metal. Again, the addition of electrons changes the chemistry as we move across the chart to argon, number 18, which

---

*The number in small type at the bottom of each block is the *atomic weight* of the element in atomic mass units (u). This will be important for our discussion in Chapter 18.

The Periodic Chart of the Chemical Elements

1 H 1.01	2 He 4.00																2 He 4.00
3 Li 6.94	4 Be 9.01											5 B 10.81	6 C 12.01	7 N 14.01	8 O 16.00	9 F 19.00	10 Ne 20.18
11 Na 22.99	12 Mg 24.30											13 Al 26.98	14 Si 28.09	15 P 30.97	16 S 32.06	17 Cl 35.45	18 Ar 39.95
19 K 39.10	20 Ca 40.08	21 Sc 44.96	22 Ti 47.90	23 V 50.94	24 Cr 51.10	25 Mn 54.94	26 Fe 55.85	27 Co 58.93	28 Ni 58.71	29 Cu 63.55	30 Zn 65.38	31 Ga 69.72	32 Ge 72.59	33 As 74.92	34 Se 78.96	35 Br 79.90	36 Kr 83.80
37 Rb 85.47	38 Sr 87.62	39 Y 88.90	40 Zr 91.22	41 Nb 92.91	42 Mo 95.94	43 Tc 98.91	44 Ru 101.07	45 Rh 102.90	46 Pd 106.40	47 Ag 107.87	48 Cd 112.41	49 In 114.82	50 Sn 118.69	51 Sb 121.75	52 Te 127.60	53 I 126.90	54 Xe 131.30
55 Cs 132.90	56 Ba 137.33	*	72 Hf 178.49	73 Ta 180.95	74 W 183.85	75 Re 186.20	76 Os 190.20	77 Ir 192.22	78 Pt 195.09	79 Au 196.97	80 Hg 200.59	81 Tl 204.37	82 Pb 207.20	83 Bi 208.98	84 Po (209)	85 At (210)	86 Rn (222)
87 Fr (223)	88 Ra 226.02	**	104	105	106												

*	57 La 138.91	58 Ce 140.12	59 Pr 140.91	60 Nd 144.24	61 Pm (145)	62 Sm 150.40	63 Eu 151.96	64 Gd 157.25	65 Tb 159.92	66 Dy 162.50	67 Ho 164.93	68 Er 167.26	69 Tm 168.93	70 Yb 173.04	71 Lu 174.97
**	89 Ac (227)	90 Th 232.04	91 Pa 231.03	92 U 238.03	93 Np 237.05	94 Pu (244)	95 Am (243)	96 Cm (247)	97 Bk (247)	98 Cf (251)	99 Es (254)	100 Fm (257)	101 Md (258)	102 No (259)	103 Lw (260)

FIGURE 15.1. Periodic Table.

is an inert gas like neon, followed by potassium, number 19, another active metal similar to lithium and sodium. So we see the pattern. The chemistry changes as we move across the table with the addition of electrons, but there is a periodic repetition of the chemical behavior as electrons are added. There are families of elements represented by the columns in the chart. All of the elements in the first column are called *alkali metals*. In the last column, we have all inert gases. All elements in the next to last column are very active nonmetals called *halogens*. The questions we want to answer here are:

- Why does the chemistry change with increasing atomic number?
- Why is the chemical behavior of the elements periodic?

15.1 The Many-Electron Atom

To get a correct description of an atom, quantum mechanics is required. Quantum mechanics is mathematically complicated and quite abstract. For our purpose, we can gain sufficient insight into atomic structure by a heuristic extension of Bohr's model of the one-electron atom and by borrowing one or two simple ideas from quantum mechanics. This procedure will be adequate to answer the questions we posed above regarding atomic spectra and the chemical behavior of the elements. We should keep in mind that this approach is vastly oversimplified and not strictly correct.

According to Bohr's model, in the ground state of hydrogen the electron is in the lowest available energy state as indicated by the $\times$ in Fig. 15.2. Following the gravitational analogue, it tries to get as close to the center of the nucleus as it can. This means the $n = 1$ state. If we add another unit of charge in the nucleus to make helium, where does the second electron go? It also seeks the lowest energy possible—the $n = 1$ state again. Presumably, if we continue this process to make heavier and heavier atoms in their ground states, the added electron each time

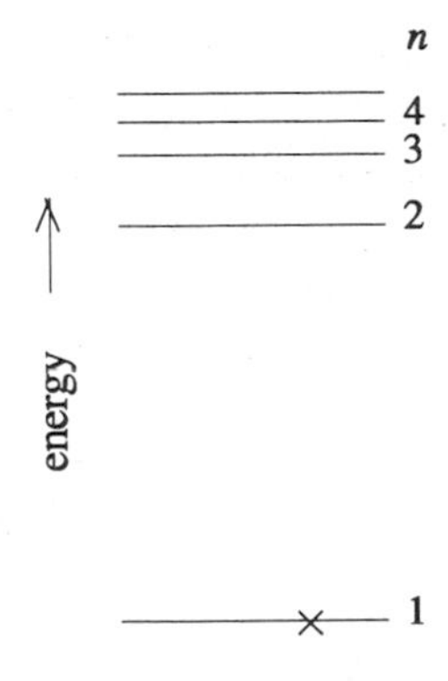

FIGURE 15.2. The ground state of hydrogen.

goes into the $n = 1$ state—the lowest possible. However, that does not happen. There is a very important principle from quantum mechanics that comes into play here—*the exclusion principle*.

First, we state the principle as it applies to electrons in an atom and then try to see what it means.

The exclusion principle: *No more than one electron can occupy a given state in an atom.**

The *state* of an electron is given by specifying a complete set of physical quantities (e.g., energy, momentum, etc.) necessary to define it uniquely. For our purpose, we consider the state of an electron in an atom to be completely determined by its total energy and its angular momentum. The rules of quantum mechanics require that these quantities be discrete. Taking into account only the Coulomb attraction between the electron and the nucleus, the total energy of the electron in the nth orbit is $E_n = -\text{constant}/n^2$, where n is a positive integer. An electron in an atom has two kinds of angular momentum: orbital angular momentum due to its motion around the nucleus and spin, which is an intrinsic property of the electron like mass or electric charge. Both of these are quantized, having discrete values of magnitude *and* direction. For small systems on the scale of atoms, molecules, and so on, angular momentum is conveniently measured in units of $h/2\pi$, where h is Planck's constant.

Thus, the state of an electron in an atom is specified by five quantities:

$$\text{energy:} \quad E_n = -\frac{\text{constant}}{n^2} \qquad n = 1, 2, 3, \ldots .$$

$$\text{orbital angular momentum:} \quad L = \sqrt{l(l+1)}\,\frac{h}{2\pi} \qquad l = 0, 1, 2, \ldots, n-1.$$

$$z\text{-component of orb. ang. mom.:} \quad L_z = m_l\,\frac{h}{2\pi} \qquad -l \le m_l \le l.$$

$$\text{spin angular momentum:} \quad S = \sqrt{s(s+1)}\,\frac{h}{2\pi} \qquad s = \tfrac{1}{2}.$$

$$z\text{-component of spin:} \quad S_z = m_s\,\frac{h}{2\pi} \qquad -s \le m_s \le s.$$

Because these quantities are all quantized, the state of an electron in an atom can be specified by five numbers (actually, four since s is always equal to $\frac{1}{2}$): n, l, m_l, s, and m_s. These numbers are called *quantum numbers*. The quantum numbers m_l and m_s are often called *projection* quantum numbers, since they represent the projection of the corresponding angular momentum vector onto the z-axis. These projections are quantized and m_l and m_s can take on only values separated by integer steps in the specified ranges. For example, if $l = 2$, m_l can have any one of the values $-2, -1, 0, +1, +2$. Since $s = \frac{1}{2}$, m_s is restricted

*The principle is more general than this statement implies, as we shall see in subsequent chapters.

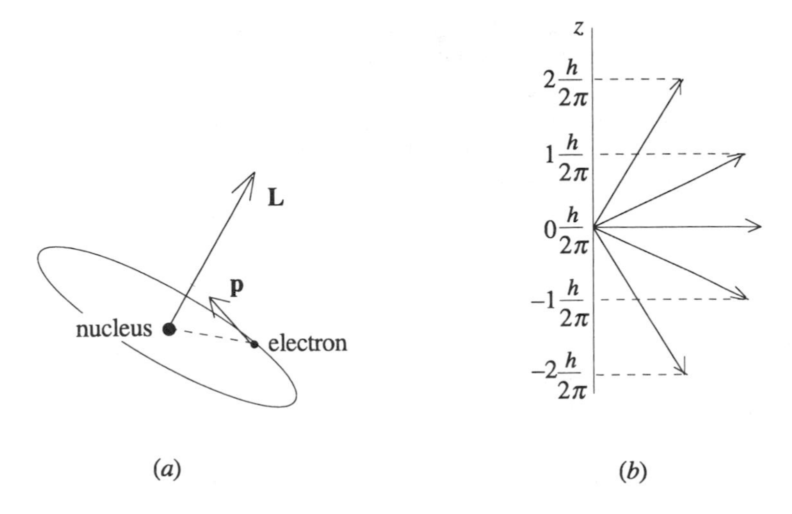

FIGURE 15.3. Orbital angular momentum projections for $l = 2$.

to one of two possible values, $m_s = \pm\frac{1}{2}$. The physical implication of this is that the direction of the angular momentum is allowed only certain orientations in space. To illustrate this, let's consider an electron traveling around the nucleus with an orbital angular momentum corresponding to $l = 2$. This is represented qualitatively in Fig. 15.3*a*. The direction of the orbital angular momentum vector **L** is perpendicular to the plane of the orbit with linear momentum **p** directed as shown. If the electron orbits in the opposite direction (i.e., reverse the direction of **p**), **L** will be in the direction opposite to that shown. With $l = 2$, the magnitude of the orbital angular momentum is $L = \sqrt{6}\,h/2\pi$. The allowed values of the projection of **L** onto the z-axis according to the rule above are illustrated in Fig. 15.3*b*.

It is convenient to represent each l-value by a letter of the alphabet according to the scheme given in Table 15.1.

According to Coulomb's law, the electrons in an atom repel each other electrically. This tends to counteract some of the electric attraction from the nucleus, reducing the binding of the electrons to the nucleus. Because of the exclusion principle, some electrons in a many electron atom will have to go into higher-energy orbits. Qualitatively speaking, the higher the energy of an electron, the farther it is from the nucleus. Therefore, the electrons closer to the nucleus will feel a stronger

Table 15.1. Spectroscopic Notation

l-value	0	1	2	3	4	5	6
symbol	s	p	d	f	g	h	i

effect of the electric charge of the nucleus than those farther away from the nucleus. For those farther out, some of the nuclear charge is effectively cancelled out by the negative charge of the inner electrons. As a result of this *screening* effect, the energies of the electron orbits in a many electron atom are shifted according to the rule: for a given n-value, an electron with a higher l-value (larger angular momentum) will have a larger orbit and will be less tightly bound (higher-energy) than an electron with the same n-value, but smaller l-value.

Rule for electron orbits: *For a given n, states with higher l lie higher in energy.*

The effect of this electron-electron interaction is illustrated in Fig 15.4, where we have used the spectroscopic notation to label the energy levels.

Atomic electrons also experience other interactions (e.g., magnetic interactions) that also affect their energies such that states with the same n and l, but different m_l or m_s will have different energies. For our purpose, we shall neglect these smaller effects.

In our approximation, states with the same n and l, but different combinations of m_l and m_s will have the same energy. For example, an electron in a p-state has $l = 1$ with three possible values for m_l $(-1, 0, +1)$ and for each of these there are two possible values of m_s $(\pm\frac{1}{2})$. Thus, each energy level in Fig. 15.4 labeled by np, represents up to six electrons in *different* states with this same energy. The maximum number of electrons that an energy level can accomodate consistent with the exclusion principle is called the *occupation number*. Occupation numbers for the energy levels in Fig. 15.4 are given in parentheses.

Let's apply our rules to the lithium atom. To make the atom in its ground state, we put the electrons in the lowest available energy levels. Lithium has three orbital electrons and three units of electric charge on the nucleus. The exclusion principle allows us to put two electrons in the 1s level, but the third one must go in the 2s

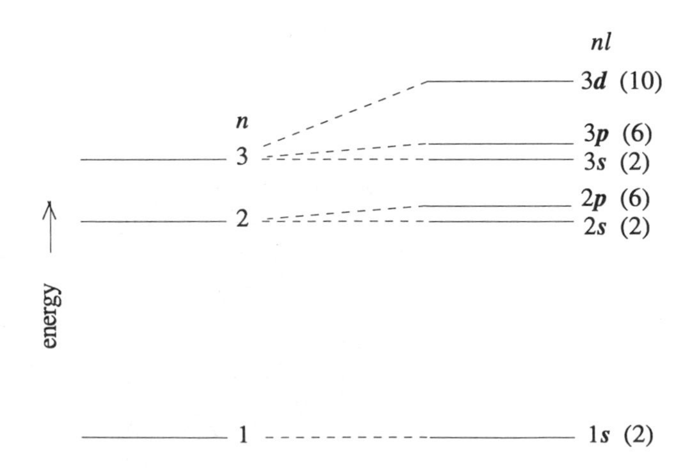

FIGURE 15.4. Energy splitting of orbital angular momentum states.

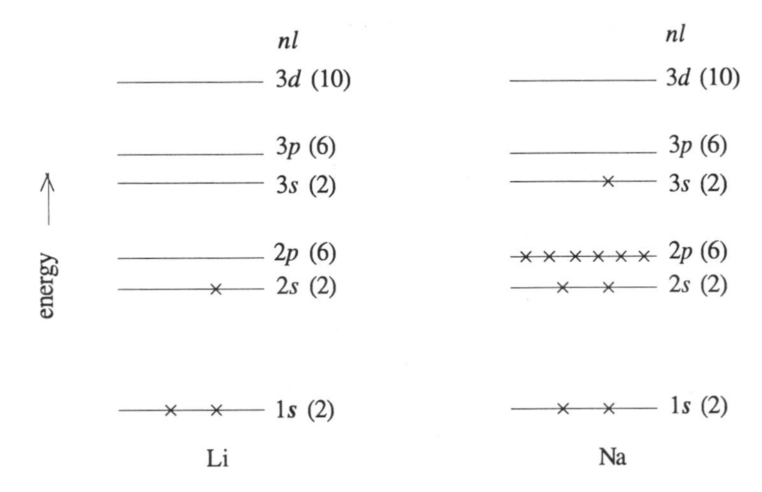

FIGURE 15.5. Electron configurations of lithium and sodium.

level. This is represented schematically in Fig. 15.5. To make the next element, beryllium, we add another electron to the $2s$ level. Boron with five electrons has the fifth one in the $2p$ level, and so on.

15.2 Atomic Spectra

If ordinary table salt (sodium chloride) is introduced into a gas flame, the flame is colored with the characteristic yellow light from sodium atoms. This light, which has a wavelength of 589 nm, is emitted by electrons returning to the ground state from the first excited state. The electron configuration of the ground state of sodium is indicated schematically on the right-hand energy-level diagram in Fig. 15.5. In a simple interpretation of this process, we consider the emission of the light to arise from changes in the motion of only the outermost electron in the sodium atom. The hot flame excites the least tightly bound electron from the $3s$ level to the $3p$ level. When this electron returns to the $3s$ level, it emits a 589-nm photon corresponding to the energy difference between these two states of the atom. This is a very simple example involving a single electron. More complicated transitions can also occur. Because of the quantized nature of the energy levels in the atom, the transitions between these levels are discrete—only photons of certain wavelengths can be emitted from a given atom. Atomic spectra must be discrete.

The actual energy difference (hence the wavelength of the emitted photon) between the two states for the electron in the example for sodium is determined by its interactions with all of the other electrons in the atom. Thus, energy differences between electron states will be different for atoms with different numbers of electrons. Because of this, atoms with different numbers of electrons (different elements) will not emit the same set of photons. Each chemical element emits a unique spectrum of radiation.

15.3 The Periodic Table of the Elements

It is the atomic electrons that are responsible for chemical reactions between atoms. For example, consider a carbon atom and an oxygen atom interacting to form a molecule in carbon monoxide gas. The interaction is represented schematically in Fig. 15.6. As the two atoms begin to penetrate each other's electron cloud, their nuclear electric charges are no longer completely shielded by their electron clouds. The positively charged nuclei will repel each other. Because of this repulsion, it is only the electrons on the outer fringes of the atoms that get close enough together to participate strongly in the interaction of the two atoms.

As the outer electron configuration changes with the addition of another electron consistent with the exclusion principle, the chemical behavior of the atom will change accordingly. As we move across the table of chemical elements (Fig. 15.1) increasing the number of electrons by one at each step, we expect the chemistry to change. We can now understand qualitatively why the periodic behavior of the chemistry mentioned at the beginning of this chapter occurs. Atoms with the same outer electron configurations have similar chemical properties. We noted that lithium and sodium are both very active metals with similar chemistry. We see from Fig. 15.5 that their outer electron configurations are similar. Both have a single electron in the outer orbital and that electron is in an s-state in both cases—$2s$ in lithium and $3s$ in sodium. Potassium, element 19, also an active metal like lithium and sodium, has a single outer electron in the $4s$-level. Similar statements follow for the other families of elements. In the next to last column, the halogens (fluorine, chlorine, etc.) all have five outer electrons (one short of a full complement) in a p level. The last column in Fig. 15.1 contains the inert gases (neon, argon, etc.), with completely filled p levels for their outer electron configurations.

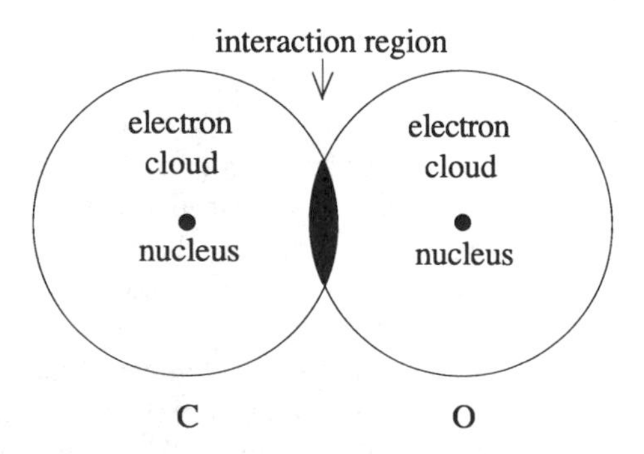

FIGURE 15.6. Carbon monoxide molecule.

Exercises

15.1. The elements lithium (Li), sodium (Na), and potassium (K) have similar chemical properties. Use our model of the many-electron atom to explain clearly why this is so.

15.2. The elements Ne, Ar, and Kr are all chemically inert gases. Use the model to explain why this is so.

15.3. Draw energy-level diagrams similar to those in Fig. 15.5 and show the electron configuration for the ground state of carbon (C) and of silicon (Si). Use your results to explain clearly why carbon and silicon have similar chemistry.

15.4. Draw energy-level diagrams similar to those in Fig. 15.5 and show the electron configuration for the ground state of aluminum (Al) and of gallium (Ga). Use your results to explain clearly why aluminum and gallium have similar chemistry.

15.5. Compare the electric force with the gravitational force that a gold nucleus exerts on an alpha particle. Which is greater? How many times greater?

15.6. Draw energy-level diagrams similar to those in Fig. 15.5 and show the electron configurations for carbon (C) and germanium (Ge). Based on your results, what can you say about the chemistry of these two elements? Explain.

15.7. Continuing the energy-level diagram in Fig. 15.4, what is the maximum number of electrons that can occupy the $5g$ level consistent with the exclusion principle?

15.8. From Fig. 15.1, we see that barium (Ba) has chemistry similar to strontium (Sr) and the chemistry of hafnium (Hf) is similar to that of zirconium (Zr). The atomic numbers of zirconium (40) and strontium (38) differ by only two units while the difference between the atomic numbers of hafnium (72) and barium (56) is much greater. In terms of our model of the many-electron atom, explain clearly and completely why this is so.

15.9. In the ground states of atoms of the heaviest elements, electrons occupy the $5f$ states. Calculate the maximum number of $5f$ electrons allowed in a given atom by the exclusion principle.

15.10. On energy-level diagrams similar to those in Fig. 15.5, construct the ground state electron configuration for lithium (Li) and fluorine (F). Based on your result and considering that a full complement of electrons in the outermost energy level corresponds to a very stable configuration, give a qualitative explanation for the strong chemical bond of the LiF molecule.

16

The Starry Messenger

Light—electromagnetic radiation—is the messenger that brings us information from the stars. We can think of many interesting things we might want to know about a star—mass, size, chemical composition, brightness, temperature, distribution of matter (density), color, age, and energy production are some that come to mind. These characteristics are interrelated and all must be extracted from the light we get from the star.

The stars we see in the sky are in various stages in their lives. By comparing characteristics of a large number of stars, we can sort out the evolutionary tracks different stars follow in their evolution. Implicit here is that stars *do* have finite lifetimes. The sun won't shine forever. We shall see later that this is indeed so and we shall also see *why* it is so.

In this chapter, we focus on two characteristics that will be extremely useful to us in discerning the paths stars take as they evolve during the courses of their lives. These are,

- intrinsic brightness
- spectral classification

16.1 Intrinsic Brightness

Notice that it is the star's *intrinsic* brightness that we want. This means how bright it *really* is and not merely how bright it *appears* to be. This latter is called *apparent* brightness and depends not only on how bright the star actually is, but also on such things as how far away it is. If we can *see* the star, we can always measure its apparent brightness. (By "see," we mean not only what the naked eye can do, but also what falls within the power of our instruments to do.)

What do we mean by the brightness of a star anyway? What do we actually measure? In practice, it is the intensity—radiation energy incident per unit area per unit time on the detection device. The intensity diminishes with increasing distance from the source. This is easily seen in Fig. 16.1. The point source emits light uniformly in all directions. Consider the energy emitted in a small time interval

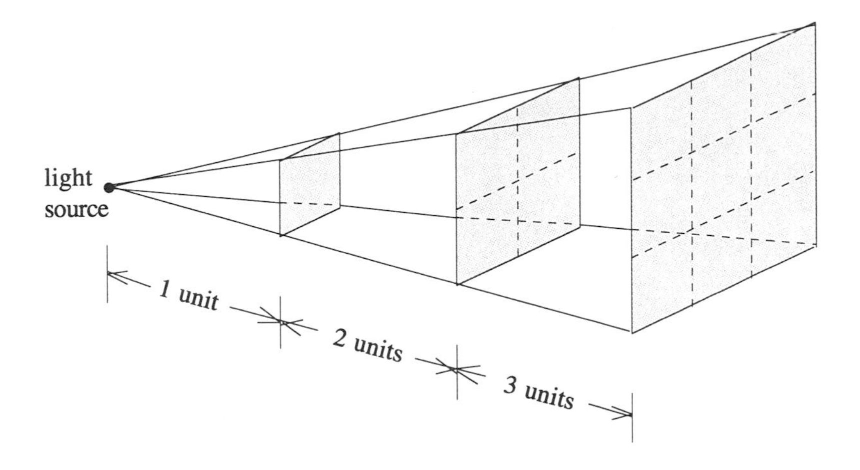

FIGURE 16.1. Inverse square law for radiation intensity.

into the rectangular cone. This energy remains in the cone, but as it propagates away from the source it spreads out over an increasingly larger area thereby reducing the intensity. Moreover, we can see from the drawing that at a distance of two units from the source, the area over which the light is spread is four times as large as the corresponding area at one unit. At a distance of three units, the area is nine times as large as at one unit. Clearly, the area increases directly with the *square* of the distance d from the source. Therefore, the intensity, or equivalently the brightness B, falls off *inversely* as the square of the distance,

$$B = \frac{C}{d^2}, \tag{16.1}$$

where C is a constant that depends on the intrinsic properties of the source (e.g., size, temperature, and so on).

EXAMPLE. A very small, 75-watt source of light radiates uniformly in all directions. Calculate the intensity of illumination on a small surface at a distance of 1.70 meter from this source. Assume that the light is incident perpendicular to the plane of the small surface.

SOLUTION. Each second, 75 joules of energy pass through the surface of any imaginary sphere centered on the source. Assume that the sphere has a radius of 1.70 meter so that a portion of its surface approximately coincides with the small surface on which the light is incident. This is illustrated in Fig. 16.2. According to the definition of intensity, we have

$$\text{intensity} = \frac{\text{energy}}{\text{time} \cdot \text{area of sphere}} = \frac{75 \text{ J/s}}{4\pi(1.70 \text{ m})^2} = 2.065 \text{ W/m}^2.$$

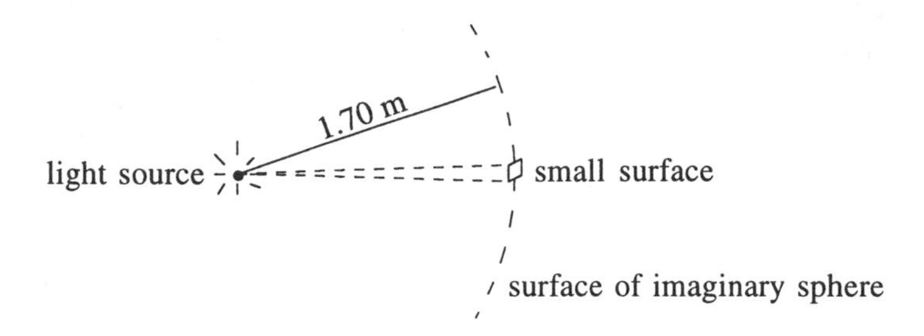

FIGURE 16.2. Radiation intensity on a small surface.

Thus, the light from the source falls on the small surface with an intensity of about 2.1 W/m^2.

16.1.1 COSMIC DISTANCES

How far away *are* the stars? The task of measuring cosmic distances is a complicated one and the larger the distance, the less certain the result. However, for a star that is relatively close to the sun, there is a direct, geometrical method of determining its distance. The idea is illustrated in Fig. 16.3. The earth orbits the sun in a nearly circular path. The mean distance from the sun to the earth is one *astronomical unit*, abbreviated AU, which is about 150,000,000 km. When the earth is at point P_A in its orbit, we see a nearby star against a background of more distant stars A. Six months later the earth has moved to point P_B from which the nearby star appears in a different background B. The change in point of observation produces an apparent shift in the position of a nearby object relative to a more remote background. This phenomenon is called *parallax*. It is familiar to us in looking at the needle on an instrument dial. For example, an automobile speedometer has a needle that is set against a fixed background of marks and numerals. To the driver, the needle appears against a background of one set of marks indicating the speed of the car. By a rider in the front passenger seat, the needle is seen with a slightly different set of marks in the background. The existence of stellar parallax was suspected long before it was first observed by F. Bessel in 1838.

When the earth is at P_A, we set our telescope to receive light from the nearby star and the distant sky beyond with the nearby star centered in the field of view. Six months later when the earth is at P_B and the telescope is set to receive light from the same point on the distant sky, it is necessary to turn the telescope through an angle 2α to center the nearby star in the field of view again. From the geometry in Fig. 16.3, we can see that to an observer at the position of the nearby star, the maximum angular separation between the earth and sun is α. This angle α is called the *parallax*. Since we know the distance from the sun to the earth, a measurement of α gives us a unique right triangle (shaded in Fig. 16.3). Using trigonometry, we can calculate the length of the side of the triangle that corresponds to the distance

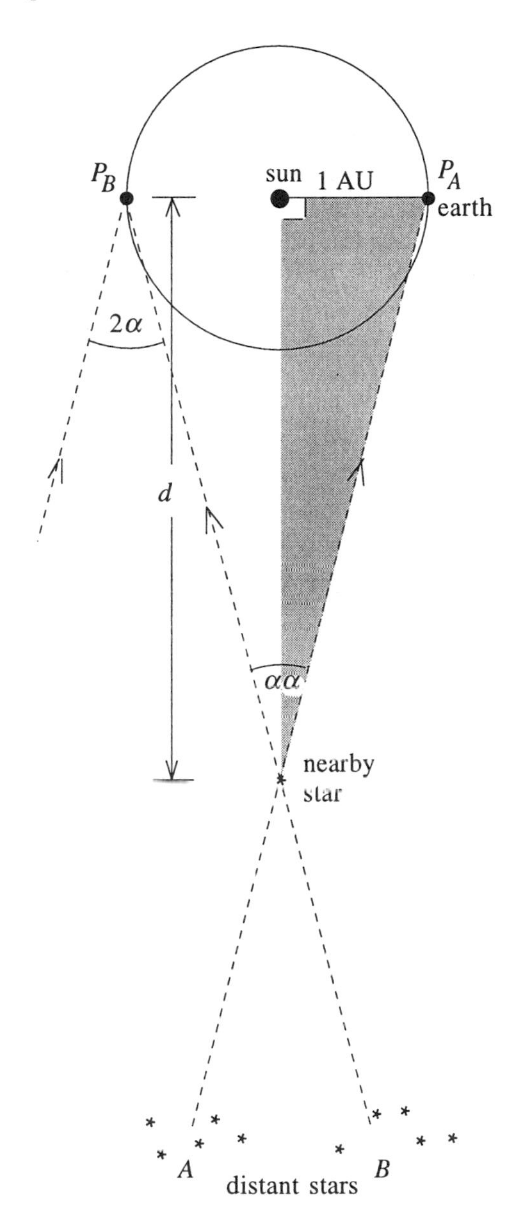

FIGURE 16.3. Determining stellar distances by trigonometric parallax.

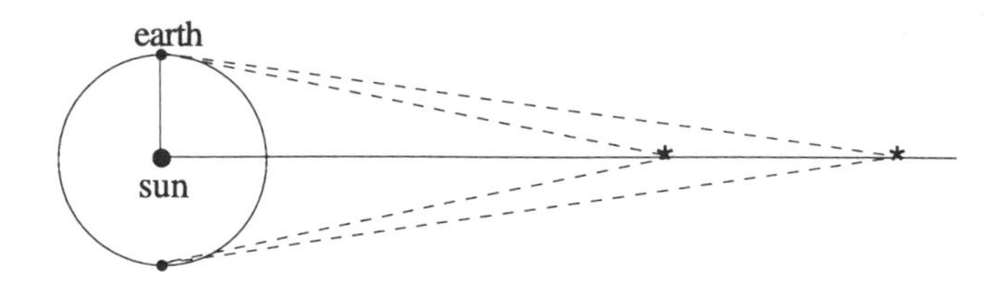

FIGURE 16.4. Parallax-distance relationship.

d from the sun to the star. This method of determining stellar distances is called *trigonometric parallax*.

From Fig. 16.4, we can see that as the distance from the solar system increases, the parallax gets smaller. Relative distances in Fig. 16.4 are grossly exaggerated. For actual stars, the parallax α is exceedingly tiny. Even for the star nearest to the sun, it is less than one second. (One second of arc is equal to one thirty-six-hundredth of a degree.) Clearly, trigonometric parallax is useful only for stars in the solar neighborhood. For more remote stars, α is so tiny our instruments are not able to measure it. Other methods are required.

The *measured* quantity is the parallax α. The distance d is *deduced* from α. It is convenient to use this relationship to define a unit for representing cosmic distances. This unit, called the *parsec* and abbreviated pc, is defined to be the distance for which the parallax is exactly one second. With reference to Fig 16.3, this means that if $\alpha = 1''$, then by definition, $d = 1$ pc. Since the star *nearest* the sun shows a parallax of less than one second of arc, its distance is greater than one parsec.

An older unit comparable to the parsec in magnitude is the *light year*. In free space, light travels with a speed of about 300,000 km/s. One light year, abbreviated ly, is defined as *the distance light travels in one year*, about 9.5×10^{12} km. One parsec is approximtely equal to 3.26 ly. Notice that the light year is a *distance* unit, not a time unit.

16.1.2 Stellar Magnitudes

The scheme that modern astronomers use to compare the brightnesses of stars is based on a classification that can be traced at least as far back as Ptolemy. In the second century A.D., Claudius Ptolemy compiled a catalogue of several hundred stars. He grouped these stars into six categories according to how bright they appeared to the naked eye. He classed the very brightest stars as *first magnitude*. The dimmest stars he could see were assigned to the sixth magnitude. This is a rather crude measurement of stellar brightness, since some stars of different brightness have the same magnitude. However, it is entirely consistent with the precision of the only instrument available to Ptolemy for making the measurement, namely his own eye. The system has been made exact and adapted for use in modern astronomical measurements.

Astronomers recognized that *on the average* Ptolemy's first magnitude stars are about one hundred times as bright as his sixth magnitude stars. To establish the scale precisely, we *define* a star of magnitude 1 to be *exactly* one hundred times as bright as a star of magnitude 6. If we make the scale uniform, then an nth magnitude star will be exactly x times as bright as a star of magnitude $n + 1$. This establishes the size of the magnitude unit. It only remains to assign a definite intensity to a specific magnitude, say magnitude zero, and the system is complete and exact.

We get an approximate value for x by recognizing that a first magnitude star is exactly x^5 times as bright as a sixth magnitude star, since each unit of magnitude difference corresponds to a factor x in brightness. But according to our definition, a first magnitude star is exactly one hundred times as bright as a sixth magnitude star. Thus,

$$x^5 = 100 \qquad \text{or} \qquad x = 2.51188\ldots.$$

So each magnitude difference in brightness corresponds to a factor of about 2.5. That is, a star of magnitude n is about 2.5 times as bright as a star of magnitude $n + 1$.

EXAMPLE. The star that appears brightest in the sky is Sirius with an apparent magnitude of -1.46. Another bright star is Regulus in the constellation Leo. The apparent magnitude of Regulus is 1.35. How many times as bright as Regulus is Sirius?

SOLUTION. Each magnitude difference is a factor of about 2.5. How many magnitudes difference do we have? The difference $m_{\text{Regulus}} - m_{\text{Sirius}} = 1.35 - (-1.46) = 2.81$ yields,

$$2.5^{2.81} = 13.1.$$

Therefore, Sirius appears about thirteen times as bright as Regulus.

We have noted that the apparent brightness B_{app} of a star depends not only on the amount of light it emits (intrinsic brightness), but also on how far away it is from the earth (where the observer is). If we could move the stars around so that they are all the same distance away from us, B_{app} would give a measure of the intrinsic or *absolute* brightness B_{abs}. Then we would be able to say that a star that *appears* to be twice as bright as another star, really *is* twice as bright, since they would be the same distance away. Of course, we can't move stars around. However, if we know the distance to a star, then, because of the inverse square law for intensity, we can determine how bright the star *would be* if we were to move it to *some other distance*. Suppose we do this for all stars whose distances we know. That is, we choose some *standard* distance d_{standard} and inquire how bright each of these stars *would be* if we moved it from its actual distance d_{actual} to the standard distance d_{standard}. We *define* the absolute brightness B_{abs} to be the apparent brightness a star would have if it were at the standard distance d_{standard}. Thus, from Eq. (16.1) we get

$$B_{\text{app}} = \frac{C}{d_{\text{actual}}^2} \qquad \text{and} \qquad B_{\text{abs}} = \frac{C}{d_{\text{standard}}^2},$$

where the constant C represents intrinsic properties of the star that do not depend on where the star is located. On dividing each side of the second equation by the corresponding sides of the first equation, we find that,

$$\frac{B_{\text{abs}}}{B_{\text{app}}} = \left(\frac{d_{\text{actual}}}{d_{\text{standard}}}\right)^2. \tag{16.2}$$

At its actual distance, a star has magnitude m called the *apparent magnitude*. The magnitude the star would have at the standard distance is called the *absolute magnitude* and is denoted by M. Let's say that a certain star is very far away so that $d_{\text{actual}} > d_{\text{standard}}$. This means that it would be brighter at the standard distance and $m > M$. How many times brighter? Each magnitude difference is a factor of about 2.5 in brightness, so at the standard distance it would be about 2.5^{m-M} times as bright as at the actual distance. Thus, from Eq. (16.2) we obtain,

$$\left(\frac{d_{\text{actual}}}{d_{\text{standard}}}\right)^2 = 2.5^{m-M}.$$

In arriving at this result, we have assumed that the observed brightness of a star depends only on its intrinsic brightness and on how far away it is. We have ignored any effects due to stuff lying between us and the star—stuff that might further diminish the observed brightness.

The only thing that remains is to choose a value for d_{standard}. By convention, this is taken to be 10 pc. Finally, we have

$$\left(\frac{d}{10 \text{ pc}}\right)^2 = 2.5^{m-M}, \tag{16.3}$$

where d is the actual distance to the star. We shall make extensive use of Eq. (16.3) in the sequel.

EXAMPLE. Betelgeuse is a very bright, red star in the constellation of Orion. The apparent magnitude of Betelgeuse is measured at 0.41. The absolute magnitude is known to be −5.50. From these data, estimate the distance to Betelgeuse.

SOLUTION. From Eq. (16.3), we have

$$\left(\frac{d}{10 \text{ pc}}\right)^2 = 2.5^{0.41-(-5.50)},$$

which yields $d = 150$ pc for the distance to Betelgeuse.

EXAMPLE. By trigonometric parallax, we find the distance to Aldebaran, a bright star in Taurus, to be 53 ly. The apparent magnitude of Aldebaran is 0.86. Use these data to determine the absolute magnitude of Aldebaran.

SOLUTION. For Eq. (16.3), we have

$$\left(\frac{d}{32.6\text{ ly}}\right)^2 = \left(\frac{53\text{ ly}}{32.6\text{ ly}}\right)^2 = 2.5^{0.86-M}.$$

Taking the logarithms of both sides yields,

$$\log\left[\left(\frac{53}{32.6}\right)^2\right] = (0.86 - M)\log 2.5.$$

Solving for M, we get

$$M = 0.86 - \frac{2\log(53/32.6)}{\log 2.5} = -0.20.$$

the absolute magnitude of Aldebaran is -0.20.

16.2 Stellar Spectra

For a typical star like the sun, the visible spectrum consists of a continuum of colors from red to violet interrupted by many dark lines. Energy is produced in the hot core of the star by means of nuclear fusion, which we shall consider in some detail later. This energy undergoes many transformations in working its way out to a relatively thin, opaque layer of the star called the *photosphere*. The photosphere absorbs the energy from the interior of the star and reradiates it consistent with the radiation laws. It is in the photosphere that the continuous spectrum originates.

16.2.1 Absorption Spectra

The dark line spectrum, known as the *absorption spectrum*, is caused by absorption in the atmosphere of the star of certain wavelengths of light coming from the photosphere. To see how this works, let us consider a hydrogen atom in the atmosphere of the star as indicated in Fig. 16.5. Suppose the star is hot enough that an appreciable fraction of the hydrogen atoms are in the excited state with $n = 2$. The $n = 3$ orbit is 1.892 eV higher in energy. If the electron absorbs a photon of just this energy (wavelength 656 nm), the photon disappears and the atom is left in the $n = 3$ excited state.* This is illustrated in Fig. 16.6.

The photosphere emits light (photons) of all wavelengths, including 656 nm. Due to the absorption in the stellar atmosphere, the light that enters our telescope from the photosphere will be deficient in photons of wavelength 656 nm compared to neighboring wavelengths. This will show up as a dark line—reduced intensity—on the continuous background of radiation from the photosphere. Dark lines will also appear in the green, blue, and violet parts of the spectrum corresponding to

---

*Compare this process with the photoelectric effect where an electron absorbs a whole bundle of energy or none.

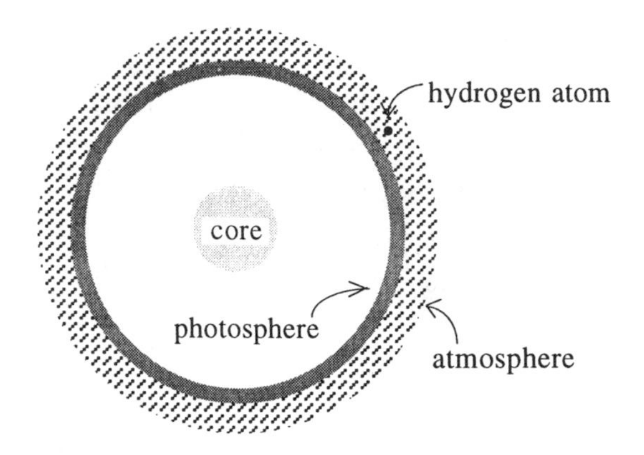

FIGURE 16.5. A model of a star.

absorption of photons by hydrogen in transitions from $n = 2$ to $n = 4, 5, 6, \ldots$ orbits. Similarly, other elements in the stellar atmosphere reveal their presence by absorption lines at wavelengths characteristic of their spectra.

Notice that to see hydrogen absorption lines in the *visible* spectrum of a star, a significant number of hydrogen atoms in the atmosphere must be already excited to the $n = 2$ orbit. Whether this is the situation in a given star is determined by the temperature in the region of the photosphere and stellar atmosphere. If the star is too cool, the hydrogen atom will most probably be in the ground state with the electron in the $n = 1$ orbit where it can only absorb ultraviolet radiation. Also we know from Planck's law that the color of the star (photosphere) is an indicator of

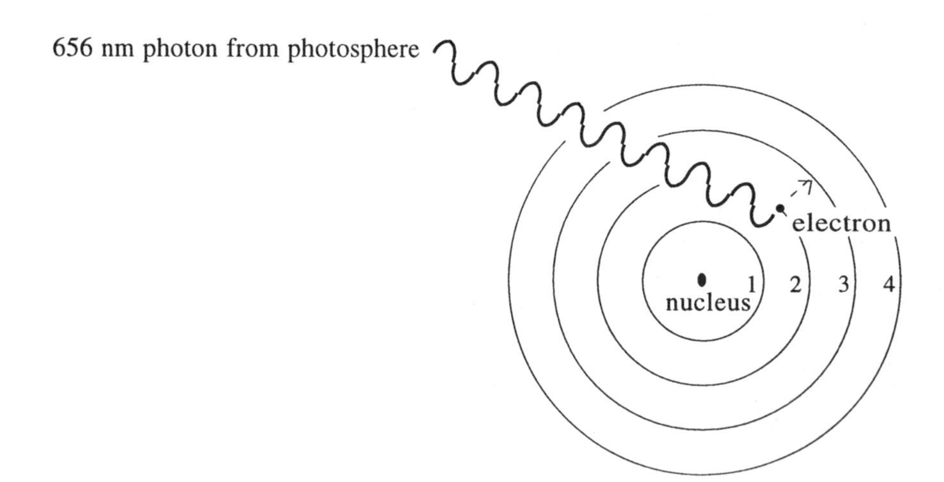

FIGURE 16.6. Excited hydrogen atom in stellar atmosphere.

its temperature. Stars with different photosphere temperatures are distinguished by different colors and by different absorption spectra.

16.2.2 Spectral Classification of Stars

It is useful to classify stars according to their spectra. Listed in Table 16.1 are the seven principal classifications with nominal temperature ranges and some salient features of the spectra.

Table 16.1. Classification of Stellar Spectra

spectral type	maximum temperature	spectral lines
O	60,000 K	singly ionized helium; weak hydrogen
B	30,000 K	neutral helium; strong hydrogen
A	10,000 K	strongest hydrogen
F	7,500 K	weak hydrogen; ionized metals
G	6,000 K	very strong singly ionized calcium
K	5,000 K	neutral metals
M	3,500 K	molecules; very weak hydrogen

Each of these spectral types is subdivided into ten subclasses denoted by single digits (e.g., in order of *decreasing* temperature, . . ., B8, B9, A0, A1,. . ., A9, F0, F1,. . .).

Typical spectra for each type are displayed in Fig. 16.7. We see that hydrogen absorption lines are weak in the spectra of the relatively cool K and M-types as expected from our discussion above. Figure 16.7 also shows that these lines are weak in the spectra of O-types. In this case, it is because these stars are so hot, much of the hydrogen in the stellar atmosphere is ionized. There is no electron on the atom to absorb a photon from the photosphere.

16.2.3 Continuous Spectrum of the Photosphere

The photosphere is opaque, absorbing energy from the interior of the star and reemitting it in a continuous spectrum of radiation characteristic of the star. In stars like the sun, the mechanism responsible for this is provided by negatively charged hydrogen ions. Let's see how this works.

A hydrogen atom can bind an extra electron to form a negative ion. The process whereby this is accomplished is illustrated in Fig. 16.8. In addition to the discrete set of bound states accessible to an electron in a hydrogen atom, there exists a set of unbound, or free-particle, states that the electron can occupy. Since an unbound electron can leave the atom with any kinetic energy whatsoever, these free-particle states form a continuum—the energy states are not discrete. A neutral hydrogen

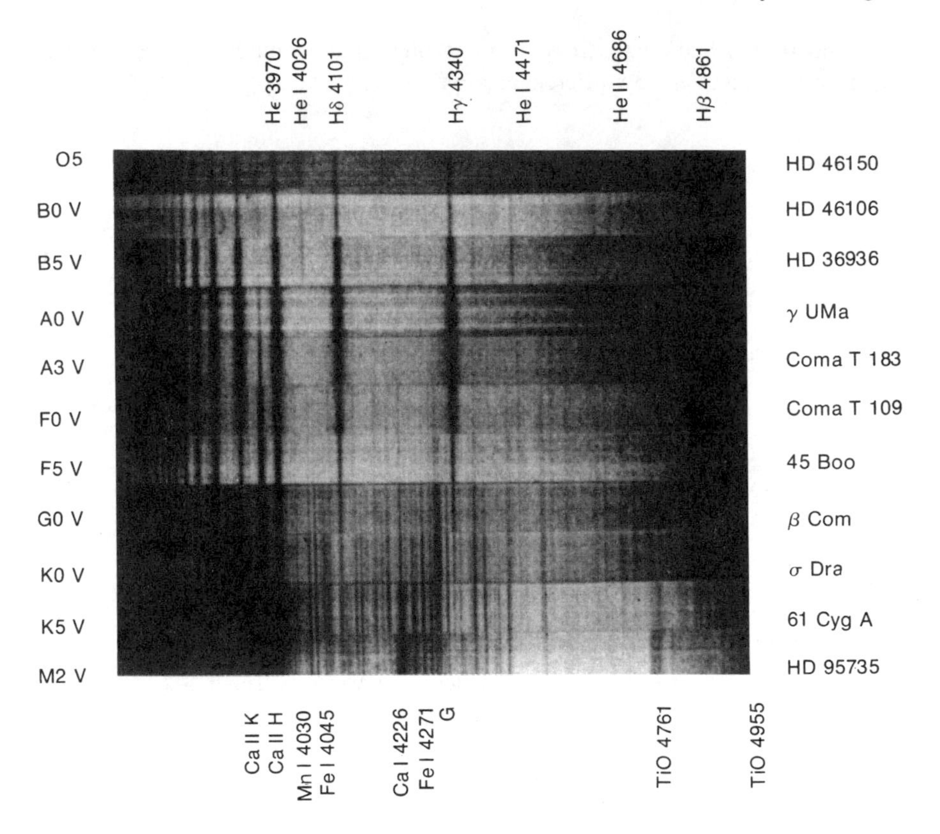

FIGURE 16.7. Spectra of stars of different types. (Courtesy of National Optical Astronomy Observatories.)

atom with its electron in the ground state may capture an additional electron from any state in this continuum. In dropping from the continuum state into a bound state, the captured electron loses energy in the form of a photon. Because the electron can fall from a continuum state with *any* energy into a bound state, the spectrum of photon energies emitted in these transitions is continuous.

Many atoms in the hot photosphere of the sun are ionized, therefore many free electrons exist in the photosphere. There is also an abundance of atomic hydrogen in the photosphere. These hydrogen atoms can capture free electrons by the process described above, producing a continuous spectrum of light from the photosphere. By the inverse process, a negative hydrogen ion produced in this way can be dissociated by absorbing a photon to produce a free electron and a neutral hydrogen atom. The ejected electron is a free electron. This means that to leave the atom, the electron may absorb a photon of *any* energy greater than its

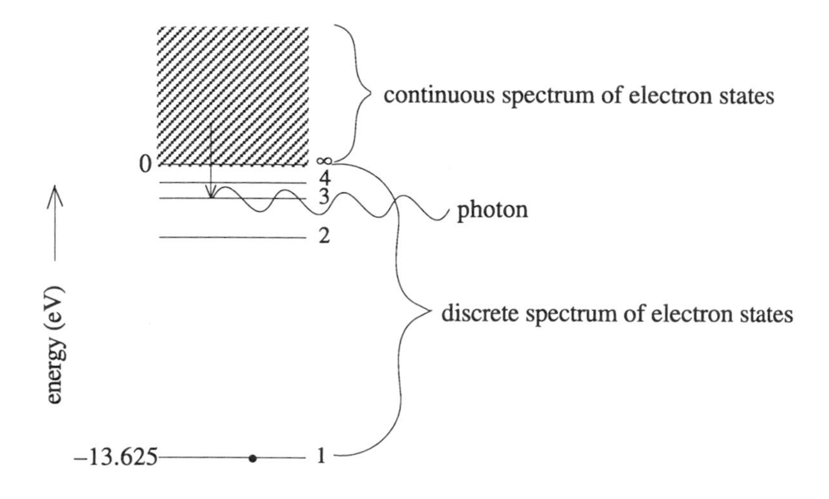

FIGURE 16.8. Energy spectrum for electron in hydrogen atom.

binding energy. Therefore, the negative hydrogen ions in the photosphere absorb a continuous spectrum of radiation making the photosphere opaque. Electromagnetic energy from the interior of the sun is absorbed in the photosphere and is reradiated as a continuous spectrum with a distribution characteristic of a hot body at a temperature of about 5,500 K.

16.3 The Hertzsprung–Russell Diagram

Suppose we measure the apparent magnitudes of all stars in the solar neighborhood whose distances we can determine. Using Eq. (16.3), we can calculate the absolute magnitude for each of these stars. The spectral class of each is obtained from its spectrum. Now we make a plot with absolute magnitude on the vertical axis (bright at the top, dim at the bottom) and spectral class on the horizontal axis (hot to the left, cool to the right) as shown in Fig. 16.9. Each data point on the plot represents one of the stars. A plot of this type is called a *Hertzsprung–Russell diagram*, abbreviated H-R diagram, after the two astronomers who, independently early in the 20th century, were the first to present stellar data in this form.

Notice that the stars are not uniformly distributed on the plot. They fall generally into four major groups. Most of the points lie in a narrow band that extends diagonally from the upper left (bright, hot blue stars) to the lower right (dim, cool red stars). This band is called the *main sequence* and represents normal or ordinary stars that are in the prime of life. There is another region of the H-R diagram representing cool, red stars that are much brighter than main-sequence stars of the same spectral class. These are the *red giants*. Another group of extremely bright stars of all spectral classes is spread across the top of the diagram. Stars in this group are called *supergiants*. Finally, there are a few very dim stars represented by

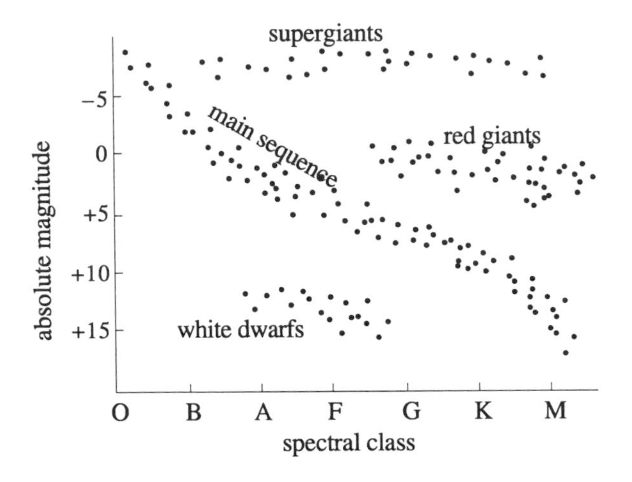

FIGURE 16.9. Hertzsprung–Russell diagram.

points below the main-sequence. Their temperatures generally place them in the B to F range of spectral types. They are known as *white dwarfs*.

The H-R diagram is extremely useful for a number of reasons. The stars represented on the diagram are in various stages of evolution, so we can trace out paths on the H-R diagram that typical stars of various masses would follow in the course of their lives. We can also apply the H-R diagram to the determination of distances to stars too far away to show any detectable parallax. For example, suppose we determine from its spectrum that a certain remote star is a main-sequence star. We can also identify its spectral class. From the H-R diagram, the absolute magnitude M of a main-sequence star of this class can be estimated. Since the star is observable, we can measure its apparent magnitude. With these data, Eq. (16.3) yields the distance to the star. This method of determining stellar distances is called *spectroscopic parallax*. It has nothing to do with the phenomenon of parallax discussed earlier.

EXAMPLE. The spectrum of a certain blue star reveals that it is a B1-type main-sequence star. Its apparent magnitude is 2.24. Estimate the distance to this star in light years.

SOLUTION. From Fig. 16.9, we find that for a main-sequence star of this type, the absolute magnitude is approximately -4. With the values $m = 2.24$ and $M = -4$, Eq. (16.3) yields $d = 174$ pc $= 568$ ly. The star is about 570 ly away.

A variation of the H-R diagram is particularly useful for representing the stars in a remote cluster. The sizes of these clusters are small compared to their distances from the earth, so to a good approximation we can regard all of the stars in a given cluster as being the same distance away from us. Since they are all at the

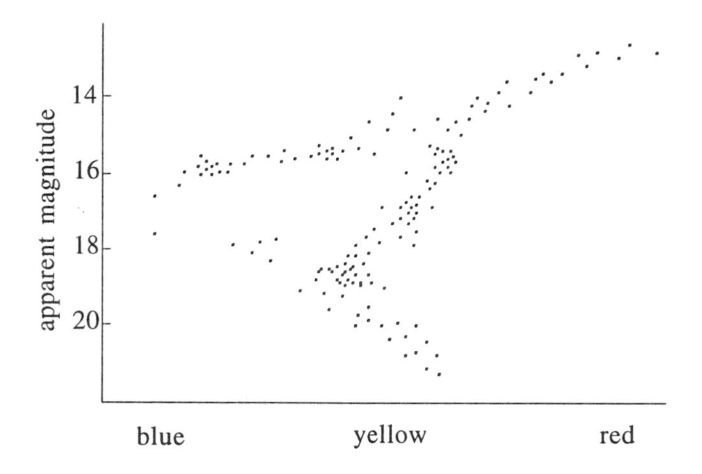

FIGURE 16.10. Color-magnitude diagram for a globular cluster.

same distance, their apparent magnitudes are indicative of their *relative* absolute magnitudes. For example, a star that *appears* twice as bright as another star in the same cluster *is* twice as bright.

In this variation, the color (temperature) is plotted on the horizontal axis and the apparent magnitude on the vertical axis as shown in Fig. 16.10 for a typical globular cluster. This kind of plot is known as a *color-magnitude diagram*.

The distance to a cluster can be deduced by comparing the color-magnitude diagram for the cluster with a standard H-R plot of field stars. First, we identify the spectral class on the H-R diagram with the corresponding color on the cluster diagram and then overlay the cluster diagram on the H-R diagram so that their main-sequences coincide. Then, from the vertical axes the difference between apparent and absolute magnitudes, $m - M$, can be read directly and the distance to the cluster obtained from Eq. (16.3).

Comparison of a cluster diagram with a standard H-R diagram also provides us with information about the *age* of the cluster, as we shall see in Chapter 19.

16.4 Binary Stars

Most of the stars in the galaxy are members of multiple star systems with the components bound together by gravity and orbiting about a common center of mass. Quite often, there are only two members in the group in which case it is called a *binary* star system. It has become customary to classify a binary according to the way in which its binary character is revealed. For example, if a binary is near enough to the earth, the two components may be seen directly in a telescope. When such direct visual resolution of the components is possible, the system is called a *visual* binary.

A binary system may also manifest its double character through its spectrum. To understand how this comes about, let's consider the simple example of two stars of equal mass orbiting about their common center of mass with the earth lying in the plane of the orbit. We further assume that the entire system is moving directly away from the earth. This is illustrated in Fig. 16.11. A portion of the absorption spectrum of the star is shown along with a corresponding emission spectrum observed in a laboratory on earth. Because of the Doppler effect, the absorprion spectrum of the star is shifted toward longer wavelength (a *redshift* due to the star's motion away from the observer). If the velocity v of the light source (star) relative to the observer (earth) is small compared to the speed of light c, the Doppler shift $\Delta\lambda = \lambda_{\text{observed}} - \lambda$ is given by

$$\frac{\Delta\lambda}{\lambda} = \frac{v}{c}, \tag{16.4}$$

where λ is the laboratory (unshifted) wavelength of the light. Clearly, if $\lambda_{\text{observed}} < \lambda$ (blueshift), then $v < 0$, implying that the source and observer are moving *toward* each other. If $\lambda_{\text{observed}} > \lambda$ (redshift), the source and observer are moving apart.

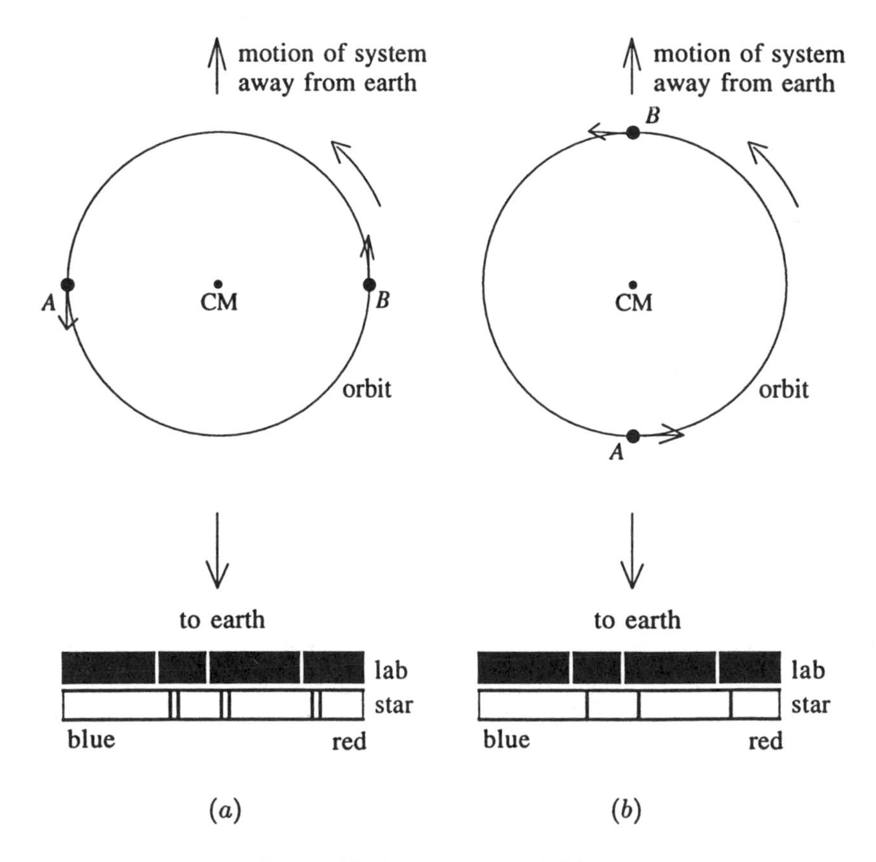

FIGURE 16.11. A spectroscopic binary.

In a binary system, the orbital motion also contributes to the Doppler effect. At times, one star may be moving away from the earth faster than the other. With the two stars positioned as in Fig. 16.11a, star B is receding from the earth with greater speed than star A. Thus, the spectrum of star B shows a slightly greater redshift than star A. This results in a splitting of the spectral lines into two slightly displaced components, one from each star. As the stars orbit around to the positions shown in Fig. 16.11b, their orbital motions have no component along the line of sight. In this case, both stars have the same recession velocity and the Doppler shift is the same for both. So, as the stars orbit, the double spectrum of Fig. 16.11a collapses to the single spectrum of Fig. 16.11b. Stellar systems that reveal their binary character in this way are called *spectroscopic* binaries.

If one of the stars is very faint, only a single-line spectrum may be seen. However, as the bright component moves on its orbit, its spectrum will shift toward the red as its orbital motion carries it away from the earth and toward the blue when its orbital motion is toward the earth. This oscillation of the spectral lines about the Doppler shifted spectrum due solely to the motion of the entire system relative to the earth is also a signature for a spectroscopic binary.

The orbital plane of a binary system may be oriented relative to the earth such that one star passes in front of the other as seen from earth. As the stars eclipse each other, the total brightness of the system diminishes. For example, Fig. 16.12 shows a hot star A and a cooler star B with their orbital plane nearly along the line of sight.

In Fig. 16.12a, light is received from both stars. In Fig. 16.12b, the cooler star is behind the hotter one, so that the earth based observer only receives light from the hotter star. Hence, the total apparent brightness of the system is reduced because no light is arriving from the cooler star. As they continue to orbit, light from the cooler star is no longer obscured by the hotter one and the total apparent brightness of the system is restored. Still later, the cooler star passes in front of the hotter one. In this eclipse, the reduction in brightness is greater than for the previous one because it is the more intense light from the hotter star that is being blocked. Even though the system may be too far away for the components to be resolved visually, these fluctuations in brightness signal the presence of a double star system.

Continued measurement of the apparent brightness of the system over several months yields a light curve as illustrated in Fig. 16.13, which shows periodic dips in the brightness as the eclipses occur. The smaller dips correspond to the cooler star

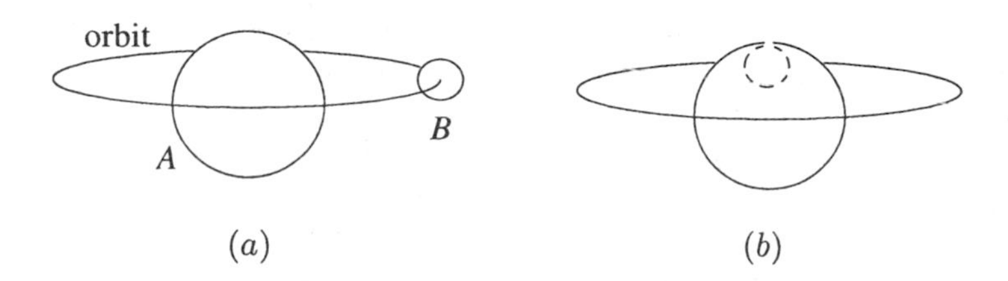

FIGURE 16.12. An eclipsing binary system.

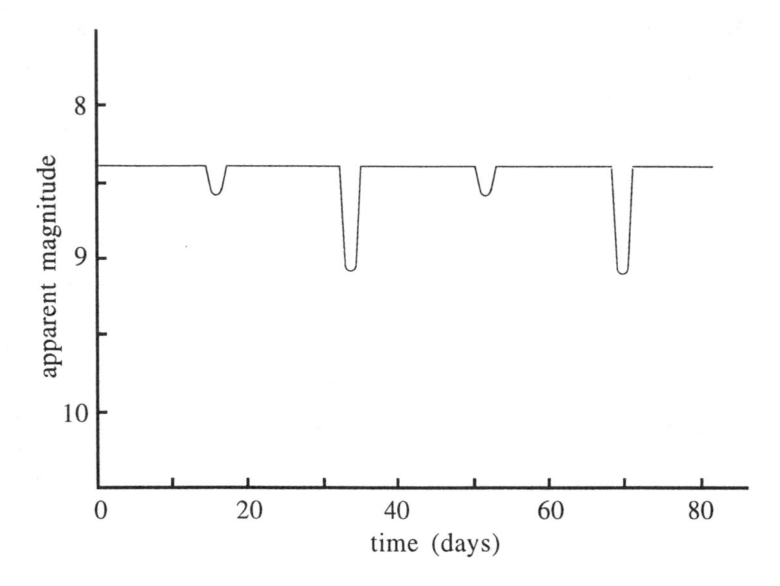

FIGURE 16.13. Simulated light curve for an eclipsing binary system.

being eclipsed by the hotter and the larger ones occur when the cooler star passes in front of the hotter one. A careful analysis of the light curve yields information about the inclination of the orbit. Coupling this with spectroscopic data, one can obtain the masses of the two stars. A measurement of the duration of each eclipse together with the orbital speed of each of the binary components may lead to a determination of the diameter of each of the stars in the system. Perhaps, the best known eclipsing binary is Algol in the constellation of Perseus.

Sometimes a relatively bright star will be paired with a very dim companion. If the system is near enough to the earth, its *proper motion* (the component of the motion perpendicular to the line of sight) can be observed. That is, by observing over times of the order of years or decades, the star can be seen to move relative to the more remote background stars. The center of mass of the system traces a linear path in the sky. However, since only the bright component is visible and it is orbiting about the center of mass, the observed path will oscillate about the actual path of the center of mass. The primary (brighter) star will appear to follow a wobbly path in the sky as illustrated in Fig. 16.14. A system that manifests its double nature in this way is called an *astrometric* binary.

Sirius, the star that appears brightest in the sky, is an astrometric binary with the primary being an A-type star somewhat larger than the sun and its companion a very hot, dim star about the size of the earth.

Clearly, a given system may be identified as binary by more than one of these means. For example, Sirius is both a visual binary and an astrometric binary.

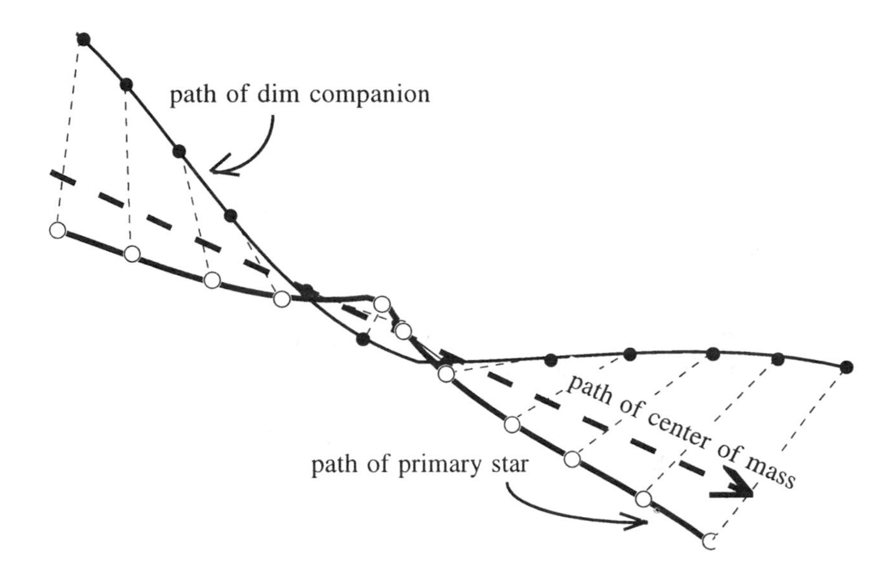

FIGURE 16.14. Simulated proper motion of an astrometric binary.

Exercises

16.1. Make a drawing representing a model of the sun. Clearly label the photosphere and the atmosphere. Make a second drawing representing Bohr's model of the hydrogen atom and showing the allowed electron orbits. Use these two drawings together to explain clearly and completely how the absorption (dark line) spectrum of a star like the sun is produced.

16.2. A tiny, 100-watt light source radiates uniformly in all directions. Calculate the intensity of radiation on a small surface at a distance of 45 centimeters from the source. The direction of propagation of the light is perpendicular to the surface on which the light falls.

Answer: (39 W/m^2)

16.3. An incandescent light bulb emits about four percent of its radiated energy as visible light. What is the intensity of visible light at a distance of four meters from a bare, 100-W light bulb?

Answer: (0.020 W/m^2)

16.4. The intensity of radiation from a certain star is measured at 1.9×10^{-9} W/m^2. The star's distance from the earth is 143 ly. Calculate the total power radiated by this star.

Answer: (4.4×10^{28} W)

16.5. If the total radiative power of the sun is 3.9×10^{26} W, estimate the amount of solar energy received by the earth each second. Treat the earth as a disk that intercepts the radiation from the sun.

Answer: (1.8×10^{17} J)

16.6. The intensity of radiation from the sun measured by instruments on board a spacecraft orbiting the earth at an altitude of 600 km is 1.4 kW/m^2. Use this information to estimate the total radiative power of the sun.

Answer: (3.9×10^{26} W.)

16.7. A very small, 7-W source emits light in all directions uniformly. How much energy from this source is absorbed each hour by a black, circular disk of diameter 3.22 mm that is 89.5 cm from the source and oriented so that the plane of the disk is perpendicular to the direction of incidence of the light falling on it.

Answer: (0.0204 J)

16.8. The earth receives energy from the sun at a rate of about 1.8×10^{17} W. If the inhabitants of the earth were required to pay for this energy at the rate electric companies charge (about 7.2¢ per kilowatt·hour), what would be the approximate daily energy bill for the earth?

Answer: ($300,000,000,000,000)

16.9. The apparent magnitude of the bright star Altair is +0.77. By trigonometric parallax, Altair's distance is determined to be 16.5 light years. Find the absolute magnitude of Altair.

Answer: (+2.2)

16.10. Two stars have the same absolute magnitude. One is fifteen times as far away as the other. How many times brighter does the nearer one appear to be? What is the difference in apparent magnitudes of these two stars?

Answer: (225; 5.91)

16.11. Spica is an ordinary, bright B1-type star in the constellation of Virgo. Its apparent magnitude is +0.91. From the data available, estimate the distance to Spica.

Answer: (80 pc)

16.12. Antares is an M1-type star in the constellation of Scorpius. It is about 130 pc from the sun and has an apparent magnitude of +0.92. Use these data and Fig. 16.9 to obtain a description of Antares.

16.13. Two main-sequence stars (let's call them Enlil and Pharpar) are known to be at distances of 183 ly and 68 ly, respectively. The apparent magnitude of Enlil is observed to be +0.54. For Pharpar, it is −0.43. From the data, determine which is the hotter star, Enlil or Pharpar?

16.14. A K0 star with absolute magnitude +6.0 is known to be at a distance of 8.6 pc from the sun. A B5 star 104 pc away has an absolute magnitude of −2.0. Which star appears brighter to an observer on earth? How many times as bright?

Answer: (10.4)

16.15. Hadar and Deneb are two of the brightest stars in the sky. Data for these two stars are given in the table. From the data, determine which star is intrinsically the brighter. How many times as bright?

Answer: (8.4 times)

star	apparent magnitude	distance (pc)
Deneb	+1.25	500
Hadar	+0.63	130

16.16. Two red stars, P and Q, are observed in the sky. P is seen to have an apparent magnitude of +16.9 at a distance of 1,522 light years from the sun. The distance to Q is 368 light years and its apparent magnitude is measured at +5.0. One of these stars is a normal star and the other is a giant. By means of calculations, determine which is the giant.

16.17. Shaula is a normal, B1 star in Scorpius with an apparent magnitude of +1.6. Estimate the distance to Shaula.

Answer: (85 pc)

16.18. The radiative power of a distant main-sequence star is 6.2×10^{27} W. By comparison with the sun (radiative power = 3.9×10^{26} W), determine the absolute magnitude and the spectral class of this star. Explain clearly and completely your reasoning.

Answer: (+1.5; B6)

16.19. The apparent magnitude of a certain star known to be just like the sun is observed to be +10.7. Estimate the distance to this star.

Answer: (170 pc)

16.20. The distance to a particular main-sequence star in the solar neighborhood is determined by trigonometric parallax to be 11.1 ly. The intensity of radiation from this star is measured at 1.6×10^{-10} W/m^2. Compare this star with the sun to determine its absolute magnitude, its apparent magnitude, and its spectral class. Explain clearly your reasoning.

Answer: (+7.6; +5.3; K0)

16.21. When seen in the sky, two main-sequence stars appear equally bright. The spectral class of one is B4. The other is F9. Which star is farther away from us? Approximately how many times farther away?

Answer: (20)

16.22. After the triple-star system α Centauri, the next nearest star to the sun is a dim, red dwarf known as Barnard's star. It's apparent magnitude is +9.5. Some observational evidence exists that suggests that this star may have one or more large planets orbiting around it. Barnard's star is near enough for its distance to be determined by trigonometric parallax. It is about six light years from the sun. From the data, determine the absolute magnitude of Barnard's star.

Answer: (+13)

16.23. Vega is a main-sequence star with an apparent magnitude of $+0.03$. By trigonometric parallax, it is found that Vega is 26 light years away. Merak, one of the bright stars in the Big Dipper, is also a main-sequence star of the same spectral class as Vega. The apparent magnitude of Merak is $+2.40$. From these data, estimate the distance to Merak. Explain clearly your reasoning.

Answer: (77 ly)

16.24. The sun is a yellow, main-sequence star of spectral class G2. Calculate the apparent magnitude of the sun. Explain clearly and completely your reasoning.

Answer: (-27)

16.25. Arneb is an F0 star in the constellation of Lepus. It is about 280 pc away from the sun and has an apparent magnitude of $+0.70$. Use these data and the H-R-diagram to obtain as complete a description of Arneb as possible.

16.26. Explain clearly and completely why absorption lines of singly-ionized helium are prominent only in O-type stars.

16.27. The stars α Centauri and Denebola are both main-sequence stars. Their distances have been measured to be 4.3 ly and 43 ly, respectively. The apparent magnitude of α Centauri is observed to be -0.01. For Denebola, the apparent magnitude is $+1.2$. Determine which star, if either, is hotter. Explain clearly and completely your reasoning in arriving at your answer.

16.28. What characteristics of the light from an M-type star show that it is cooler than a G-type?

16.29. A comparison of the main-sequence in a cluster color-magnitude diagram with that in the standard H-R diagram shows that the apparent magnitude of the stars on the cluster main-sequence are seven magnitudes larger than the absolute magnitudes of stars in corresponding positions on the standard H-R diagram. How far away is the cluster?

16.30. From the spectrum of an eclipsing binary, it is found that the orbits are circular with a period of 5.03 yr. From the earth, the orbital plane is seen edge-on. The orbital velocity of one of these stars is determined to be 7.73 km/s and the other 17.38 km/s. Find the mass of each star.

Answer: (0.93 $M_\odot$; 2.08 $M_\odot$)

16.31. The two stars A and B of an eclipsing binary system describe circular orbits about their common center of mass with a period of 88.2 days. The system is oriented such that the orbital plane is viewed edge-on as seen from the earth. Hydrogen lines are observed in the spectra of both components. Data for three of these lines are given in the table. The maximum and minimum observed values of the wavelengths are given for each star. From the Doppler shifts evident in the data, it is clear that the stars have different orbital speeds, hence different masses. From the data, determine how fast and in which direction along the line of sight the entire system is moving relative to the earth. Determine the orbital speed of each star and the distance between the two. From your results, calculate the mass of each star.

	star A		star B	
laboratory (nm)	minimum (nm)	maximum (nm)	minimum (nm)	maximum (nm)
656.273	656.202	656.532	656.332	656.402
486.133	486.080	486.325	486.177	486.228
434.047	434.000	434.218	434.086	434.132

16.32. The optical spectrum of the element sodium is characterized by a strong yellow line with a wavelength of about 589 nm. It is this radiation that gives the sodium vapor lamps commonly used for parking lot illumination their yellowish color. Careful laboratory measurements show that this line is actually split into two components separated by about 0.597 nm. These two lines are shown in the figure as a laboratory emission spectrum. Also shown are the sodium absorption lines observed in the spectrum of a spectroscopic binary in which only light from the brighter star can be detected. The figure shows the spectrum with the shortest (minimum) wavelengths observed for the two lines and the longest (maximum) values observed. From the data, estimate how fast and in what direction the system is moving relative to the earth along the line of sight. Explain clearly and completely your reasoning.

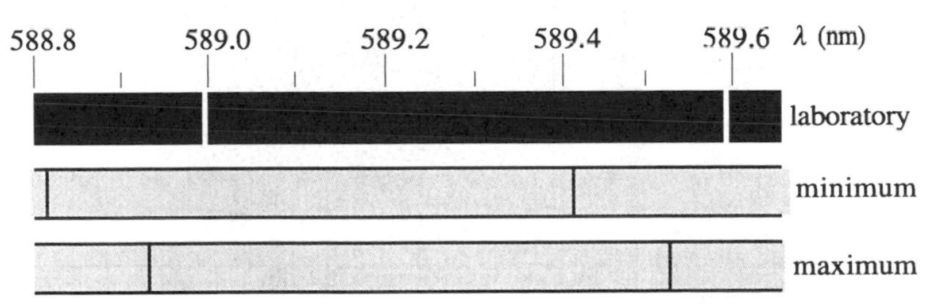

17

The Sun Is a Gas

The sun is a ball of gas—a special kind of gas to be sure. It is so hot in some regions in the interior that the atoms have come apart completely to form a state of matter known as a *plasma*—an *electrically charged* gas. To understand the physical processes that cause stars, like the sun, to shine and to understand how and why stars evolve, we must learn something about the behavior of gases. A quantity that is quite useful in describing the state of a fluid (gas or liquid) is the *pressure*. We define the average pressure $\bar{P}$ on a small area ΔA in a fluid as $(\Delta F)_{\perp}$, the component of the force exerted perpendicular to the area, divided by the area,

$$\bar{P} = \frac{(\Delta F)_{\perp}}{\Delta A}. \tag{17.1}$$

As the area around a given point becomes very small, this ratio approaches the value for the pressure at the point. Note that the pressure is a *scalar* quantity. To make this notion of pressure intuitively clear, let's consider a liquid (e.g., water) in an open continer as illustrated in Fig. 17.1. We focus our attention on a small,

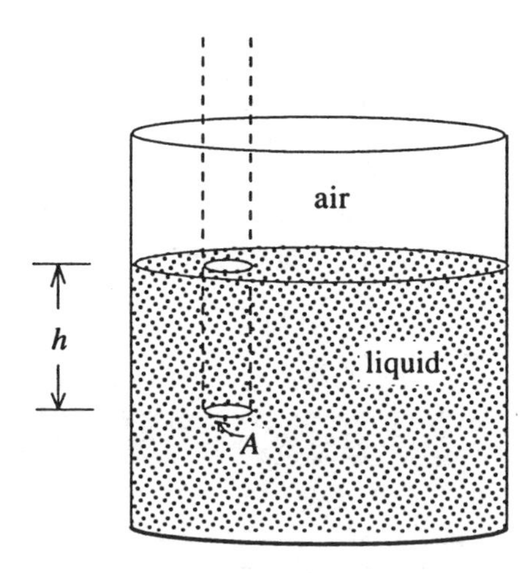

FIGURE 17.1. Pressure in a fluid.

horizontal circular area A at a depth h in the liquid. The force exerted on this area is due to the weight of the fluid directly above it, that is, a fluid column of cross-sectional area A consisting of a cylindrical volume of liquid of height h and a cylindrical column of air extending from the surface of the liquid all the way to the top of the atmosphere.

From Eq. (17.1), we find that

$$\begin{aligned} P &= \frac{\text{weight of liquid column} + \text{weight of air column}}{\text{area of column}} \\ &= \frac{(\rho h A)g}{A} + P_{\text{atm}} = \rho h g + P_{\text{atm}} \end{aligned}$$

where ρ is the density of the liquid, g is the acceleration due to gravity, and P_{atm} is the pressure of the atmosphere at the top of the liquid.

17.1 Boyle's Law

Let's begin with a very simple gas—the air in the room. Suppose we trap a small volume of air in a sealed glass cylinder fitted with an airtight piston. By pushing the piston in or drawing it out, we can change the volume that the gas occupies. Since the system is tightly sealed, we assume no air leaks in or out.

To measure the pressure of the gas trapped in our glass cylinder, a pressure gauge is connected to the cylinder by means of a small, flexible tube as illustrated in Fig. 17.2. The volume of gas is obtained from the inside diameter of the cylinder and the distance from the end of the cylinder to the piston.

Let us attempt a graphical analysis of the data collected from these measurements. First, we make the straightforward plot of pressure versus volume shown

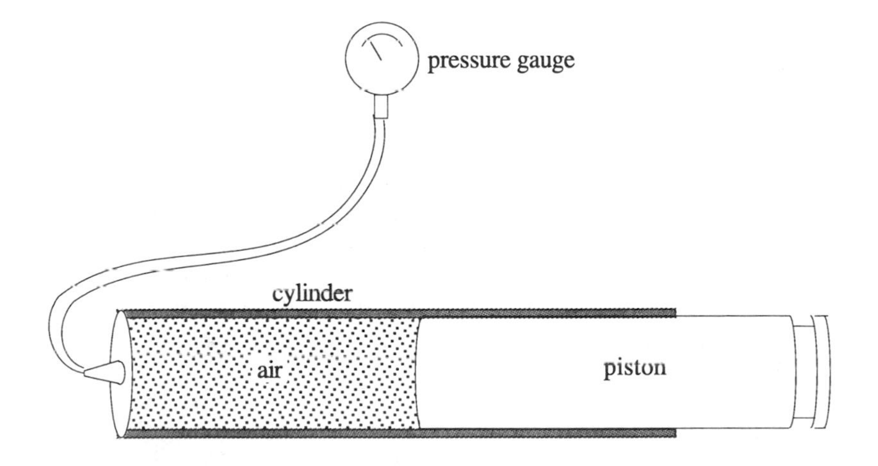

FIGURE 17.2. Measurements of pressure and volume of air.

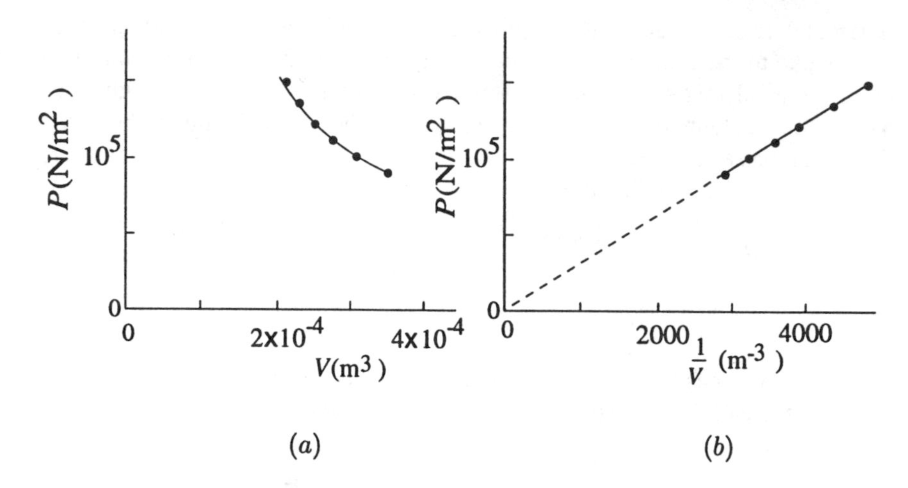

FIGURE 17.3. Pressure–volume relationship.

in Fig. 17.3*a*. Clearly, the graph is not linear. However, we can see that there appears to be some kind of inverse relationship between P and V (as V increases, P decreases). As a first guess at the correct relationship, we try a simple inverse proportionality and plot the pressure versus the *reciprocal* of the volume ($\frac{1}{V}$) as shown in Fig. 17.3*b*. The data plotted in this way are seen to be consistent with a straight line, which extends through the origin of the graph, the point (0,0). This brings us to the empirical relationship

$$PV = \text{constant}, \tag{17.2}$$

known as *Boyle's law*. Under the conditions we have described, air behaves in a manner consistent with this law and so do many other gases.

17.2 A Molecular Model of the Gas

Now let's see if we can understand this relation in terms of a microscopic model in which we picture the gas as a collection of tiny, structureless particles ("molecules") in random motion. We assume that the molecules have no size (point particles), they all have the same mass m, and they interact with one another and with the walls of the container only through elastic collisions. Because they have no structure, they have no internal excitations. To simplify the geometry, we consider the gas to consist of N of these molecules confined to a rectangular box of dimensions L_x, L_y, and L_z as illustrated in Fig 17.4.

The pressure the gas exerts on the walls of the box arises from the individual molecules striking the walls and bouncing off. The average force a molecule exerts on a given wall is, by Newton's third law of motion, equal to the force the wall exerts on the molecule. According to Newton's second law, this is just the momentum

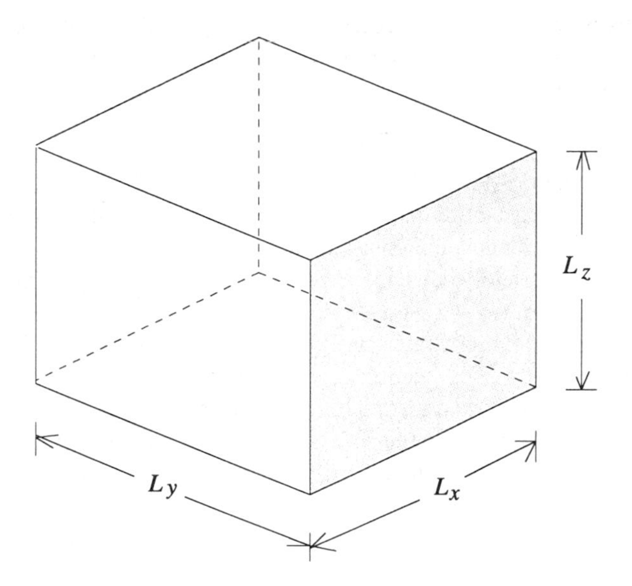

FIGURE 17.4. Gas molecules confined to a rectangular box.

change suffered by the particle in collision with the wall divided by the average time interval between collisions with this wall.

For definiteness, we take the shaded wall in Fig. 17.4 to calculate the pressure,

$$P = \frac{\text{(number of molecules)} \cdot \text{(average force per molecule)}}{\text{area of shaded wall}}$$

$$= \frac{N(\overline{\Delta(mv_y)/\Delta t})}{L_x L_z} = \frac{N(\overline{2mv_y)/(2L_y/v_y})}{L_x L_z} = \frac{N(m\overline{v_y^2})}{L_x L_y L_z}. \tag{17.3}$$

Note that the magnitude of the velocity $\mathbf{v}$ is related to the squares of its rectangular components by

$$v = \sqrt{v_x^2 + v_y^2 + v_z^2}.$$

Also, since all directions look the same to a molecule (it can't tell up from down, left from right, etc.), the average values of the squares of the components of the velocity are equal,

$$\overline{v_x^2} = \overline{v_y^2} = \overline{v_z^2}.$$

Thus,

$$\overline{v_y^2} = \tfrac{1}{3}\overline{v^2}.$$

On substituting this result into Eq. (17.3), we get

$$PV = \tfrac{2}{3} N(\tfrac{1}{2} m\overline{v^2}), \tag{17.4}$$

where we have recognized that the volume $V = L_x L_y L_z$. We have also inserted a factor of $\frac{1}{2}$ within the parentheses and a factor 2 outside. The quantity in parentheses is seen to be the average kinetic energy per molecule,

$$\overline{KE}_{\text{mol}} = \tfrac{1}{2} m\overline{v^2}.$$

We see that Eq. (17.4) is consistent with Boyle's law, Eq. (17.2), provided the right-hand side is constant. One factor on the right-hand side is N, the number of molecules in the container. Consistent with our assumption that the container has no leaks, N is certainly constant. What about the factor in parentheses on the right-hand side of Eq. (17.4)? Should that remain constant? Let's examine more carefully the experiment illustrated in Fig. 17.2 which led us to Eq. (17.2). When the piston moves in, it pushes the molecules into a smaller volume—that is, it exerts a *force* on the gas and *displaces* it. The piston does work on the gas. Therefore, the energy of the gas, represented by $\overline{KE}_{\text{mol}}$, should increase. Accordingly, the right-hand side of Eq. (17.4) should increase, not remain constant as Eq. (17.2) implies. Where is the discrepancy?

The resolution lies in recognizing that the walls of the container (glass cylinder) also contain molecules in constant motion. These molecules are in contact with the gas inside the cylinder *and*, through the glass, with the air molecules in the room. The molecules in the walls absorb the extra energy acquired by the gas in compression and share this energy with the very large number of molecules in the room. Very quickly—much more quickly than we can notice—room, cylinder, and gas all come to equilibrium with the same average kinetic energy per molecule. The system of air molecules in the room is so large that it can absorb the tiny amount of extra energy coming from compression of the gas with no perceptible change in the average kinetic energy of its molecules. Because, in equilibrium, the gas molecules inside the cylinder will have the same average kinetic energy per molecule, the gas shows no noticeable change in $\overline{KE}_{\text{mol}}$ either. Hence, under the conditions we have described, the right-hand side of Eq. (17.4) should remain constant in accord with Eq. (17.2)

Notice that the quantities P and V on the left-hand side of Eq. (17.4) are *macroscopic* quantities associated with the bulk properties of the gas—quantities we can measure directly with instruments like pressure gauges, meter sticks, and so on. On the right-hand side, we have the quantities N and $\overline{KE}_{\text{mol}}$ that are more related to the *microscopic* description of the gas and suggest measurements that involve counting molecules or somehow finding the average kinetic energy of a molecule in the gas.

17.3 The Ideal Gas Law

For a completely macroscopic description, we need to introduce new parameters in place of N and $\overline{KE}_{\text{mol}}$ that will allow us to measure the amount of gas and the average energy of the particles with laboratory instruments that change in an observable way as the state of the gas changes. An obvious macroscopic alternative to N is simply the total mass of the gas, which we could obtain by weighing the container twice—first evacuated, and then filled with the gas. The difference between these measurements would tell us how much gas we had in the container. In practice, for gases there is a more convenient quantity that measures the amount of gas. We shall return to this shortly.

For the moment, let's turn our attention to $\overline{KE}_{\text{mol}}$. We look for a probe to insert into the gas that will change in a regular and observable way when the average kinetic energy of the molecules in the gas changes. We can construct such a device that will work under the conditions we described provided the amount of gas is large enough so that the probe itself does not change $\overline{KE}_{\text{mol}}$ by any observable amount. The device, illustrated in Fig. 17.5(*a*), consists of a long, thin sealed glass tube containing a supply of mercury (a liquid under our conditions) in a small bulb at one end of the tube. We insert the bulb into the gas, sealing it in such a way that molecules cannot enter or escape from the container.

In equilibrium, the glass bulb, the mercury, and the gas all have the same $\overline{KE}_{\text{mol}}$. The mercury atoms are jiggling around in constant motion, causing them to maintain a certain average separation from their nearest neighbors. If we increase $\overline{KE}_{\text{mol}}$ (e.g., by compressing the gas), the mercury atoms will jiggle more rapidly and increase the average distance between neighboring atoms. This causes the mercury to expand and rise in the tube. The height of the mercury column is an indicator of $\overline{KE}_{\text{mol}}$. Such a device, appropriately calibrated, is called a *thermometer*. The macroscopic quantity that corresponds to the observable change in the length of the mercury column when $\overline{KE}_{\text{mol}}$ for the gas changes is the *temperature*.

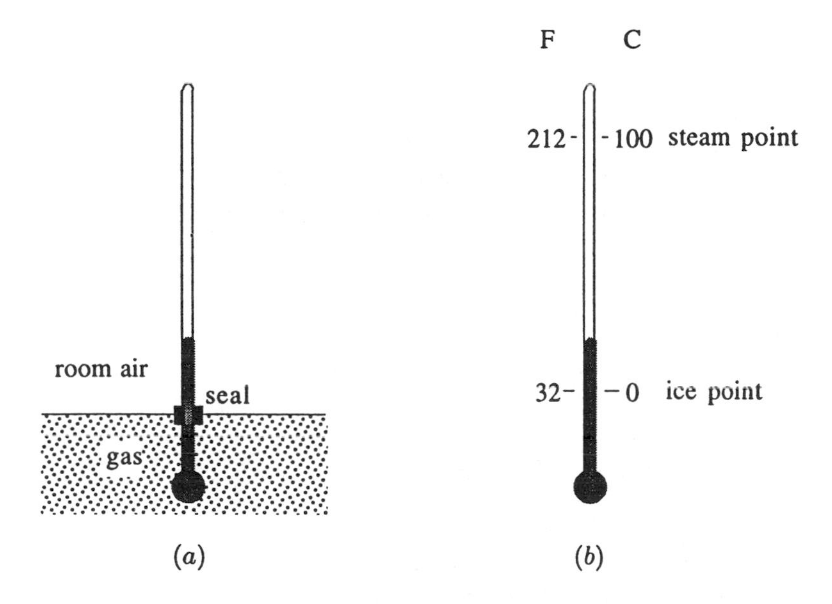

FIGURE 17.5. Mercury column for temperature measurement.

To a good approximation, the length of the mercury column expands uniformly with increasing $\overline{KE}_{\text{mol}}$. So we can calibrate the glass tube by defining a temperature scale according to the following procedure:*

1. Immerse the mercury bulb in a bath of ice and water in equilibrium and mark the position of the top of the mercury column (*ice point*).
2. Immerse the mercury bulb in a mixture of water and steam in equilibrium and mark the new position of the top of the mercury column (*steam point*).
3. Subdivide the interval between the two marks on the glass tube into equal subintervals. The number of subintervals between the ice point and the steam point determines the size of the temperature unit (*degree*) and the number assigned to *one* point (e.g., ice point) determines the temperature scale uniquely.

There are two systems in general use. Both are defined in Fig. 17.5*b*. On the Fahrenheit (F) scale, there are 180 divisions (degrees) between ice point and steam point with the ice point arbitrarily set at 32. The Celsius (C) scale has 100 units between ice point and steam point with the ice point defined at 0. We also use another temperature unit called the *kelvin*, abbreviated K, which has the same magnitude as the Celsius degree.

The exact connection between the temperature and the average kinetic energy per molecule is still at our disposal. We choose a simple, linear relationship with absolute† temperature T defined by

$$\overline{KE}_{\text{mol}} = \tfrac{3}{2}kT, \tag{17.5}$$

where the constant parameter k is known as *Boltzmann's constant*. With this definition, Eq. (17.4) now reads

$$PV = NkT, \tag{17.6}$$

which is called the *ideal gas law*.

The ideal gas law describes a gas that is *ideal* in the sense that it consists of structureless, point particles that interact only through elastic collisions. No real gas is made up of such particles.‡ Nevertheless, over certain ranges of temperature and pressure, the behavior of some real gases (e.g., air) is represented by Eq. (17.6).

17.4 Absolute Temperature

Suppose we have a metal sphere filled with a fixed volume of air, which behaves like an ideal gas over certain ranges of temperatures and pressures. We immerse the sphere in a water bath. In equilibrium, the temperature of the air in the sphere is the

---

*In practice, this procedure is carried out at a specific pressure.

†The significance of the adjective *absolute* will become clear shortly.

‡For example, the molecules in such a gas cannot bind to form liquids and solids as real substances do at low temperature and high pressure.

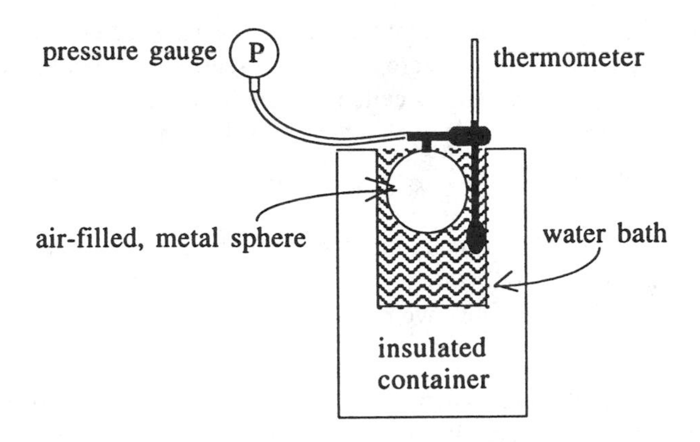

FIGURE 17.6. Pressure–temperature measurement of fixed volume of air.

same as the temperature of the surrounding water. The experimental arrangement is illustrated in Fig. 17.6. The gauge displays the pressure of the air in the sphere. By changing the temperature of the water bath, we can obtain a relationship between the pressure and the temperature of the fixed volume of air trapped inside the sphere. The data are shown in Fig. 17.7.

In this experiment, we have used a Celsius thermometer to measure the temperature. From the graph, we can see that when the temperature is 0 °C, the pressure is about 0.94×10^5 N/m^2 (clearly, *not* zero). Since the amount of gas N and the volume V are constant, it is clear from Eq. (17.6) that at zero pressure ($P = 0$), we must have zero temperature ($T = 0$). The graph shows that $P = 0$ corresponds to a Celsius temperature of −273°C. To obtain a temperature consistent

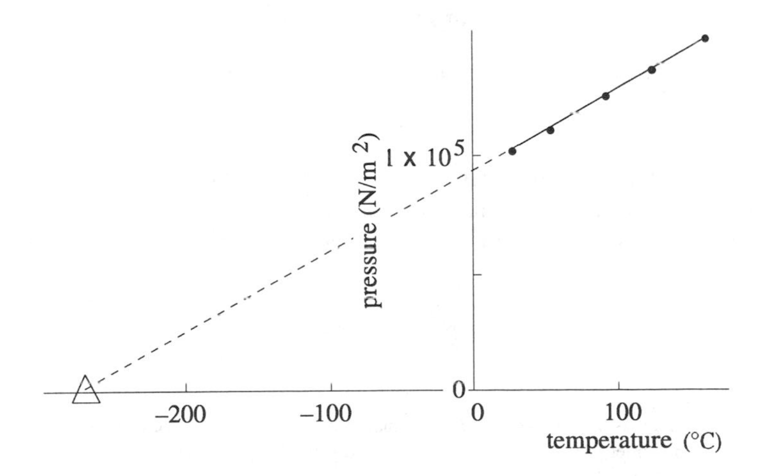

FIGURE 17.7. Temperature corresponding to zero pressure of an ideal gas.

with Eq. (17.6), we define a temperature scale with a unit (degree) the same size as the Celsius degree, but with the zero on the new scale corresponding to $-273°$C. This *absolute temperature* scale is called the *Kelvin* scale.

The Kelvin temperature K is related to the Celsius temperature C by

$$K = C + 273.$$

In the form, Eq. (17.6), the equation of state for the gas still contains the microscopic quantity N. Let us now find an alternative which represents the amount of gas macroscopically. To do this, we introduce a unit called the *mole* defined in the following way:

> One mole of a substance is defined to be the amount of the substance whose mass in grams is numerically equal to its molecular weight in atomic mass units* (u).

The molecular weights of substances are determined by chemical means and are extensively tablulated in handbooks. A nitrogen molecule, for example, consists of two nitrogen atoms each of which has a mass of approximately 14 u. Thus, one mole of nitrogen has a mass of approximately 28 grams—a macroscopically measureable amount. So we can specify the amount of gas we have in terms of the number of particles N (microscopic) or the number of moles n (macroscopic). To clarify the connection between these two representations, let us calculate in two different ways the mass of one mole of any substance.

$$\begin{aligned}\text{mass of one mole} &= (\text{number of molecules per mole}) \cdot (\text{mass per molecule}) \\ &= N_0 A \cdot (1.66 \times 10^{-24}\ \text{g}),\end{aligned}$$

where N_0 is the number of molecules in one mole and A is the molecular weight of the substance in u. But according to our definition of a mole, we also have

$$\text{mass of one mole} = A\ \text{g}.$$

Setting the two expressions equal and solving for N_0 , we get

$$N_0 = 6.02 \times 10^{23},$$

which is the number of molecules in one mole† of *any substance*. So if we know the number of moles, we can immediately calculate the number of molecules, regardless of the substance. With $N = nN_0$, the ideal gas law, Eq. (17.6), can now be written,

$$PV = nRT, \tag{17.7}$$

---

*The atomic mass unit, u, is defined to be equal to one-twelfth of the mass of an atom of a certain isotope of carbon, namely, ^{12}C, which is about 1.66054×10^{-27} kg. We learn more about isotopes in Chapter 18.

†The quantity N_0 is called *Avogadro's number.*

where we have combined k and N_0 in a single constant parameter, $R \equiv kN_0$, called the *universal gas constant*.[‡] This form of the ideal gas law contains only macroscopic quantities.

EXAMPLE. A metal cylinder contains 15.2 moles of methane gas. Estimate the mass of the gas in this cylinder.

SOLUTION. First, we find the molecular weight of methane, CH_4. The mass of the carbon atom is 12 u and each of the four hydrogen atoms has a mass of about 1 u. Therefore, the mass of the methane molecule is 16 u. By definition, the mass of one mole of methane is 16 grams.

$$\text{mass of gas} = (\text{number of moles}) \times (\text{mass per mole})$$

$$\text{mass of gas} = (15.2 \text{ moles}) \times (.016 \text{ kg/mole}) = 0.243 \text{ kg}.$$

Thus, the cylinder contains about 0.243 kg of methane.

EXAMPLE. A metal tank has a volume of 91.6 liters. Oxygen gas (O_2) is pumped into the tank until the pressure reaches 1.41×10^6 N/m^2 at a temperature of 24.3°C. A leak develops. After a week, the pressure has dropped to 1.02×10^6 N/m^2 and the temperature has climbed to 28.7°C. Find the amount of gas (in kilograms) that leaked out.

SOLUTION. We use Eq. (17.7) to calculate the number of moles n_i initially in the tank,

$$(1.41 \times 10^6 \text{ N/m}^2)(0.0916 \text{ m}^3) = n_i \, (8.314 \text{ J/mole} \cdot \text{K})(297.3 \text{ K}),$$

from which we get $n_i = 52.2$ moles. The number of moles n_f in the tank at the end of the week is also obtained from Eq. (17.7),

$$(1.02 \times 10^6 \text{ N/m}^2)(0.0916 \text{ m}^3) = n_f \, (8.314 \text{ J/mole} \cdot \text{K})(301.7 \text{ K}),$$

which yields $n_f = 37.2$ moles. The amount of gas that leaked out during the week is $n_i - n_f = 15.0$ moles. The molecular weight of O_2 is about 32. Therefore,

$$\text{amount of } O_2 \text{ leaking out} = (15.0 \text{ moles}) \times (.032 \text{ kg/mole}) = 0.48 \text{ kg}.$$

About 0.48 kg of oxygen leaked out of the tank during the week.

Exercises

17.1. In terms of a molecular picture, account for the temperature and the pressure of a confined gas. Use the molecular model to explain clearly how these two quantities are related to each other.

[‡]In mks units, $R = 8.314$ J/mole·K

17.2. Why does the pressure of a gas enclosed in a rigid-walled container increase as the temperature increases? (Two reasons.)

17.3. On average which, if either, moves faster in air—oxygen molecules or nitrogen molecules? Explain clearly.

17.4. By means of a piston, air at atmospheric pressure is trapped in the end of a cylinder as shown in Fig. 17.2. The length of the air column in the cylinder is 83.2 mm. The piston is pushed 13.7 mm futher into the cylinder. When equilibrium is reached after compression, what is the pressure of the air in the cylinder?

Answer: (1.21×10^5 N/m^2)

17.5. A cubical metal box 50 cm in length is filled with nitrogen gas, sealed, and placed in a room where the temperature is 24°C. Nitrogen is a diatomic molecule. A gauge on the box indicates that the absolute pressure inside the box is 6.0×10^5 N/m^2. Calculate the approximate number of nitrogen molecules inside the box. What is the mass (in kilograms) of the nitrogen in the box?

Answer: (1.83×10^{25} molecules; 0.851 kg)

17.6. Identical amounts of a certain gas (which behaves like an ideal gas) are heated separately. The pressure of each is measured at various temperatures. The results are shown graphically below. *From the data*, what can you say about system X relative to system Y? Explain clearly your reasoning.

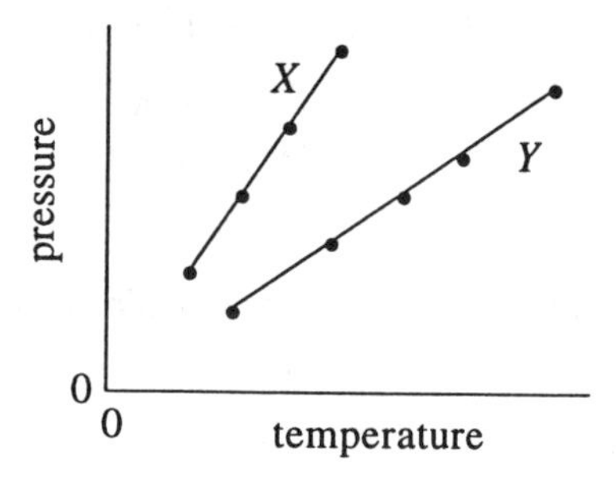

17.7. The amount of a given substance can be represented by the mass, by the number or moles, or by the number of molecules of the substance. By converting from one of these representations to another, complete the table below.

substance	formula	mass (kg)	number of moles	number of molecules
argon	Ar			2.189×10^{24}
flourine	F_2	21.772		
oxygen	O_2		140.3	
water	H_2O		948.2	
ozone	O_3			3.619×10^{25}
xenon	Xe		28.7	
methane	CH_4	7.635		

17.8. A steel tank sits in a room where the temperature is 22°C. It contains 41.8 liters of argon gas at an absolute pressure of 2.39×10^7 N/m^2. Calculate the mass (in kilograms) of the gas in the tank. Argon is a monatomic molecule.

Answer: (16.3 kg)

17.9. Pure oxygen at 0°C is enclosed in a rigid metal cylinder having a volume of 12 liters. If the absolute pressure in the cylinder is 2.0×10^6 N/m^2, how many moles of oxygen are inside the cylinder? How much pressure would the oxygen exert if the temperature were increased to 73°C?

Answer: (10.6 moles; 2.53×10^6 N/m^2)

17.10. Helium is pumped into a balloon until the pressure in the balloon is 1.98×10^6 N/m^2 and the diameter of the (spherical) balloon is 41.3 cm. The balloon is in a room where the temperature is 14°C. Later in the day, the room has warmed to 31°C and the diameter of the balloon is measured to be 41.7 cm. What is the pressure in the balloon then?

Answer: (2.04×10^6 N/m^2)

17.11. A metal cylinder contains 51.8 moles of chlorine gas (Cl_2). Estimate the mass of gas in the cylinder.

Answer: (3.67 kg)

17.12. The pressure of a certain gas is measured at various temperatures. The gas behaves like an ideal gas. The data are presented in the graph at the top of the following page. From the slope of the graph, determime the particle density of the gas.

Answer: (2.9×10^{25} molecules/m^3)

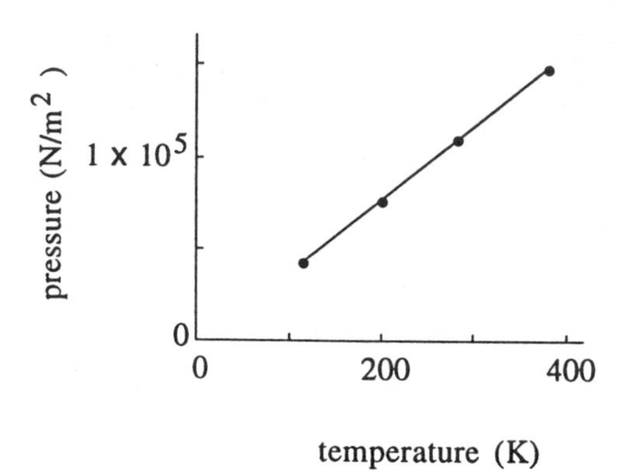

17.13. An automobile tire pressure gauge reads the *difference* between the absolute pressure in a tire and the atmospheric pressure. What would a tire pressure gauge read when applied to a *flat* tire? Early in the morning when the temperature is 8°C, the gauge pressure in an automobile tire is 1.85×10^5 N/m^2. Later in the day, the temperature has climbed to 22°C. What will the pressure gauge read then when applied to the same tire?

Answer: (1.99×10^5 N/m^2)

17.14. A steel cylinder contains 19.3 moles of the monatomic gas xenon at a pressure of 4.73×10^5 N/m^2 and temperature of 21°C. The gas behaves like an ideal gas. The cylinder has a slow leak. Fifty days later the temperature is up to 29°C, but the pressure is down to 2.09×10^5 N/m^2. How many kilograms of gas leaked out during the fifty days?

Answer: (1.44 kg)

17.15. An ideal gas at 19.3°C occupies a volume of 64.3 liters at 5.17×10^5 N/m^2. At a later time, the pressure is measured to be 2.26×10^5 N/m^2 at 27.6°C and the volume has expanded to 75.6 liters. Make a quantitative comparison of the initial and final states of the gas and write a sentence describing what has happened to the gas in the time between the two observations.

17.16. By means of a moveable piston, a small amount of gas at atmospheric pressure is trapped in the end of a metal (circular) cylinder as illustrated in the drawing. The cylinder is in thermal equilibrium with the atmosphere in the laboratory where the temperature is 22.3°C. Calculate the final pressure of the gas, if the piston is pulled 28.3 mm to the left of its initial position.

Answer: (9.14×10^4 N/m^2)

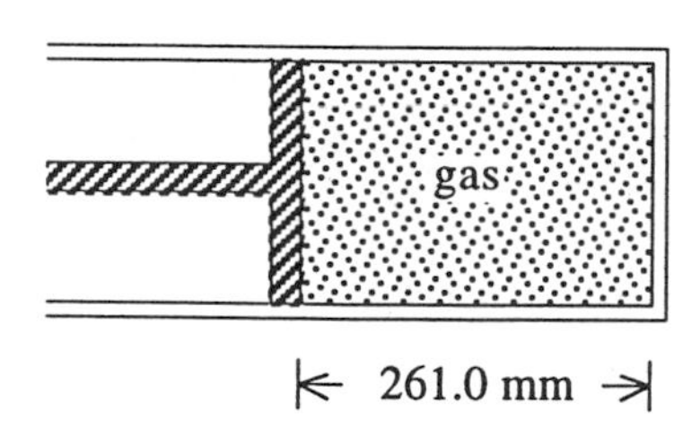

17.17. A metal tank is equipped with a relief valve. When the pressure of the gas in the tank exceeds 8.0×10^5 N/m^2, the valve opens and releases gas until the pressure in the tank drops again to the value at which the valve is set. On a given day, the valve starts to release gas when the temperature reaches 21°C. Later in the day, the temperature is up to 26°C. What fraction of the gas escapes during this time?

Answer: (1.7%)

17.18. The water in a cylindrical water tank is 8 m deep. The tank sits vertically on top of a tower with the bottom of the tank 17 m above the ground. An air bubble with a diameter of 1.30 mm forms on the bottom of the tank. The bubble rises toward the top of the tank. What is the diameter of the bubble just as it reaches the surface of the water?

Answer: (1.57 mm)

17.19. The pressure of a quantity of methane gas (CH_4) at 31.7°C is 3.183×10^5 N/m^2. Assuming that methane behaves like an ideal gas, calculate the density (in kg/m^3) of this gas.

Answer: (2.01 kg/m^3)

17.20. A cylindrical tank 123.8 cm long has a diameter of 38.5 cm. The tank is filled with 2.67 kg of fluorine gas (F_2) at 20.6°C. Calculate the pressure of this gas.

Answer: (1.19×10^6 N/m^2)

17.21. A steel tank is installed in a laboratory where the temperature is 23°C. It contains 1,393 liters of hydrogen gas (H_2) at an absolute pressure of 4.72×10^6 N/m^2. Calculate the mass of the gas in the tank. Later in the day, the temperature in the laboratory has increased by seven Celsius degrees. What is the pressure of the gas then?

Answer: (5.38 kg; 4.83×10^6 N/m^2)

17.22. A large glass jar sits in a laboratory where the temperature is 22.8°C. To create a very good vacuum, air is pumped out of the jar until the pressure is 10^{-12} N/m^2. Calculate the number of air molecules per unit volume remaining in the evacuated jar. Estimate the average distance between adjacent molecules in this vacuum.

Answer: (245 air molecules per cubic centimeter; 0.16 cm)

17.23. The pressure gauge on an air compressor tank reads 6.891×10^5 N/m^2 and the temperature is 31.3°C. Later, when the temperature is 23.9°C, the valve is opened briefly and some of the air is released. The pressure gauge then shows 4.121×10^5 N/m^2. Assume that atmospheric pressure remains constant at 1.034×10^5 N/m^2. What fractiion of the air escaped while the valve was open?

Answer: (33%)

17.24. The density of matter in the space between the galaxies is about one hydrogen atom per cubic meter at a temperature of about −270°C. Calculate the pressure of this intergalactic gas.

17.25. In a room where the temperature is 23.3°C, a small balloon is filled with pure helium gas until the pressure reaches 2.713×10^5 N/m^2 and the diameter of the balloon is 1.332 m. Some time later, the temperature in the room has dropped to 18.1°C, the diameter of the balloon has shrunk to 1.281 m, and the pressure of the helium is down to 2.109×10^5 N/m^2. Some helium leaked out between the two observations. How much? Express your answer in number of moles, in number of molecules, and in kilograms.

Answer: (40.4 moles; 2.43×10^{25} molecules; 0.16 kg)

17.26. A balloon contains a small amount of gas at a pressure of 3.193×10^5 N/m^2 and a temperature of 21.8°C. Later, the size of the balloon has increased by 3 percent and the pressure of the gas has risen to 3.292×10^5 N/m^2. Find the Celsius temperature of the gas at the later time. Assume no gas leaked out.

17.27. The temperature in the central core of the sun is approximately 1.5×10^7 K. The density of hydrogen in this region is about 4.5×10^{31} atoms/m^3. Assuming that the matter in the sun behaves like an ideal gas, estimate the contribution of protons (hydrogen nuclei) to the pressure of the gas in the solar core.

Answer: (10^{16} N/m^2)

18

The Sun Is a Nuclear Furnace

Why does the sun shine? A simple answer is, "Because of nuclear fusion." Although there is some sense in which this is a correct answer, it is at best an incomplete one. Other factors contribute to bring about conditions in stars under which nuclear fusion can take place. Still other factors are responsible for converting the energy produced by fusion into forms more directly perceptible to our senses—forms such as light and heat radiation.

In any event, nuclear fusion does play a central role in energy production in stars, so it is appropriate that we learn something about the structure of atomic nuclei and how they interact with each other to produce energy.

18.1 The Structure of the Atomic Nucleus

First, we list some general characteristics of atomic nuclei.

- The nucleus of an atom carries a positive electric charge.
- The nucleus accounts for most of the mass of the atom.
- The diameter of the atom is about 100,000 times larger than that of the nucleus.

We present a simple model of the nucleus that will explain those features of nuclear structure and behavior that will concern us in our study here. In this model, the atomic nucleus is a bound system consisting of two kinds of particles, *neutrons* and *protons*. In Table 18.1, some properties of these particles are compared to those of the electron.

Table 18.1. The Constituents of Atoms

particle	mass	electric charge
proton	1,836	$+e$
neutron	1,837	0
electron	1	$-e$

We can account for many of the properties of matter by assuming that these are all we have in the world—electrons, protons, and neutrons. In this model, the atom has a tiny nucleus consisting of neutrons and protons with the electrons distributed over the rest of the volume of the atom. In the normal state, matter is electrically neutral. The identity of a chemical element is characterized by the *atomic number* Z, the number of electrons in a neutral atom of the element (e.g., $Z = 6$ for carbon, 8 for oxygen, and so on). To maintain the electrically neutral character, we see from Table 18.1 that there must be exactly Z protons in the nucleus. The mass of a proton is roughly one atomic mass unit u. Looking at the atomic weights given in the Periodic Chart of the Elements in Fig. 15.1, it is clear that Z protons and Z electrons cannot account for all of the mass of an atom. To make up the rest of the mass, we put some neutrons into the nucleus. For example, carbon ($Z = 6$) requires six protons in the nucleus, but Fig. 15.1 shows a mass of about 12 u. Six neutrons in the nucleus with the six protons will bring the mass to about the right value. This kind of carbon nucleus is represented schematically in Fig. 18.1*a*.

To identify a particular nucleus (or the atom to which it belongs) we have introduced the notation $^{A}\mathrm{X}$, where X stands for the chemical symbol of the element (see Fig. 15.1) and A denotes the total number of particles (neutrons and protons together) in the nucleus. The superscript A is called the *mass number*.* Notice that the mass number is, by definition, an integer, whereas the atomic weight is not. There are at least three important reasons why the atomic weight is not a whole number, as we shall soon see.

We can put another neutron into this nucleus without changing the chemistry—it is still carbon with six electrons and six protons. The additional neutron increases the atomic weight by about one u. We now have $^{13}\mathrm{C}$ as illustrated in Fig. 18.1*b*. For a given element, atoms with different numbers of neutrons are called *isotopes*

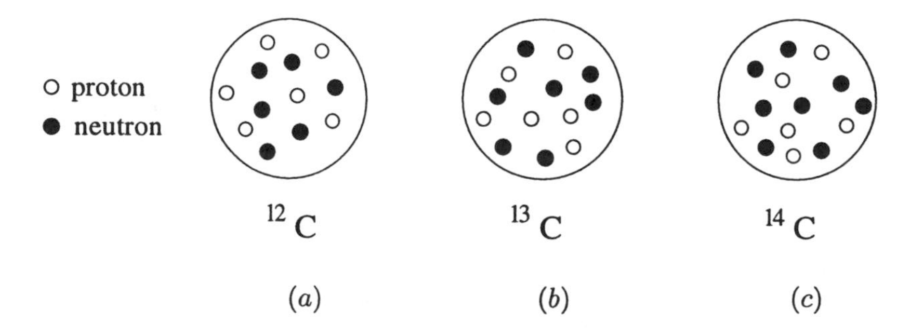

FIGURE 18.1. Nuclei of carbon isotopes.

*The letter A has a different meaning here from its use in Chapter 17 where it stands for molecular weight.

of the element. For example, ^{12}C is the mass-twelve isotope of carbon, ^{13}C is the mass-thirteen isotope of carbon, and so on.

In Fig. 18.1*c*, we have ^{14}C. But now we have gone too far. There are too many neutrons for the number of protons in ^{14}C. It is unstable and will decay by a radioactive process called *beta decay*, in which an electron (β-particle) is emitted from the nucleus, leaving behind an isotope of nitrogen ^{14}N. We represent this process symbolically by

$$^{14}\mathrm{C} \xrightarrow{\beta^-} {}^{14}\mathrm{N}.$$

Let's see if we can understand how this instability arises.

18.1.1 Nuclear Stability

First, why do neutrons and protons stick together to make a nucleus in the first place? So far, we have talked about only two fundamental forces—the force of gravity and the electromagnetic force. However, neither of these will do the job here. The positive electric charges of the protons cause them to repel one another. The neutrons do not interact electrically, since they carry no electric charge. Both particles have some magnetism, but it is far too weak to hold the nucleus together. The gravitational force is the weakest of all fundamental forces and for particles with such tiny masses as neutrons and protons, it is completely negligible compared to the electric repulsion of the protons. So we need a completely new kind of attractive force strong enough to overcome this repulsion and bind neutrons and protons tightly together in the nucleus. This force is called the *strong* (or *nuclear*) force and has the following features:

- It is the strongest of all the fundamental forces.
- It is effective only over a very short distance between interacting particles.

Since the electric repulsion of the protons is disruptive, why not build a nucleus using neutrons alone? The neutrons attract each other through the strong force, but do not have the disadvantage of electric repulsion, since they carry no electric charge. The answer to this question will help us to understand why ^{14}C is unstable and radioactive β-decay occurs.

The neutron and proton each carries one-half unit of intrinsic angular momentum; therefore both are subject to the exclusion principle. Analogous to the way electrons occupy discrete states in the atom, neutrons and protons occupy discrete states in the nucleus. In the ground state, these particles drop into the lowest energy states available, consistent with the exclusion principle. In Fig. 18.2, we illustrate schematically how this would look for a twelve-particle nucleus. For illustrative purposes, we assume that there are four distinct single-particle states for each energy level in the nucleus.

The system consisting of twelve neutrons is depicted in Fig. 18.2*a*. In Fig. 18.2*b*, we show the configuration for a twelve-particle nucleus of six neutrons and six protons. We can think of building the nucleus one nucleon at a time. First, we drop

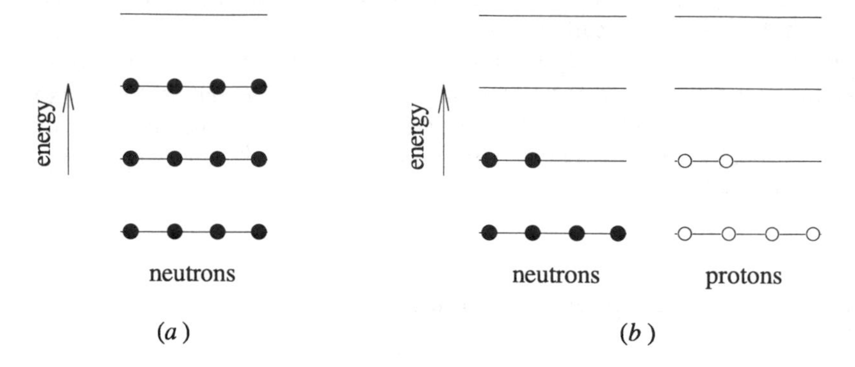

FIGURE 18.2. Nuclear stability and the exclusion principle.

in six neutrons. If we add six more neutrons, as in Fig. 18.2*a*, these must go into higher-energy states because of the exclusion principle. However, if we add six protons, as in Fig. 18.2*b*, we can put them into the same low-energy states that the neutrons occupy, because the exclusion principle applies separately to neutrons and protons. Thus, with equal numbers of neutrons and protons we get a lower-energy (more stable) configuration for the twelve-particle nucleus than if we built it with neutrons alone. A stable nucleus requires a proper balance between neutrons and protons. In the case of carbon, ^{12}C and ^{13}C are stable, but in ^{14}C there are too many neutrons (8) for the number of protons (6).

A free neutron is unstable. It is heavier than a proton and will spontaneously decay according to

$$n \longrightarrow p + e + \bar{\nu}.$$

A *fourth* fundamental force called the *weak force* is responsible for this process in which a neutron n disappears with the creation of a proton p, an electron e, and an elusive little particle called a *neutrino* ν (actually, an *antineutrino* $\bar{\nu}$).* When there are too many neutrons for the number of protons in the nucleus, as in ^{14}C, one of the neutrons will undergo this process with electron and antineutrino leaving the nucleus and the proton remaining behind in place of the neutron that decayed. The residual nucleus now has seven protons and seven neutrons, that is, ^{14}N. This is represented pictorially in Fig. 18.3.

The neutron–proton imbalance can also take place in the other direction with too many protons for the number of neutrons. In this case, the process

$$p \longrightarrow n + e^{+} + \nu \tag{18.1}$$

converts one of the protons into a neutron with a *positron* e^{+} and a neutrino ν emitted from the nucleus. There is not enough energy available for the reaction

*The neutrino is a spin one-half particle with no mass and no electric charge.

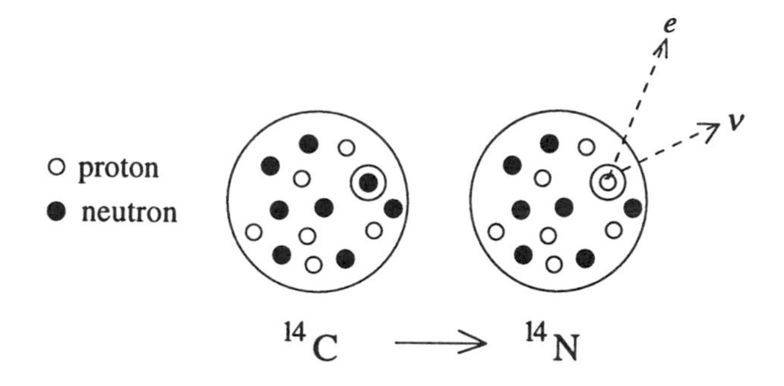

FIGURE 18.3. Radioactive beta-decay of ^{14}C.

Eq. (18.1) to occur for a free proton, but inside the nucleus a proton can "borrow" enough energy for this process to take place. For example, ^{11}C has too many protons (6) for the number of neutrons (5), so we have

$$^{11}\mathrm{C} \xrightarrow{\beta^+} {}^{11}\mathrm{B},$$

where the β^+ indicates that a positron e^+ is emitted.

So for a stable nucleus, because of the exclusion principle, we require a proper balance between the number of neutrons and the number of protons. Notice that for the lighter elements, we have approximately equal numbers of neutrons and protons.

If we examine the ratio of neutrons to protons in the most stable isotopes, we find that this ratio increases with atomic number. For example, the ratio is given for a few representative isotopes in Table 18.2. This generally holds throughout the Periodic Table. We can understand this feature from what we have learned so far about the stability of atomic nuclei. The nuclear force is strong enough to bind neutrons and protons in a stable nucleus in spite of the electric repulsion of the protons. The nuclear force has a very short range. Because of this, a given nucleon will interact through the nuclear force only with its nearest neighbors. On the other hand, the electric force is long-range and each proton repels every other proton in the nucleus. So, as atomic number increases, the number of electric repulsions between protons increases much more rapidly than the number of proton–nucleon nuclear bonds.

As we have seen, the exclusion principle tends to favor equal numbers of neutrons and protons in a stable nucleus. But the increasingly disruptive effect of the protons makes it more energetically favorable (lower-energy) to add in a few extra neutrons, which bind to other nucleons without repelling them electrically. A stable isotope represents a proper balance between the exclusion principle's call for equal numbers of neutrons and protons and the Coulomb repulsion favoring more neutrons.

For isotopes of elements heavier than lead, the Coulomb effect becomes so disruptive and the nucleons so loosely bound as a result, that the system can reach

Table 18.2. Ratio of Neutrons to Protons in Nuclei

isotope	$\frac{\text{neutron number}}{\text{proton number}}$
^{12}C	1.00
^{32}S	1.00
^{63}Cu	1.17
^{121}Sn	1.42
^{208}Pb	1.54
^{238}U	1.59

a lower-energy configuration by emitting an α-particle. This is the radioactive process we call α-decay. For example, the mass of a ^{238}U nucleus is greater than the combined masses of a ^{4}He nucleus and a ^{234}Th nucleus. This uranium isotope will spontaneously decay, yielding two products. The energy equivalent of the mass difference between the original uranium nucleus and the products appears as kinetic energy of the α-particle (^{4}He) and the ^{234}Th nucleus. Symbolically, for this process we write

$$^{238}\text{U} \xrightarrow{\alpha} {}^{234}\text{Th}.$$

Sometimes, following radioactive α- or β-decay, the residual nucleus will be in an excited state that will decay to the ground state by photon (γ-ray) emission. For example, the α-decay of ^{211}Bi produces an isotope of thallium, which may be in an excited state that subsequently decays to the ground state according to

$$^{207}\text{Tl} \xrightarrow{\gamma} {}^{207}\text{Tl}.$$

18.1.2 Natural Radioactivity

Bismuth, element number 83, has one stable isotope, ^{209}Bi. All known isotopes of elements with more than 83 protons are radioactive. Some naturally occuring elements have relatively short half-lives.

The *half-life* of a radioactive isotope is the time required for half of a sample of the pure isotope to decay. For example, ^{14}C has a half-life of 5,730 years. This means that if we start with a gram of pure ^{14}C today, 5,730 years later the mass of ^{14}C remaining will be one-half a gram. At the end of 11,460 years, one-fourth of a gram will remain. The sample will contain one-eighth of a gram of ^{14}C 17,190 years from now, and so on.

The most stable isotope of radium—the naturally occuring element discovered by Marie Curie—has a half-life of only 1,620 years. The earth is about 4.5 billion years old. Why is there still a measurable amount of radium still around in the earth? Why, after all this time, hasn't it decayed to an undetectable level? The answer is that isotopes of some heavier elements have half-lives that are comparable with the age of the earth. For example, ^{238}U has a half-life of about 4.5 billion years. This means that roughly one-half of the primordial uranium ^{238}U present when the earth was formed is still around. It disintegrates through a series of alpha and beta emissions until it reaches a stable configuration as ^{206}Pb. One of the intermediate stages in this process is ^{226}Ra. So, the radium that occurs naturally in the environment today was once uranium that is on its way to becoming lead.

18.1.3 Nuclear Binding

In Table 18.3, we give the masses (in atomic units) of some selected isotopes. These are *atomic* masses, that is, the mass of the neutral atom consisting of nucleus and orbital electrons. It is far too difficult to remove all of the electrons from an atom and measure directly the mass of a bare nucleus. In practice, we use the masses of neutral atoms—which can be measured quite precisely—and subtract out any contributions from the electron masses. (In most nuclear processes we consider, the electron masses will automatically cancel out.) The electromagnetic binding of electrons to the nucleus is completely negligible compared to the binding of the nucleons in the nucleus.

We can construct all of the isotopes from ^{1}H atoms (each consisting of one proton and one electron) and neutrons. For example, a ^{4}He atom consists of two electrons, two protons, and two neutrons. The total mass of these constituents is obtained from

$$\begin{aligned} 2M_{^1\mathrm{H}} &= 2(1.007825\ \mathrm{u}) = 2.015650\ \mathrm{u} \\ 2M_n &= 2(1.008665\ \mathrm{u}) = \underline{2.017330\ \mathrm{u}} \\ \text{total mass of constituents} &= 4.032980\ \mathrm{u} \end{aligned}$$

But according to Table 18.3, the total mass of ^{4}He is only 4.002603 u. What happened to the rest of the mass? The missing mass accounts for the binding of the system. Clearly, the 4.002603 u in ^{4}He is insufficient to make two ^{1}H atoms and two neutrons. So the energy equivalent ($E = mc^2$ again) of this mass deficiency (0.030377 u) is the binding energy (28.30 MeV). The binding energy is just the amount of energy required to separate a ^{4}He atom into four particles—two ^{1}H atoms and two neutrons.

Clearly, an atom holds together as a bound system because it does not possess enough total mass to make the separate hydrogen atoms and neutrons that are its constituents.

An atom of any isotope ^{A}X consists of Z ^{1}H atoms and $A - Z$ neutrons. The nuclear binding energy for this isotope is

$$B = \left[Z M_{^1\mathrm{H}} + (A - Z) M_n - M_{^A\mathrm{X}} \right] c^2.$$

Table 18.3. Atomic Masses of Selected Isotopes[a]

isotope	u	isotope	u	isotope	u
n	1.008665	^{40}Ca	39.962589	^{116}Sn	115.901775
^{1}H	1.007825	^{42}Ca	41.958625	^{117}Sn	116.902959
^{2}H	2.014102	^{43}Ca	42.958780	^{118}Sn	117.901606
^{3}H	3.016050	^{44}Ca	43.955490	^{119}Sn	118.903314
^{3}He	3.016030	^{46}Ca	45.953689	^{120}Sn	119.902199
^{4}He	4.002603	^{58}Mn	57.940260	^{121}Sn	120.904227
^{6}Li	6.015125	^{54}Fe	53.939617	^{122}Sn	121.903442
^{7}Li	7.016004	^{56}Fe	55.934937	^{124}Sn	123.905272
^{12}C	12.000000	^{57}Fe	56.935398	^{150}Sm	149.917276
^{13}C	13.003354	^{58}Fe	57.933282	^{183}Ta	182.951470
^{14}C	14.003242	^{58}Co	57.935761	^{183}W	182.950324
^{14}N	14.003074	^{58}Ni	57.935342	^{197}Au	196.966541
^{16}O	15.994915	^{75}As	74.921597	^{200}Hg	199.968327
^{17}O	16.999133	^{90}Rb	89.914820	^{201}Hg	200.970308
^{20}F	19.999987	^{90}Sr	89.907747	^{202}Hg	201.970642
^{20}Ne	19.992440	^{90}Y	89.907163	^{206}Pb	205.974468
^{23}Ne	22.994473	^{90}Zr	89.904700	^{207}Pb	206.975903
^{23}Na	22.989771	^{92}Zr	91.905031	^{208}Pb	207.976650
^{23}Mg	22.994125	^{107}Ag	106.905094	^{209}Pb	208.981082
^{32}P	31.973909	^{109}Ag	108.904756	^{209}Bi	208.980394
^{32}S	31.972074	^{112}Sn	111.904835	^{226}Ra	225.974640
^{39}K	38.963710	^{114}Sn	113.902773	^{235}U	235.043915
^{39}Ca	38.970691	^{115}Sn	114.903346	^{238}U	238.050770

[a] Reprinted from J. H. E. Mattauch, W. Thiele, and A. H. Wapstra, *Nuclear Physics* **67** (1965) 1, with kind permission of Elsevier Science, Amsterdam.

We define the atomic mass unit to be *exactly* one-twelfth of the mass of a ^{12}C atom. Expressed in other units, we have

$$1 \text{ u} = 1.66056 \times 10^{-27} \text{ kg} = 931.50 \text{ MeV}/c^2. \tag{18.2}$$

In general, the binding energy increases with mass number A. For comparing the relative binding of different isotopes, it is convenient to use B/A, which is the *binding energy per nucleon*. We use the term *nucleon* for a particle that is either a neutron or a proton. The general behavior of B/A as a function of A is shown in Fig. 18.4 along with specific values for selected isotopes.

EXAMPLE. Calculate the average binding energy per nucleon for the phosphorous isotope ^{32}P.

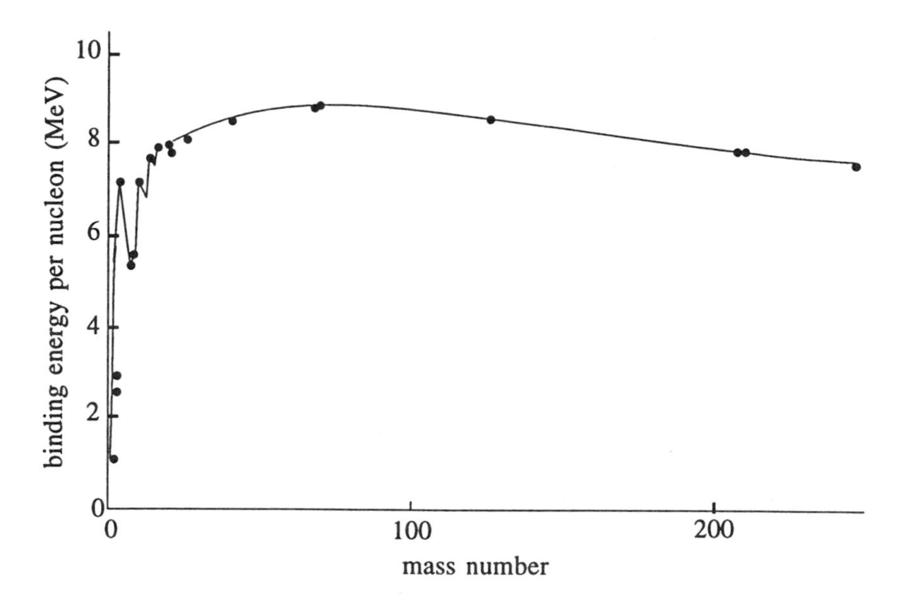

FIGURE 18.4. Average binding energy per nucleon.

SOLUTION. From Table 15.1, we see that phosphorous is element number 15. This tells us that 15 of the 32 particles in the ^{32}P nucleus are protons and the remaining 17 are neutrons. Using the masses for the neutron, the hydrogen atom, and the ^{32}P atom given in Table 18.3, we get

$$\text{mass of constituents of } ^{32}\text{P} = 15 \times (1.007825 \text{ u}) + 17 \times (1.008665 \text{ u}) = 32.264680 \text{ u}$$

$$\text{mass of } ^{32}\text{P atom} = 31.973909 \text{ u}$$

The mass difference is

$$\text{mass of constituents of } ^{32}\text{P} - \text{mass of } ^{32}\text{P atom} = 0.290771 \text{ u},$$

which represents the binding of ^{32}P. From Eq. (18.2), we find that the energy equivalent of this mass difference is,

$$\text{total binding energy of } ^{32}\text{P} = 270.853 \text{ MeV}.$$

Hence,

$$\frac{B}{A} = \frac{270.853 \text{ MeV}}{32 \text{ nucleons}} = 8.464 \text{ MeV/nucleon}.$$

We obtain an average binding energy per nucleon of 8.464 MeV for ^{32}P.

18.1.4 Why Are the Atomic Masses Not Integers?

Neutrons and ^{1}H atoms have nearly the same mass—1 u each. Since the atoms of all of the elements can be made from whole numbers of these two constitutents, why are not all of the atomic masses given in Table 15.1 integers (whole numbers)? Of course, some small difference is expected because the masses of the ^{1}H atom and the neutron are not *exactly* equal. But this cannot account for such large departures from whole numbers as chlorine's atomic mass of 35.45 u. Does this imply that the nucleus of a chlorine atom contains a fraction of a neutron or a proton? Let us see.

The atomic masses given in Fig. 15.1 are determined by chemical means. Natural chlorine is a mixture of two isotopes, ^{35}Cl and ^{37}Cl, with atomic masses of 34.968851 u and 36.965898 u, respectively. The chemical behavior is the same for both, so the chemically determined mass of chlorine is a weighted average of the two isopopes.

Knowing that there are only two isotopes, we can calculate the relative abundance in nature of each isotope. Let x denote the fraction of natural chlorine atoms represented by ^{35}Cl. Taking a weighted average of the two isotopes, we have

$$x(34.968851 \text{ u}) + (1 - x)(36.965898 \text{ u}) = 35.45 \text{ u},$$

from which we obtain $x = 0.759$. Thus, in a sample of natural chlorine, 75.9% of the atoms are ^{35}Cl and 24.1% are ^{37}Cl atoms.

Some elements have only one stable isotope. Beryllium and gold are examples. Even for these elements, the atomic mass in atomic units is not exactly a whole number. Some of the mass of the constituent neutrons and ^{1}H atoms has been released in binding the nuclear particles together. So, the nuclear binding energy also accounts in part for the discrepancy between atomic masses expressed in atomic units and integers.

18.2 Nuclear Reactions

We noted in Chapter 15 that only the outer atomic electrons participate directly in chemical reactions because the positively charged nuclei of the interacting atoms repel each other. The faster two nuclei move toward each other, the closer they approach, before the Coulomb repulsion can push them apart. If the relative motion is great enough, the nuclei can get close enough together for the short-range nuclear force to become effective. When this happens, a *nuclear reaction* can take place. For example, suppose we direct a beam of α-particles with energies of several MeV onto a beryllium target and observe the products emerging from this reaction. We represent the process symbolically by

$$\alpha + {}^9\text{Be} \longrightarrow ({}^{13}\text{C})^* .$$

The α-particle (^{4}He) interacts with the ^{9}Be nucleus to form a compound system with $2 + 4 = 6$ protons and $2 + 5 = 7$ neutrons. This looks like some form of

^{13}C, but with far too much energy (mass)—implied by the asterisk—for the ground state of ^{13}C. Because of this excess energy, the compound system will decay to a lower-energy configuration by one of a variety of channels. Some of these are,

$$\begin{aligned}({}^{13}\mathrm{C})^* &\longrightarrow {}^{12}\mathrm{C} + n\,,\\ &\longrightarrow {}^{12}\mathrm{B} + p\,,\\ &\longrightarrow {}^{13}\mathrm{C} + \gamma\,,\\ &\longrightarrow {}^{9}\mathrm{Be} + \alpha\,,\\ &\longrightarrow \cdots \text{ and so on,}\end{aligned}$$

Most often in writing a nuclear reaction, we omit the compound system and write only the particles in the initial and final states of the system. For example, for the above reaction,

$$\alpha + {}^{9}\mathrm{Be} \longrightarrow {}^{12}\mathrm{C} + n\,.$$

In writing a nuclear reaction in this way, two conditions must be met:

- The total electric charge to the left of the arrow must be equal to the total electric charge to the right of the arrow.
- The total number of neutrons and protons together on the left of the arrow must be equal to the total number of neutrons and protons on the right of the arrow.

EXAMPLE. Complete the following nuclear reaction,

$${}^{6}\mathrm{Li} + {}^{28}\mathrm{Si} \longrightarrow {}^{31}\mathrm{P} + \quad .$$

SOLUTION. The + sign on the right-hand side tells us that we need one more particle to balance this reaction. From Table 15.1, we find that there are $3 + 14 = 17$ units of charge on the left-hand side. With 15 units of charge on the phosphorous nucleus, we need two more units on the right-hand side to balance the charge. So the missing particle is some isotope of helium. Looking at the mass, there are $6 + 28 = 34$ units on the left and the phosphorous isotope on the right has 31 units. We need 3 more units of mass on the right. Therefore, we conclude that the missing particle is a ^{3}He nucleus.

18.2.1 FISSION

Since neutrons are electrically neutral, these particles—even very slow ones—can penetrate an atomic nucleus and interact with it through the nuclear force. When this occurs in certain very heavy isotopes, the compound system is highly unstable and will split into two pieces of comparable size accompanied by the emission of several neutrons. Because there are two large pieces among the reaction products, this type of nuclear reaction is called *fission*. The uranium isotope ^{235}U has this property. A typical fission reaction is,

$$n + {}^{235}\mathrm{U} \longrightarrow {}^{114}\mathrm{Rh} + {}^{119}\mathrm{Ag} + 3\mathrm{n}\,.$$

There are several features of fission that are worthy of note. These are:

- There are two large fragments among the reaction products.
- These fission fragments are highly radioactive.
- Energy is released.
- The reaction is initiated by slow neutrons.
- Several neutrons are among the reaction products.

We see from Fig. 18.4 that the binding energy per particle for ^{235}U is relatively low compared to that of the products, rhodium and silver. This is a general feature of fission reactions. The products are much more tightly bound than the original fissioning isotope. Some of the mass difference appears as kinetic energy of the reaction products.

The ratio of neutrons to protons in ^{235}U is much greater than the neutron/proton ratio for stable isotopes of rhodium and silver. As a result, the fragments ^{114}Rh and ^{119}Ag have far too many neutrons for the number of protons. These isotopes undergo a series of radioactive beta decays until they reach a stable balance between neutrons and protons.

Because the nuclear force is so much stronger than the electromagnetic force, the energy released in a typical nuclear fission reaction is many times greater than the energy released in a typical chemical reaction. The fact that fission is initiated by a neutron and several neutrons are produced in the process suggests that a self-sustaining chain reaction might be possible. The implications of this notion were recognized immediately, when nuclear fission was discovered in Hitler's Germany in 1938.

18.2.2 Fusion

Fission induced by slow neutrons is a nuclear reaction in which a loosely bound, heavy nucleus splits into two more tightly bound nuclei accompanied by a release of energy. Conversely, it is possible to combine two lightweight, loosely bound nuclei to form a single, more tightly bound nucleus again with energy released. This kind of nuclear reaction is called *fusion*. For example, Fig. 18.4 shows that ^{2}H with $B/A = 1.11$ MeV/nucleon is a very loosely bound nucleus compared to ^{3}He with 2.57 MeV/nucleon or ^{3}H with 2.83 MeV/nucleon. There is therefore a net release of energy in the following two fusion reactions:

$${}^{2}\mathrm{H} + {}^{2}\mathrm{H} \longrightarrow {}^{3}\mathrm{He} + n$$

$${}^{2}\mathrm{H} + {}^{2}\mathrm{H} \longrightarrow {}^{3}\mathrm{H} + p.$$

Atomic nuclei carry positive electric charge and repel each other through the long-range Coulomb force. For two nuclei to get close enough together for the short-range nuclear force to become effective and produce a heavier nucleus, they must approach each other with very high relative velocities. In practice, this means that for nuclear fusion to occur in a hot gas, the temperature must be exceedingly high—tens of millions of degrees. Such temperatures exist naturally in the interiors of stars like the sun. The sun is a nuclear furnace.

Exercises

18.1. Neutrons and protons have mass and therefore exert an attractive gravitational force on each other. Explain clearly and completely why this force cannot be the force responsible for binding neutrons and protons together to form a stable nucleus.

18.2. When the isotope ^{214}Po emits a β-particle, it transforms into a new element. What is the new element? What is its mass number? What is its atomic (proton) number? Suppose this isotope emits an α-particle instead of a β-particle. What is the new element in this case?

18.3. Complete the radioactive decays

$$^{211}\mathrm{Po} \xrightarrow{\alpha} \qquad ^{190}\mathrm{Au} \xrightarrow{\beta^+} \qquad ^{138}\mathrm{Cs} \xrightarrow{\beta^-} \qquad ^{234}\mathrm{U} \xrightarrow{\alpha}$$

$$^{131}\mathrm{Xe} \xrightarrow{\gamma} \qquad ^{40}\mathrm{K} \xrightarrow{\beta^-} \qquad ^{6}\mathrm{He} \xrightarrow{\beta^-} \qquad ^{89}\mathrm{Zr} \xrightarrow{\beta^+}$$

18.4. An isotope of strontium ^{90}Sr is a fairly long-lived (28-year half life) radioactive byproduct of nuclear fission that has found its way into the food chain through dairy products. Write the formula for β^--decay of this isotope and calculate the energy released when it decays.

18.5. Calculate the energy released in the ordinary radiaoactive β-decay of the isotopes ^{20}F, ^{32}P, and ^{14}C.

Answer: (7.03 MeV; 1.71 MeV; 0.16 MeV)

18.6. Calculate the energy released in radioactive β^--decay of each of the isotopes ^{3}H and ^{23}Ne.

18.7. Calculate the average binding energy per nucleon for ^{56}Fe.

Answer: (8.79 MeV)

18.8. Calculate the average binding energy per nucleon for each of the tin isotopes ^{112}Sn, ^{116}Sn, ^{120}Sn, and ^{124}Sn. Which of these isotopes is most strongly bound? Explain.

18.9. A sulfur atom is assembled from sixteen hydrogen atoms and sixteen neutrons. Calculate the energy released in assembly and the average binding energy per nucleon.

Answer: (272 MeV; 8.49 MeV)

18.10. Calculate the average binding energy (in MeV) per particle for each of the isotopes ^{58}Mn. ^{58}Fe, ^{58}Co, and ^{58}Ni. Calculate the energy released in ordinary β-decay of ^{58}Mn. The energy released in β^+-decay of ^{58}Co is

1.29 MeV. By means of a calculation, verify this result. Explain clearly your reasoning. (NOTE: It is important to remember that it is *atomic* masses of *neutral* atoms that we use to calculate *nuclear* binding energies. Hence, the number of electrons must be the same in initial and final states so that the electron contributions cancel. This is automatically achieved in β^--decay but *not* in β^+-decay.)

18.11. Calculate the average binding energy per nucleon for the uranium isotopes ^{235}U and ^{238}U.

Answer: (7.591 MeV; 7.570 MeV)

18.12. The mass-39 isotope of potassium (K) is stable. This isotope is produced in the β^--decay of a radioactive isotope in which the energy released is 0.565 MeV. Identify this radioactive isotope and calculate its mass and its average binding energy per nucleon.

18.13. The average binding energy per nucleon for ^{16}N is 7.374 MeV. Calculate the atomic mass of this isotope. Use your result to calculate the energy released in radioactive β^--decay of this isotope. Give the symbolic representation of this radioactive decay.

Answer: (16.00610 u; 10.4 MeV)

18.14. In ^{209}Bi, the average binding energy per nucleon is 7.848 MeV. Calculate the atomic mass of this isotope. From your result, calculate the energy released in β^--decay of ^{209}Pb. Express this decay process symbolically.

Answer: (208.98042 u; 0.61 MeV)

18.15. The mass of a ^{234}Th atom is 234.04358 u. Calculate the average nuclear binding energy per nucleon for this isotope. Calculate the energy released in the radioactive alpha decay of ^{238}U.

Answer: (7.597 Mev/nucleon; 4.3 MeV)

18.16. The nuclei of the isotopes ^{23}Na and ^{23}Mg are called "mirror" nuclei, because the proton number of each is equal to the neutron number of the other. Calculate the difference in binding energy (in MeV) for these two nuclei. Which is the more tightly bound system? Give a physical reason why you would expect it to be more tightly bound.

18.17. Calculate the energy released in radioactive β^+-decay of ^{39}Ca. (See note on Exercise 18.10.)

Answer: (5.48 MeV)

18.18. There are five stable isotopes of the element calcium: ^{40}Ca, ^{42}Ca, ^{43}Ca, ^{44}Ca, and ^{46}Ca. Calculate the average binding energy per nucleon for each of these. Which has the most tightly bound nucleus? Explain.

18.19. Each of the isotopes ^{16}O, ^{32}S, ^{75}As, ^{109}Ag, ^{150}Sm, ^{197}Au, and ^{209}Bi is stable. Calculate the average binding energy per nucleon for each of these.

Based on your results, write a general statement regarding the relationship between the average nuclear binding energy per nucleon and the mass number for stable isotopes.

18.20. All isotopes of radium are radioactive. The longest-lived isotope is ^{226}Ra, an α-particle emitter with a half-life of 1,620 years. Write the decay formula and calculate the energy released in the radioactive decay of this isotope.

18.21. Natural lithium is a mixture of two isotopes, ^{6}Li and ^{7}Li. Calculate the natural relative abundance of each of these isotopes.

Answer: (7.6% and 92.4%)

18.22. There are two stable isotopes of boron. One has mass 10.01294 u and the other 11.00931 u. The chemically determined mass of natural boron is 10.811 u. In a sample of natural boron, what fraction of the atoms will be ^{10}B atoms?

18.23. What fraction of the *mass* of a sample of natural chlorine is the ^{35}Cl isotope?

Answer: (74.9%)

18.24. Complete the following nuclear processes:

$$\alpha + {}^{90}Zr \longrightarrow \quad + p \qquad {}^{12}C + {}^{16}O \longrightarrow \quad + \gamma$$
$$p + {}^{37}Cl \longrightarrow {}^{34}S + \qquad {}^{7}Li + {}^{10}B \longrightarrow {}^{14}C +$$
$$+ {}^{28}Si \longrightarrow {}^{55}Fe + {}^{3}He \qquad {}^{2}H + {}^{123}Sb \longrightarrow \quad + {}^{3}He$$
$$n + {}^{99}Ru \longrightarrow \quad + \alpha \qquad p + {}^{11}B \longrightarrow \quad + n$$
$${}^{12}C + {}^{28}Si \longrightarrow {}^{40}Ca + \qquad \alpha + {}^{65}Cu \longrightarrow {}^{68}Zn \quad +$$

18.25. Calculate the energy released in each of the nuclear reactions

$${}^{3}H + {}^{14}N \longrightarrow {}^{16}O + n \qquad \text{and} \qquad {}^{3}He + {}^{14}N \longrightarrow {}^{16}O + p.$$

Answer: (14.48 MeV; 15.24 MeV)

18.26. The nucleus of an atom of ^{27}Al is struck by an α-particle. The products of the ensuing nuclear reaction are a proton and the nucleus of another atom. Write the complete equation for this nuclear process.

18.27. No element has a stable isotope with mass number equal to eight. When a proton is incident on a ^{7}Li nucleus at rest, two α-particles are produced. By how much does the combined kinetic energy of the α-particles exceed the kinetic energy of the incoming proton?

18.28. A nuclear reaction occurs when a ^{3}He nucleus strikes a ^{88}Sr target nucleus. One of the two reaction products is a proton. Write the complete equation for this nuclear reaction. The product nucleus is radioactive and decays by ordinary beta decay. Write the nuclear radioactive equation for the decay process. Calculate the energy released in the radioactive decay.

18.29. In the fission of ^{235}U by slow neutrons, an average energy of about 200 MeV is released in each fission. The energy released in the Hiroshima nuclear explosion in August of 1945 was about 8×10^{13} J (roughly equivalent to the explosive energy of 20,000 tons of TNT). Estimate the mass of ^{235}U that fissioned in the Hiroshima bomb. If 18% of the ^{235}U fissioned, estimate the mass of ^{235}U used in constructing the bomb.

Answer: (0.97 kg; 5.4 kg)

18.30. Explain clearly why fission fragments (nuclei produced in the fission of a heavy element) are usually highly radioactive.

18.31. There is one stable isotope of fluorine, ^{19}F. When a beam of slow neutrons falls on a sample of natural fluorine, β^- particles are detected. Write the two nuclear equations that represent the complete process of formation and decay of the radioactive nucleus. Calculate the energy released in the radioactive decay.

Answer: (7.03 MeV)

18.32. Why does nuclear fusion require such high temperatures (tens of millions of degrees)? Do not omit any relevant fact in your explanation.

18.33. The fusion of two ^{2}H nuclei proceeds by either of the reactions

$$^2\mathrm{H} + {}^2\mathrm{H} \longrightarrow \begin{cases} {}^3\mathrm{He} + n \\ {}^3\mathrm{H} \;\; + p \end{cases}.$$

Calculate the energy released in each reaction.

Answer: (3.27 MeV; 4.03 MeV)

18.34. Given that no stable isotope with mass number equal to five exists, complete the fusion reaction,

$$^2\mathrm{H} + {}^3\mathrm{H} \longrightarrow \quad + $$

Calculate the energy gained from the reduction in mass in this reaction.

19

No More to Wonder What You Are. . .

19.1 The Evolution of a Star Like the Sun

Now we are ready to apply the physics we have learned toward understanding the structure and evolution of stars. Our approach to this will be to track the progress of a star like the sun on a standard H-R diagram as it passes through the various stages in its evolution. The sun is a relatively light weight star. Heavier stars will follow a somewhat different course, but the physical processes are essentially the same. For this purpose, we use the diagram in Fig. 19.1. The range of temperatures is quite large, therefore, it will be convenient to use a logarithmic temperature scale. For reference, we show the main-sequence on the diagram.

As noted in Chapter 9, a star begins its life as a local condensation of matter in a huge cloud of interstellar gas. The Trifid Nebula, shown in Fig. 19.2, is an

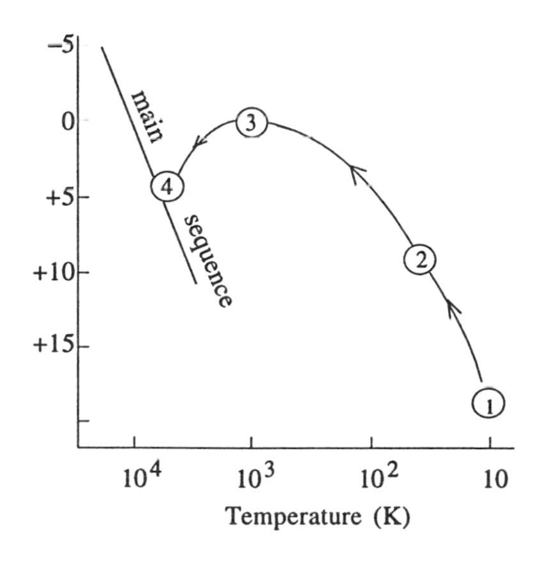

FIGURE 19.1. Pre-main-sequence track of a star like the sun.

FIGURE 19.2. The Trifid Nebula in Sagittarius. (Courtesy of UCO/Lick Observatory.)

example. Parts of this nebula are made luminous by atoms, especially hydrogen, absorbing ultraviolet radiation from hot stars in the interior of the cloud. An ultraviolet photon can ionize a hydrogen atom. When the ejected electron recombines with a hydrogen or other ion in the cloud, radiation is emitted. This radiation includes visible light, making the cloud luminous. The dark lanes are caused by concentrations of dust grains that are especially effective in scattering visible light and thereby prevent the light from escaping the nebula, causing these regions to appear dark against the surrounding luminous part of the cloud. Balls of cold gas formed within a nebula like this can condense under gravity and give birth to new stars.

The gas in such a globule is quite cold at this point. The atoms—mostly hydrogen—are electrically neutral and far apart. Gravity is the only force through which the particles can interact. In this stage, the star would be represented by a point at (1) in Fig. 19.1. Many such condensations occur in the gas, giving birth to a large number of stars in the same cloud. We refer to the collapsing gas in one of these globules as a *stellar nebula*. The slightly larger gravity of a given condensation will draw in matter from the surrounding region, further increasing the gravity. As the matter in this region continues to collapse by gravity into a smaller and smaller volume, its temperature rises and it gets brighter—gravitational potential energy is converted to thermal energy. During this phase, the "star" moves up (brighter) and to the left (hotter) on the H-R diagram—path (2) in Fig. 19.1. When the temperature gets hot enough, the atoms will collide with enough energy to knock electrons from the atoms. Particles in the gas are now electrically charged and interact with the radiation. The collapsing stellar nebula becomes opaque. As a result, it begins to get dimmer. At the same time, it continues to collapse under gravity, getting hotter. The "star" moves down (dimmer) and to the left (hotter) on the diagram at (3) in Fig. 19.1.

Eventually, the temperature in the central core of the nebula will become high enough—several million degrees—for nuclear fusion to occur at an appreciable rate. Remember that these high temperatures are necessary for nuclear fusion to take place because of the two principal forces involved. It is the strong force that is responsible for binding the interacting nuclei in nuclear fusion. But the nuclei all have positive electric charges that cause them to repel each other. (Like charges repel.) Even though the nuclear force is much stronger than the electric repulsion, it acts only over very short distances. On the other hand, the Coulomb (electric) repulsion is a long-range force. Hence, two nuclei repel each other even when they are separated by a large distance. For this reason, two positively charged nuclei must approach each other very fast to get close enough together for the short-range nuclear force to become effective and fuse them to form a heavier nucleus. This means that the kinetic energies of the colliding particles must be quite large. We see from Eq. (17.5) that for the particles in the stellar interior to have sufficiently high kinetic energies, the temperature there must be correspondingly high.

Hydrogen accounts for roughly 74 % (by mass) of the matter in the cloud. But this is the mass-one isotope ^{1}H. The conversion of this isotope of hydrogen to helium by nuclear fusion takes place through the three-step process known as the *proton–proton chain*,

$$\begin{aligned} {}^1\mathrm{H} + {}^1\mathrm{H} &\longrightarrow {}^2\mathrm{H} + e^+ + \nu \\ {}^1\mathrm{H} + {}^2\mathrm{H} &\longrightarrow {}^3\mathrm{He} + \gamma \\ {}^3\mathrm{He} + {}^3\mathrm{He} &\longrightarrow {}^4\mathrm{He} + {}^1\mathrm{H} + {}^1\mathrm{H} \end{aligned} \tag{19.1}$$

19.1.1 Life on the Main Sequence

Gravity tends to pack the matter more densely in the core, making it hotter and driving up the reaction rate for the process in Eq. (19.1). Energy released in the gravitational collapse and in nuclear fusion appears as kinetic energy of the particles of matter—i.e., as thermal energy—which increases the pressure of the gas at each level of the star. When the pressure at each level is such that it can support the weight of all the matter that lies above it, the star is said to be in *hydrostatic equilibrium*. In this stage, the star would be represented by a point on the main-sequence of the standard

H-R diagram—at (4) in Fig. 19.1. This is the normal state of a star in the "prime of life" with equilibrium maintained by energy produced in the core through nuclear fusion balancing the effect of gravity tending to collapse the system.

In the three-step process in Eq. (19.1), four hydrogen atoms disappear from the universe and are replaced by a single helium atom with the evolution of energy. A couple of neutrinos are also produced, but because of their weak interaction with matter, they promptly leave the star altogether. Since the positron e^+ is the antiparticle of the electron e^-, it will quickly find an electron and these two will annihilate each other, producing two photons. This is an example of the mutual annihilation of matter and antimatter. The net effect of the three-step process represented by Eq. (19.1) is summarized by

$$4p \longrightarrow \alpha + 2e^+ + 2\nu, \tag{19.2}$$

where the α-particle denotes the ^{4}He nucleus.

Four particles in the star are replaced by one. We see from Eq. (17.6) that the pressure of a gas depends on the temperature T and the *number* of particles per unit volume N/V—not on the *mass* of the particles. The decrease in the number of particles in the core resulting from nuclear fusion will contribute to a decrease in pressure. This will lead to internal changes that will cause the star to take up slightly different positions on the main-sequence during its lifetime.

In heavier stars, another process in which carbon functions as a catalyst plays a significant role in converting protons to helium nuclei. This is the carbon–nitrogen–oxygen (CNO) cycle. The principal steps in this process are

$$\begin{aligned}
{}^1\text{H} + {}^{12}\text{C} &\longrightarrow {}^{13}\text{N} + \gamma \\
{}^{13}\text{N} &\longrightarrow {}^{13}\text{C} + e^+ + \nu \\
{}^1\text{H} + {}^{13}\text{C} &\longrightarrow {}^{14}\text{N} + \gamma \\
{}^1\text{H} + {}^{14}\text{N} &\longrightarrow {}^{15}\text{O} + \gamma \\
{}^{15}\text{O} &\longrightarrow {}^{15}\text{N} + e^+ + \nu \\
{}^1\text{H} + {}^{15}\text{N} &\longrightarrow {}^{12}\text{C} + {}^4\text{He}.
\end{aligned} \tag{19.3}$$

We see that isotopes of carbon, nitrogen, and oxygen appear in the intermediate steps. This process is also summarized by Eq. (19.2). That is, four protons are converted into an α-particle, two positrons, two neutrinos, and energy. Notice that ^{12}C appears in the initial state of the first step and in the final state of the last step. The carbon is not consumed, hence its catalytic role.

Because the carbon nucleus has a larger electric charge ($6e$) than hydrogen or helium, the electric repulsion is much greater in the first step in Eq. (19.3) than in any of the steps in Eq. (19.1). Therefore, higher temperatures ($> 15,000,000$ K) are required for the CNO cycle than for the proton–proton chain. For this reason, the CNO cycle accounts for only about 2% of the energy production in the sun. In more massive stars ($\gtrsim 1.1\ \text{M}_\odot$), it plays a dominant role.

A star's career on the main-sequence is determined principally by its mass. In this stage, the energy to provide the pressure required to balance gravity is supplied by nuclear fusion. A very heavy star must produce energy at such a prodigious rate that it exhausts its supply of nuclear fuel in a relatively short time. Such stars have typical lifetimes of only a few million years as main-sequence stars. On the other hand, a very low mass star has such small energy production requirements to maintain stability that its fuel reserves last for tens or hundreds of billions of years—longer than the age of the universe itself.

Are there limits to the masses main-sequence stars can have? The answer is yes. If a protostar has too little mass (less than about 8% of the solar mass), the gravity is too weak to raise the central temperature high enough for hydrogen fusion to occur. There is also an upper limit to the mass of a main-sequence star. In practice, this turns out to be in the neighborhood of 100 $M_{\odot}$.

The very massive stars are the hottest and produce the most energy. These are the bright, hot, blue, short-lived stars on the upper end of the main-sequence. The lightweight stars have low temperatures and generate the least energy. They are the dim, cool, red, long-lived stars at the lower end of the main-sequence. A star's position on the main-sequence is an indicator of its mass and its lifetime.

19.1.2 Red Giants

What happens when the nuclear fuel runs low after a substantial fraction of the hydrogen has been converted to helium? When this occurs, energy production in the core drops and gravitational collapse begins again. Some gravitational potential energy of the particles in the core is turned into kinetic energy, thereby raising the temperature. As the core gets hotter, so does a shell of hydrogen around the core. When this shell becomes hot enough, nuclear fusion can occur there, too.

For the rest of the life of the star, we can trace its evolution on a standard H-R diagram as shown in Fig. 19.3. With the depletion of the nuclear fuel in the core as

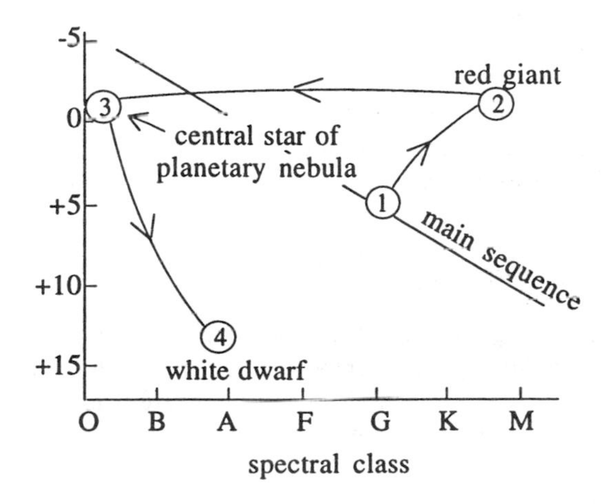

FIGURE 19.3. Post-main-sequence track of a star like the sun.

the star ages, its position on Fig. 19.3 moves from (1) up and to the right to the region of the red giants (2). Continuing collapse of the core releases gravitational potential energy driving the temperature—hence the nuclear fusion rate—still higher in the hydrogen shell around the core. Energy is produced so prodigiously the size of the star must increase for the star to be able to radiate away this huge amount of energy. However, the photosphere, which emits the continuous spectrum of light that is radiated away into space, is relatively cool. The star is a very bright, cool star—a red giant.

The stellar models predict that while the star is in the red giant phase, it undergoes a series of internal changes. During this time, the core temperature rises to around 100,000,000 K, which is high enough for the helium nuclei in the core to combine to form carbon according to the two-step process,

$$\begin{aligned} {}^{4}\mathrm{He} + {}^{4}\mathrm{He} &\longleftrightarrow {}^{8}\mathrm{Be}, \\ {}^{4}\mathrm{He} + {}^{8}\mathrm{Be} &\longrightarrow {}^{12}\mathrm{C} + \gamma. \end{aligned} \tag{19.4}$$

Notice that the arrow in the first step in Eq. (19.4) points to the left and to the right indicating that the reaction goes both ways. The reason is that ^{8}Be is unstable* and quickly falls apart into two α-particles. However, in the collapsing core the temperature and density are so great that in an appreciable number of instances another ^{4}He nucleus will combine with the ^{8}Be before it can fall apart, as indicated in the second step of Eq. (19.4). This is sometimes called the *triple-alpha-process*, since three α-particles come together to make a carbon nucleus with the release of energy. This eventually leads to a carbon core. Some oxygen is also produced in the reaction,

$${}^{4}\mathrm{He} + {}^{12}\mathrm{C} \longrightarrow {}^{16}\mathrm{O} + \gamma.$$

For a star that began its career on the main-sequence with about one solar mass, this is the end of the line for nuclear reactions.

In this stage, two nuclear processes are going on simultaneously in different parts of the red giant as illustrated in Fig. 19.4. The reaction in Eq. (19.4) is turning the core into carbon and hydrogen is converted into helium in the shell around the core through the series in Eq. (19.1). As the supply of helium in the core begins to run out, the carbon core collapses under gravity, but before it can get hot enough for the carbon itself to fuse, the exclusion principle comes into play.

19.1.3 White Dwarfs

In the hot carbon core, the electrons are not bound to the atoms in states that are described by angular momenta. Under these circumstances, the states that are available to an electron are characterized by *linear* momentum and spin. Nevertheless, the exclusion principle still applies to the electrons in the core of the star. Because

---

*In fact, no element has a stable mass-8 isotope.

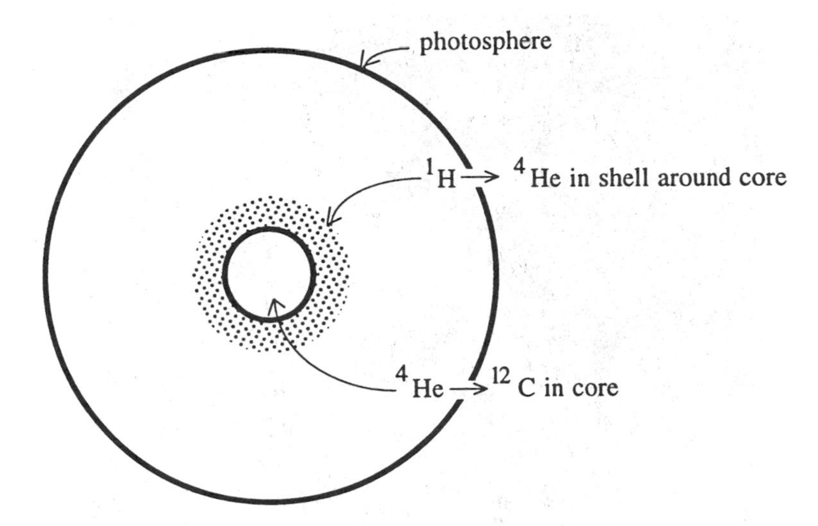

FIGURE 19.4. Schematic model of a red giant.

of this, the electrons cannot all be in the lowest-momentum state. Some electrons are forced into higher-momentum states, giving rise to an electron pressure that halts the gravitational collapse of the stellar core. For a given number of electrons, the exclusion principle requires that there is a minimum volume into which these electrons can be squeezed.

Observational evidence indicates that late in the red giant stage, the star begins to eject its outer layers into space. As this matter moves away from the star, it forms a shell of gas that is made luminous by absorbing ultraviolet radiation from the hot central core and reradiating it as visible radiation. To an observer on earth, these luminous shells of gas appear in the sky as fuzzy, extended objects called *planetary nebulae*—although they have nothing to do with planets. A classic example is the Ring Nebula in Lyra, shown in Fig. 19.5

As the spreading gas shell becomes thinner, the hot central core of the star becomes visible, as is evident in Fig. 19.5. The star's position on the H-R diagram when it reaches the planetary nebula stage in its evolution is represented by (3) in Fig. 19.3. The star has now exhausted its nuclear fuel. It is still very hot, but just a cinder in the sky steadily becoming dimmer and slowly radiating away its thermal energy into space. The star collapses under gravity as far as the exclusion principle will permit, reaching the *white dwarf* stage at (4) in Fig. 19.3.

A white dwarf is an extremely dense star in which matter of roughly the mass of the sun is squeezed by gravity into a volume approximately the size of the earth. The white dwarf is stable because gravity is balanced by the electron gas pressure resulting from the effect of the exclusion principle. Over time, a white dwarf will radiate away nearly all of its thermal energy and become a cold, dead object—a black dwarf.

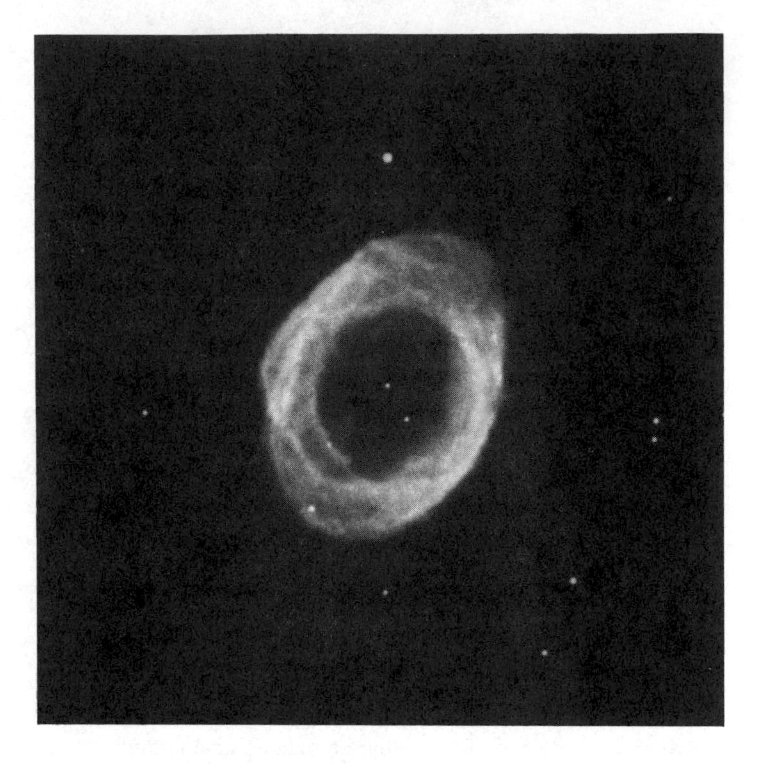

FIGURE 19.5. A planetary nebula. (Courtesy Of UCO/Lick Observatory.)

If a star in the red giant phase can eject enough matter so that its remaining mass is less than about 1.4 $M_{\odot}$, it can stabilize as a white dwarf. Heavier stars follow a different evolutionary track after they leave the main-sequence.

19.2 The Evolution of a Heavier Star

When an aging massive star reaches the stage where its core has been converted to carbon, the weight of the overlying matter is sufficient to compress the core to the point where carbon itself can fuse at temperatures around 600,000,000 K. In addition, the helium shell outside the core heats up and helium fusion begins in this shell. Moreover, a shell of hydrogen outside the helium shell becomes hot enough for nuclear fusion to occur here. This process continues in successive stages. As the core exhausts its fuel, gravity compresses it still further until the temperature rises enough to ignite another nuclear fuel source. The structure of the star is a series of layers with successively heavier elements toward the center.

We see from the B/A curve in Fig. 18.4 that as lighter elements fuse to produce heavier ones, energy is released as long as the mass number of the product is less than about 60. This is in the neighborhood of the element iron. Nuclear fusion of iron *absorbs* energy rather than releases it.

19.2.1 A Neutron Star

When a massive star has evolved to the point where a substantial fraction of its core is iron, this star has reached a critical stage. As gravity collapses the iron core, no energy is released in nuclear fusion to provide the pressure required to slow the collapse. The core is compressed until the exclusion principle provides the electron gas pressure to stabilize it. The fuel is gone in the core, but nuclear reactions are still going on outside the core. Eventually, the gravitational effects become so strong that the protons in the nuclei of the core capture the electrons to form neutrons according to

$$p + e^{-} \longrightarrow n + \nu. \tag{19.5}$$

By this process, the electron gas of the core is replaced by a neutron gas.

Neutrons also obey the exclusion principle. Because of the neutron's large mass, the minimum volume into which a neutron gas can be compressed is very much smaller than for an equal number of electrons. Therefore, the iron core quickly collapses to a few kilometers in diameter. With no energy production from the core to hold them up, the outer layers fall freely toward the center. When the matter falls onto the rigid core, it rebounds with a shock wave that blows the star apart catastrophically. The force of the explosion is enhanced by the escaping neutrinos of Eq. (19.5).

In the high-temperature blast that destroys the star, nuclear fusion produces elements heavier than iron, spewing them into the interstellar medium as seeds for new generations of stars. The brightness of this event rivals that of the entire galaxy—a single star for a brief time has the brightness of tens of billions of stars. Such an event is called a *supernova*.

Because of dust that obscures light in the plane of the galaxy, supernovae are not seen very frequently in the Milky Way. The supernova most recently seen in our own galaxy was observed by Johannes Kepler in the seventeenth century. However, since hundreds of millions of galaxies are accessible to the large optical telescopes, it is not uncommon for astronomers to observe a supernova in another galaxy. An example is seen in Fig. 19.6, which shows two views of the galaxy NGC 5253. The left photograph shows this galaxy as it appeared in 1959. The one on the right was taken thirteen years later in 1972 and clearly reveals the supernova. The brilliance of the supernova is seen to be comparable to that of the whole galaxy

The photograph in Fig. 19.7 shows the remnant of a supernova that occurred when a star in the Milky Way reached the end of its life. The event was seen on earth in the year 1054 by astronomers in the Orient, who recorded that it was bright enough to be visible even in the daytime.

In a supernova, much of the original matter of the star is blown away in the explosion. All that remains is the core—a tiny, extremely dense, *neutron star*, in

FIGURE 19.6. A supernova in a distant galaxy. (Courtesy of Palomar Observatory/Caltech.)

which gravity is held in check by the gas pressure derived from the exclusion principle by the neutrons.

19.2.2 A Black Hole

In even more massive stars (60 $M_\odot$–100 $M_\odot$), not even the neutron gas pressure is sufficient to counteract the gravity. Nothing can withstand the gravity. The star undergoes complete gravitational collapse. The final state is a *black hole* where gravity is so strong that not even light can escape.

19.2.3 Main Sequence Lifetimes

The large, massive, hot, bright, blue stars on the upper main-sequence spend much shorter times on the main-sequence than do the small, lightweight, cool, dim, red stars at the lower end of the main-sequence. If we assume that the stars in a given cluster all begin to form at about the same time from the same cloud of gas and dust, then the hot, blue stars will be the first to evolve off the main-sequence. Thus, the length of the main-sequence in a color-magnitude diagram for the cluster is an indicator of the age of the cluster. For example, the very short main sequence in

FIGURE 19.7. Supernova remnant in the constellation of Taurus. (Courtesy of Palomar Observatory/Caltech.)

Fig. 16.10 shows that the cluster represented in this diagram is very old. It is old enough for most of the stars heavier than the sun to have left the main-sequence. This is typical of globular clusters that contain mostly red stars. Open clusters, on the other hand, often have short-lived, hot, blue stars on the main-sequence, indicating that the cluster is still quite young.

19.3 The Stuff Between the Stars

We have noted that new stars are formed from gigantic clouds of intestellar dust and gas—the gas being mostly hydrogen. How do we know this? We can tell that the dust is there because it prevents visible light from reaching us from luminous sources that lie in the plane of the galaxy. But how do we know the hydrogen is there? We don't see visible light emitted from the atoms, so if the hydrogen is there it must be so cold that in most atoms the electron is in the ground state. No population of electron excited states means no emission of radiation. So how is it detected?

In making our model of the hydrogen atom in Chapter 14, we only took into account the electric attraction between the electron and the nucleus (proton). This is the dominant effect to be sure, but there are other interactions that exist as well. For example, the electron and the proton both have intrinsic magnetism. Each is a tiny magnet—the electron being the stronger of the two. This magnetism is related to the spin of each particle. The atom has slightly more energy when the electron and proton spins are aligned (parallel) than when they have opposite orientations (antiparallel). Thus, the $n = 1$ state of hydrogen actually has two components—one with a slightly higher energy than the other.

When the electron in the higher-energy state spontaneously flips its spin in making a transition to the lower-energy state, a photon with energy just equal to the energy separation between the two components is emitted, as illustrated in Fig. 19.8. Because the magnetic effect is so weak, this splitting of the $n = 1$ state is very tiny; hence the wavelength of the emitted radiation is quite long—about 21 cm. This is in the microwave region of the radio spectrum and detectable with radio telescopes. So, even very cold hydrogen gas can reveal its presence to a radio telescope by emitting this 21-cm radiation.

Interstellar hydrogen can also be detected by absorption in the microwave spectrum. This is seen in Fig. 19.9, which shows a continuum of microwave radiation incident from the left on a cloud of cold hydrogen gas. The schematic plot below the incoming radiation in Fig. 19.9 shows a uniform intensity for this radiation around 21 cm. In passing through the cloud, hydrogen atoms in the lower-energy spin state absorb 21-cm photons, flipping the spin of the electron to the higher-energy state. Thus, the number of photons transmitted through the cloud is diminished in the neighborhood of 21 cm and the graph below the radiation leaving the cloud shows a corresponding dip in intensity of the transmitted radiation around this wavelength in the spectrum.

If the hydrogen cloud is moving along the line of sight relative to the earth, the 21-cm absorption line will be shifted by the Doppler effect. By measuring this shift, an observer can determine if the cloud is moving toward or away from the earth

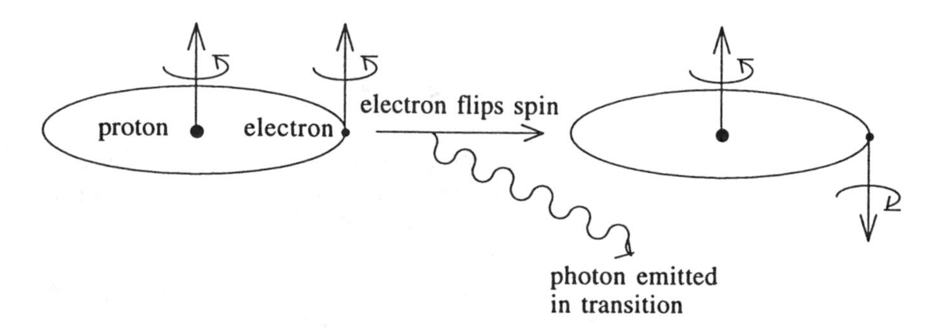

FIGURE 19.8. The origin of the 21-cm radiation from hydrogen.

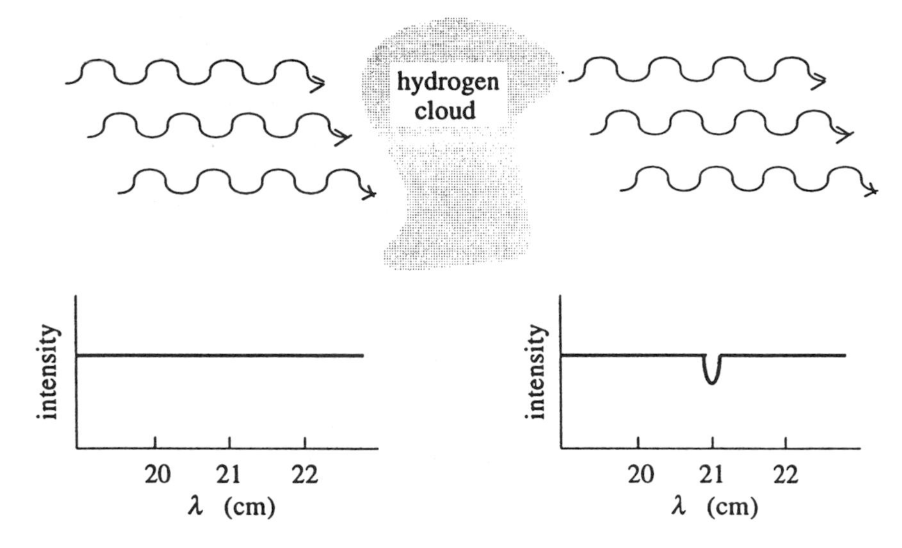

FIGURE 19.9. Absorption of radiation by interstellar hydrogen.

and how fast it is moving along the line of sight. In this way, radio astronomers have been able not only to map the distribution of the clouds of hydrogen gas in the galaxy, but also to be able to tell how fast these clouds are moving in different directions.

19.4 Solar Neutrinos

A neutrino ν appears as a reaction product in the first step in the proton–proton chain of Eq. (19.1) responsible for converting hydrogen to helium in the central core of a star like the sun. The CNO cycle of Eq. (19.3) also produces neutrinos in the core of the star. Potentially, these particles can supply information directly about nuclear processes within the stellar core.

Computer models have been developed to account for the evolution of stars. These models predict the conditions inside the star (density, temperature, chemical composition, etc.) that are required to match the observed properties of the star (size, mass, color, luminosity, etc.). This implies certain reaction rates for nuclear processes within the stellar core, hence a corresponding rate for neutrino production. In principle, the actual neutrino production rate can be deduced from earth-based measurements of the neutrino flux from the star. The measured rate provides a test of the stellar model.

Neutrinos are elementary particles that have no mass and carry no electric charge. Only the weak force is significant in the interaction of the neutrino with matter.

Because stellar matter is virtually transparent to neutrinos, a neutrino can come promptly out of the core of the star with a high probability of not interacting at all with the matter in the star.

The virtue of relying on neutrinos as carriers of information from the stellar core lies in the neutrino's very weak interaction with matter. Energy produced in the core of a star through the reactions of Eq. (19.1) undergoes many transformations in working its way out of the star to the photosphere where it is radiated away as starlight. It takes millions of years for the energy produced in the core of a star to make its effects known to us. Neutrinos, on the other hand, interact weakly with the matter in the outer layers of the star. They come out promptly. Some wing their way toward the earth at the speed of light. So, potentially, the neutrinos may bring information immediately from the interior of the star and not millions of years later as with other products of nuclear reactions.

The weakly interacting nature of the neutrino is a mixed blessing for neutrino astrophysics. Because the neutrinos interact so weakly with matter, only a few interact with the detector. This makes stellar neutrino experiments hard to do. In fact, the probability for a neutrino to interact with an earth-based detector is so small that the sun is the only star close enough to provide enough neutrino flux at the earth for any measurements to be feasible. So, in practice, only *solar* neutrinos can be studied with the present technology.

The first solar neutrino experiment could not even detect the neutrinos produced in the first step in Eq. (19.1). It relied on detection of higher-energy neutrinos from secondary reactions converting the ^{3}He produced in the second stage in Eq. (19.1) to ^{4}He. In spite of the experimental difficulties, data were obtained that indicated the observed neutrino flux from the sun was only about one-third of the prediction based on solar models. This experiment has been running for nearly three decades. Others that can detect the low-energy neutrinos from the first stage of Eq. (19.1) have also come into operation. With the accumulation of more extensive and reliable data, the discrepancy persists and appears to be real. This discrepancy constitutes the *solar neutrino problem*. To compound the problem, as confidence in the reliability of the experimental data has increased over the years, so has confidence in the standard solar models.

One attractive resolution of the problem is rooted in the fact that there are three distinct kinds of neutrinos. Only one kind is produced in the nuclear reactions in the sun and that is the only kind that can be detected by the absorption detectors on the earth. However, if neutrinos have even a tiny bit of mass, it is possible that the neutrinos are produced in nuclear reactions in the solar core in the abundance required by the solar models, but in passing from the dense core of the sun out to the low-density region of interplanetary space they are converted into a different kind of neutrino that does not interact appreciably with the ordinary matter from which detectors are made. Hence, the detectors do not count them.

Other explanations that call for new physics or departures from the standard solar models have also been offered. It is an open question and the solar neutrino problem is still with us.

Exercises

19.1. Explain clearly how energy is produced in the interiors of stars like the sun. Give a clear and complete explanation of why the stellar core must reach such high temperatures (tens of millions of degrees) before this energy production can occur. Do not omit any relevant detail.

19.2. The total radiative power of the sun is estimated at 4×10^{26} W. This energy is generated in the central core of the sun in which some of the sun's mass is converted to energy according to Einstein's mass-energy relation,

$$E = mc^2.$$

Estimate the amount of the sun's mass (in kilograms) converted to energy each year. The sun is expected to continue to produce energy at this rate for another 5 billion years. Approximately, what percent of the sun's present mass will be converted to energy in this time interval?

Answer: (1.4×10^{17} kg; 0.04 %)

19.3. The conversion of hydrogen to helium in the core of a star is sometimes referred to as "hydrogen burning." Explain clearly and completely how this differs from what we ordinarily mean by the burning of hydrogen gas in the earth's atmosphere.

19.4. The net effect of the three-step process represented in Eq. (19.1) in which four protons are converted to helium can be expressed as

$$4p \longrightarrow \alpha + 2e^+ + 2\nu,$$

where the α-particle denotes the ^{4}He nucleus. Add the appropriate number of electrons to both sides of this equation and use the atomic masses in Table 18.3 to estimate the energy released in this process.

Answer: (24.7 MeV)

19.5. The ^{3}He produced in the second step of Eq. (19.1) can be converted directly to ^{4}He through fusion with a proton in the solar core according to the reaction

$$^1\text{H} + {}^3\text{He} \longrightarrow {}^4\text{He} + e^+ + \nu.$$

Calculate the maximum energy of the neutrino produced in this reaction. HINT: The equation here represents a *nuclear* reaction. Because we use *atomic* masses, we must be sure to take into account the appropriate number of electrons in calculating reaction energies.)

Answer: (18.8 MeV)

19.6. Helium is the second most abundant element in the universe after hydrogen. The ^{4}He present in the solar core can complete the conversion of the ^{3}He produced in the second stage of Eq. (19.1) through the three-step process,

$$\begin{aligned} ^3\text{He} + {}^4\text{He} &\longrightarrow {}^7\text{Be} + \gamma, \\ e^- + {}^7\text{Be} &\longrightarrow {}^7\text{Li} + \nu, \\ ^1\text{H} + {}^7\text{Li} &\longrightarrow {}^4\text{He} + {}^4\text{He}. \end{aligned}$$

The isotope ^{7}Be has mass 7.01693 u. Calculate the energy of the neutrino produced in the second step of this process.

Answer: (0.35 MeV)

19.7. Calculate the net kinetic energy gained per reaction in the third stage of Eq. (19.1),

$$^3\text{He} + {}^3\text{He} \longrightarrow {}^4\text{He} + {}^1\text{H} + {}^1\text{H}.$$

Answer: (12.9 MeV)

19.8. Why are blue main-sequence stars brighter than red main-sequence stars? Give a clear and complete physical explanation. Include all relevant facts.

19.9. Calculate the average kinetic energy per particle for particles in the core of the sun where the temperature is 1.5×10^7 K. What is the particle speed corresponding to this (nonrelativistic) kinetic energy?

19.10. Name the three observable phases of evolution through which a low-mass star like the sun passes after it leaves the main sequence. Describe the physical characteristics of the star in each phase.

19.11. In white dwarf stars, nuclear reactions have ceased. With no energy produced from fusion to balance the gravity, why doesn't the white dwarf collapse completely under gravity?

19.12. Certain *physical* properties of a white dwarf star are responsible for the *visually observable* characteristics that give rise to the term "white dwarf." What are these physical properties? How is each related to the corresponding visual characteristic? Be very clear.

19.13. Write a paragraph describing the evolution of a star like the sun. Include in your discussion the roles played by nuclear fusion, the exclusion principle, planetary nebula, gravity, the hydrogen shell, and pressure.

19.14. Write a paragraph explaining clearly why a heavier main-sequence star eventually becomes unstable and explodes as a supernova. Include all of the key factors that lead to the instability and cause the explosion.

19.15. Explain clearly why the sun will never explode as a supernova.

19.16. What causes a star that has exhausted its fuel supply to become very dense? What stops this density increase in medium-weight stars (heavier than the sun)?

19.17. With regard to their stability, tell how a white dwarf and a neutron star are similar. Tell how they differ.

19.18. If we see a massive main-sequence star, what can we assume about its age relative to most stars? Explain clearly your reasoning.

19.19. Use words and drawings to explain clearly and completely how the 21-cm radiation from hydrogen is produced. How does this radiation differ from the radiation we normally see from hot hydrogen gas? From an atomic point of view, *why* is it different? Why is it particularly useful to astronomers in mapping the distribution of matter in our galaxy? Be very clear. Do not omit any relevant fact.

19.20. Why are neutrinos particularly valuable as carriers of information about stars compared to other kinds of stellar radiations? Be clear and complete.

19.21. Presumably, all main-sequence stars emit neutrinos in abundance. Why is it that currently only measurements on neutrinos from the sun are being carried out?

19.22. What is "the solar neutrino problem"? Give a clear and complete explanation. What are the implications for astronomy?

20

The Flight of the Galaxies

20.1 The Nebulae

On a clear, moonless night in the countryside away from city lights, with a good pair of binoculars, we can see many faint, fuzzy-looking patches of light distributed over the sky. These are the *nebulae*. They are there all the time.

When a comet approaches the sun—yet is still too far away to have an observable tail—it has the same general appearance as a nebula. For someone interested in discovering new comets, a catalogue of the nebulae would be quite useful. If a fuzzy object appears in the telescope, a quick reference to the catalogue would tell whether this is an interloper or one of these nebulous objects that are in the sky all the time. In the late eighteenth century, Charles Messier, a French astronomer and avid comet hunter, did exactly this. From a survey of the sky seen from the northern hemisphere, he made a list of the most prominent objects that might easily be confused with a comet when it first becomes visible. Messier's catalogue contains about 100 of these objects with their celestial coordinates. Astronomers today commonly refer to these objects according to their listing in Messier's catalogue. For example, the Crab Nebula, depicted in Fig. 19.7 (the remnant of the supernova observed and recorded by the Chinese in 1054) in the constellation of Taurus is designated M1, since it is the first item in Messier's list.

Messier's entries are all extended objects with a fuzzy appearance to the naked eye or as seen in a small telescope. With the increased resolution of a larger telescope, we find a variety of objects that have this cloudy appearance and are listed among the nebulae—objects such as star clusters, gas clouds, planetary nebulae, and supernova remnants.

Early in the twentieth century, evidence began to accumulate that indicated that the sun is a part of a vast system of stars, gas, and dust called the *galaxy*—from the Greek word for milk suggested by the ancient name Milky Way which is a luminous manifestation of the plane of the galaxy. Around 1920, there was a great controversy among astronomers regarding the nature of the spiral-shaped nebulae that were seen in all directions away from the Milky Way. Were these objects components of the galaxy or were they "island universes"—galaxies—in their own right? It was a question of size and distance. How big is the galaxy and how far away are the spirals? The answer was provided around the middle of the decade by Edwin Hubble, the astronomer for whom the Hubble Space Telescope is named. To follow Hubble's path to the solution to the riddle of the spirals, let us digress briefly to consider the properties of a certain class of stars called *variable stars*.

20.2 Variable Stars and Cosmic Distances

A variety of types of stars corresponding to different stages in stellar evolution vary in intrinsic brightness over time. The mechanisms responsible for the variations are different for different types of stars. In some of these stars, the variation in brightness is irregular with time. Such stars are referred to as *explosive* or *eruptive* variables. Others show regular, periodic fluctuations in brightness. These are called *periodic* or *pulsating* variables and are the ones of interest to us here.

As the term *pulsating* suggests, these stars are fluctuating in size. The outer regions of a pulsating variable are not in hydrostatic equilibrium. They swell and contract in a regular way, fluctuating about an equilibrium size. These size fluctuations of the outer layers are driven by a mechanism that involves the ionization of helium.

Two classes of pulsating variables have turned out to be extremely useful in establishing distances in the cosmos. These are the Cepheid variables and RR Lyrae variables. As seen in Fig. 20.1, all RR Lyrae variables are moderately bright with approximately the same absolute magnitude, $M \approx +0.5$. These stars are bright enough to be seen in remote clusters of stars and can be used to determine the distances to these clusters. They are quite common in globular clusters.

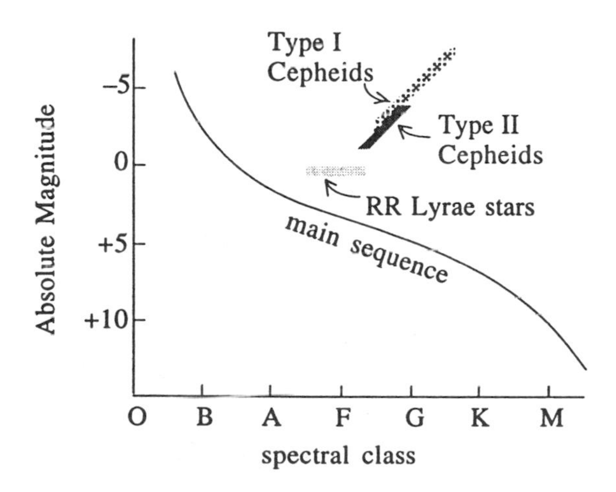

FIGURE 20.1. Pulsating variable stars.

EXAMPLE. A star in a certain cluster is identified as an RR Lyrae variable. The apparent magnitude of this star is measured at +15.2. Estimate the distance to this cluster.

SOLUTION. The magnitude indicates that the star is so far away we can consider all stars in the cluster to be the same distance from us. We only need to know the distance to one of the stars to know the distance to the cluster. From Eq. (15.3), we have

$$\left(\frac{d}{10\text{ pc}}\right)^2 = 2.5^{15.2-0.5} = 707,486,$$

from which we obtain $d = 8411$ pc. The cluster is about 8,400 pc or 27,000 ly away from us.

The Cepheid variable stars are yellow supergiants that can be seen at even greater distances than the RR Lyrae stars. The time interval over which a Cepheid changes from maximum brightness to minimum back to maximum again—the period—is correlated with the average absolute magnitude of the star. Because of its larger surface area, a large star shines brighter than a small one. At the same time, the large star will oscillate more slowly about its equilibrium size, giving rise to a longer period for its brightness fluctuation.

There are two types of Cepheid variable stars. The relationships between the brightness and period are given for both in Fig. 20.2. The average apparent magnitude and the period of a Cepheid variable are readily observed. From these data and the calibration graph in Fig. 20.2, the distance to the star can be easily calculated.

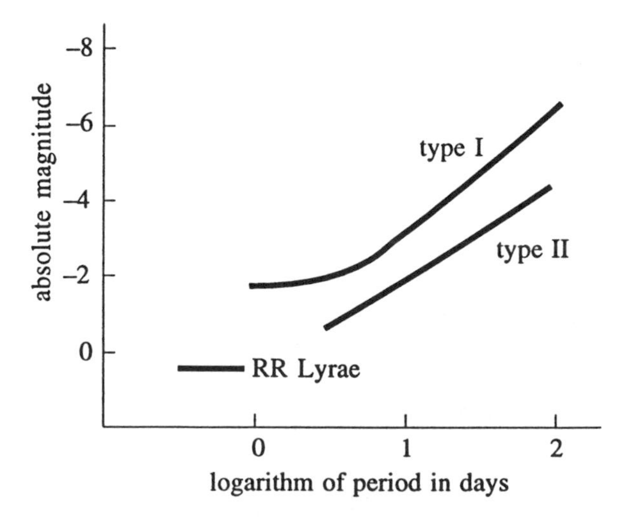

FIGURE 20.2. Absolute magnitudes of pulsating variable stars.

EXAMPLE. A bright, yellow star is identified as a type I Cepheid variable. Its average apparent magnitude is measured at +3.2. It is seen to vary in brightness over a period of about 50 days. Estimate the distance to this star.

SOLUTION. The logarithm of 50 is about 1.7. Therefore, from Fig. 20.2, we see that for a type I Cepheid with a 50-day period, the absolute magnitude is about −5.4. Substituting these magnitudes into Eq. (15.3), we get

$$\left(\frac{d}{10\text{ pc}}\right)^2 = 2.5^{3.2-(-5.4)} = 2{,}644,$$

from which we find that $d = 514$ pc. The distance to this Cepheid variable star is about 500 pc or 1,700 ly.

20.3 Hubble's Discovery

Around 1924, using the 100-inch telescope on Mt. Wilson, Hubble was able to resolve Cepheid variable stars in some of the spiral nebulae. From measurements of the light from these objects, he determined that they are well beyond the limits of the galaxy, which by then had been fairly well established. Hubble demonstrated convincingly that the spiral nebulae are galaxies like our own Milky Way.

Analyses of the light from the extragalactic nebulae had already established that the spectral lines of these objects were shifted toward longer wavelengths—a redshift—with the shift tending to be greater for the fainter ones. This is illustrated for five galaxies in Fig. 20.3. The spectrum of each galaxy in the photograph is compared with a calibration spectrum of lines from hydrogen and helium in the laboratory. The principal features of the spectrum (the broad smear in each photograph) for each galaxy are the double absorption lines produced by singly ionized calcium atoms, marked by a horizontal arrow in each photograph. These two lines in the violet part of the spectrum have unshifted (laboratory) wavelengths of 393.37 nm and 396.85 nm. The photographs clearly show that the fainter the galaxy (nearer the bottom of the figure), the greater the shift of the calcium lines to the right (to longer wavelength).

If this redshift is interpreted as a consequence of the motion of these nebulae relative to the earth, the implication is that they are all moving *away* from us with the fainter ones moving away faster (larger redshift). Furthermore, if the faintness is due to distance, then the interpretation is that the more distant nebulae are moving away from us faster than the nearer ones. If the speed of a receding source of light is small compared to the speed of light c, the recession velocity v is related to the wavelength shift $\Delta\lambda$ by Eq. (16.4)

$$v = \frac{\lambda_{\text{obs}} - \lambda}{\lambda} c \equiv \frac{\Delta\lambda}{\lambda} c, \qquad (20.1)$$

where λ_{obs} is the *observed* wavelength and λ is the *actual* (unshifted) wavelength emitted by the source.

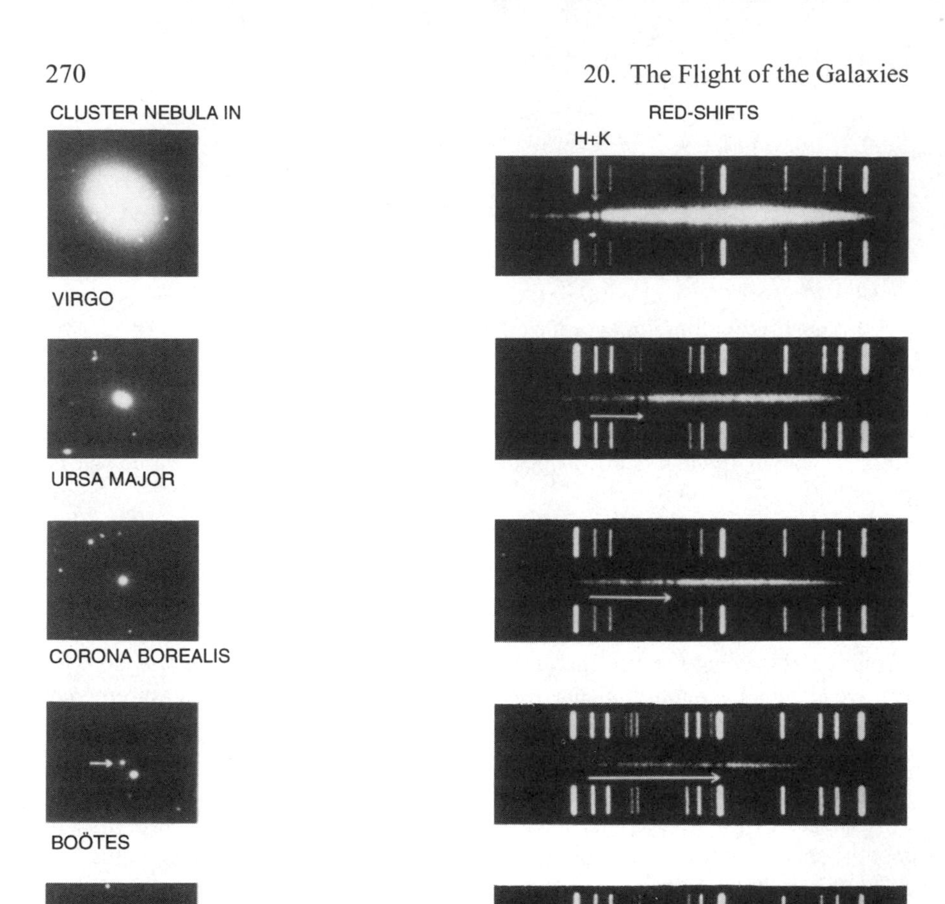

FIGURE 20.3. Spectra of five galaxies. (Courtesy of Palomar Observatory/Caltech.)

If the relative velocity of the source and observer is near the speed of light, the approximation in Eq. (20.1) relating the redshift and the recession velocity is not valid and one must use the exact expression obtained from special relativity,

$$\frac{v}{c} = \frac{\left(1 + \frac{\Delta\lambda}{\lambda}\right)^2 - 1}{\left(1 + \frac{\Delta\lambda}{\lambda}\right)^2 + 1}.$$

Following Hubble's discovery that these nebulae are galaxies, the effort accelerated to establish distances to them. By the end of the decade, Hubble had collected distance and redshift data on enough of the galaxies to claim that the distant galaxies are all receding from the earth with velocity proportional to the distance—exactly what one expects if the universe is *expanding*. This is a remarkable discovery with important cosmological significance. These results lead us to the conclusion that the galaxies are all getting farther apart as the universe expands.

The relationship between recession velocity v and distance d to the galaxy, is known as *Hubble's law*,

$$v = Hd. \tag{20.2}$$

The parameter H is called *Hubble's constant.* A graphical representation of Hubble's law is shown in Fig. 20.4. The data points do not correspond to real galaxies, but are given to illustrate a typical distribution of data used to establish Hubble's law. The value of Hubble's constant is in dispute among astronomers with different sets of data leading to different values. The range is 50–100 km/s/Mpc. In Eq. (20.2), we see that the slope of the graph in Fig. 20.4 corresponds to H and has the value 65 km/s/Mpc.

EXAMPLE. The mean wavelength of the double line from singly ionized calcium is 395 nm. This line in the spectrum of a distant galaxy is observed to have a mean wavelength of 411 nm. Use Hubble's law to estimate the distance to this galaxy.

SOLUTION. First, calculate the recession velocity from Eq. (20.1),

$$v = \frac{\Delta\lambda}{\lambda} c = \frac{411 \text{ nm} - 395 \text{ nm}}{395 \text{ nm}} 3 \times 10^5 \text{ km/s} = 1.22 \times 10^4 \text{ km/s}.$$

With this value for v and $H = 65$ km/s/Mpc, we obtain from Eq. (20.2) $d = 187$ Mpc. The distance to this galaxy is about 190 Mpc or about 600,000,000 light years.

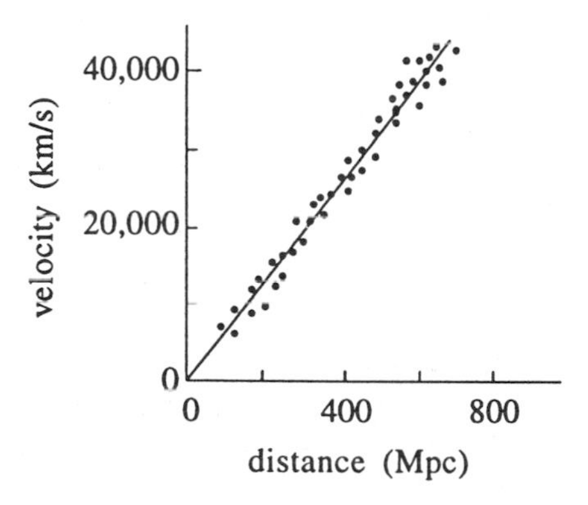

FIGURE 20.4. Hubble's law.

FIGURE 20.5. The spiral galaxy NGC 628 in Pisces. (Courtesy of Palomar Observatory/Caltech.)

FIGURE 20.6. The spiral galaxy NGC 4565 in Coma Berenices. (Courtesy of Palomar Observatory/Caltech.)

20.4 The Structure of a Spiral Galaxy

A typical spiral galaxy similar to the Milky Way is shown in Fig. 20.5. This face-on view shows how the spiral arms are arranged around the central core of the galaxy. These arms contain large quantities of gas and dust from which new stars can form. Frequently, these occur as members of a cluster of a few hundred to several thousand stars that formed about the same time from a single cloud of gas. These so-called open or galactic clusters very often contain bright, blue main-sequence stars. The presence of massive blue stars on the main-sequence indicates that the cluster is quite young, supporting the notion that new stars are being born in the spiral arms of the galactic plane.

Another spiral galaxy, also similar to the Milky Way, is depicted in Fig. 20.6. This edge-on view shows the central bulge of the galaxy, which is more densely populated with stars than the spiral arms. These stars are mostly old, red stars. Also distributed around the galaxy's halo, away from the plane of the galaxy, is another type of stellar cluster. These clusters are called globular clusters and generally comprise tens of thousands to a few million stars. A photograph of a globular cluster visible in the constellation of Hercules is shown in Fig. 20.7. No significant new star formation takes place in the globular clusters, because they contain very

FIGURE 20.7. The globular cluster M13 in Hercules. (Courtesy of Palomar Observatory/Caltech.)

little gas and dust from which to make new stars. These clusters are so old that the heavier stars have long ago evolved off the main sequence to become red giants or supergiants. The remaining main-sequence stars are the longer-lived reddish ones. Thus, globular clusters are dominated by old, red stars.

Exercises

20.1. Using the (then new) 2.5-m telescope on Mt. Wilson in the 1920s and 1930s, Edwin Hubble accumulated extensive data on the nebulae which led him to two remarkable discoveries. What were they?

20.2. The apparent magnitude of a certain RR Lyrae variable star in a globular cluster is observed to be +16.5. How far away is the cluster?

Answer: (49,700 ly)

20.3. The nebula M31, visible in the constellation Andromeda, is about two million light years away from us. It is a spiral galaxy similar to our own Milky Way. Assume that these two galaxies are gravitationally bound and follow circular orbits about a common center of mass. Estimate the time it takes for the Milky Way to make one complete orbit about this center of mass. Estimate the mass of the Milky Way from the number of stars, assuming that the average mass of a star is equal to the mass of the sun.

20.4. The double violet line (mean wavelength 395 nm) in the spectrum of singly ionized calcium is very strong in the absorption spectra of many stars. In the spectrum of a certain galaxy, these lines are observed to have a mean wavelength of 398 nm. From these data, estimate how far away the galaxy is. Explain clearly the physical ideas underlying your method of determining your result.

Answer: (35 Mpc)

20.5. In a certain galaxy, one of the stars is identified as a type I Cepheid variable. The brightness of the star is measured over a period of several months. The data are presented in the graph below. From the data, estimate the distance to the galaxy. Explain clearly and completely your reasoning.

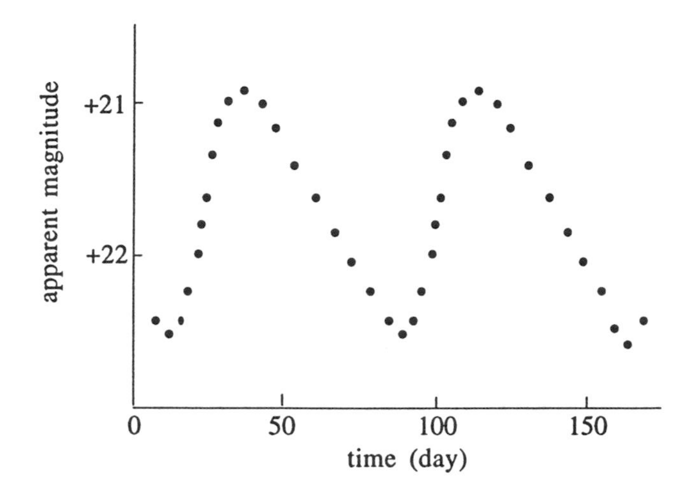

20.6. On a certain day when a type-I Cepheid variable star reaches maximum brightness, its magnitude is measured at +14.7. Continued observation shows that its brightness diminishes, reaching a minimum 12.2 days later when the magnitude is 17.2. After another 8.7 days, the star is back to maximum brightness again. How far away is this star? Could it be in the Milky Way galaxy? Explain clearly your reasoning.

Answer: (3×10^5 ly)

20.7. A galaxy lies 240,000,000 ly away. Its spectrum shows a redshift of 0.0173. From these data, estimate the value of Hubble's constant.

Answer: (70.5 km/s/Mpc)

20.8. A faint star in a distant galaxy is identified as a type I Cepheid variable. The time it takes for the brightness of the star to change from maximum to minimum and back to maximum is exactly 100 days. The mean apparent magnitude of this star is +26.2. The spectrum of the galaxy in which this star resides shows a redshift of 0.0056. From the data, obtain an estimate of Hubble's constant.

Answer: (52 km/s/Mpc)

20.9. Each of the photographs of the spectra of the five galaxies in Fig. 20.3 also shows for comparison portions of the spectra of hydrogen and helium. These unshifted (laboratory) spectral lines can be used as a reference to calibrate the photograph and obtain the shifted wavelengths of the singly ionized calcium absorption lines in the spectra of the galaxies. The laboratory wavelengths of these hydrogen and helium lines are given below. With your ruler determine from the photograph the position of each line. From these results, make a graph of *actual wavelength* (nm) versus *position on the photograph* (mm).

line	(nm)	position (mm)
———	388.8	0
———	396.5	
———	402.6	
———	410.2	
———	412.0	
———	414.3	
———	434.0	
———	438.7	
———	447.1	
———	471.3	
———	486.1	
———	492.2	
———	501.5	

The unshifted (laboratory) wavelengths of the two calcium lines are 393.3 nm and 396.8 nm. Measure with your ruler the position (in millimeters) of each of these lines in each photograph and use the graph you obtained above to determine the actual (*shifted*) wavelength in nanometers. Use these results and Hubble's law to determine the distance to each of the five galaxies (hence to the cluster of galaxies in which it lies).

20.10. A certain star is identified as a type I Cepheid variable. The average apparent magnitude is measured to be 8.4. In 1947, this star was seen to appear brightest successively on August 15, September 24, and November 3. From the data, determine whether this star is in our galaxy. Explain clearly and completely your reasoning.

20.11. Charles Messier was a French astronomer who flourished in the eighteenth century. Tell what his primary interest was and explain clearly how that interest led him to accomplish a task that proved to be extremely useful to other astronomers who had nothing to do with Messier's principal interest. Describe the task and explain clearly how it was useful to the other astronomers.

21

The Big Picture

21.1 The Cosmological Principle

In this chapter, we take a brief look at recent efforts to obtain an understanding of the universe as a whole. Cosmologists begin with a simplifying assumption about the nature of the universe on a large scale—namely, that in looking out from our vantage point in the cosmos, we see essentially the same kind of universe that an observer stationed in any other part of it, no matter how remote, would see. As far as our telescopes can reach, we see galaxies and clusters of galaxies distributed more or less the same in every direction. If we consider big enough pieces of the universe, it is homogeneous and isotropic. This feature, ascribed to the large-scale universe, is called *the cosmological principle*. It is not a proven fact, but it is a reasonable starting assumption.

For over three decades, the expansion of the universe implied by the flight of the galaxies stood as the only piece of observational evidence of cosmological significance. Just after the middle of the twentieth century, there were two general classes of cosmological theories presented by different groups of cosmologists that fit this single known feature of the cosmos. These two classes were:

- evolutionary (Big Bang)
- steady-state

The essential idea in the evolutionary cosmologies is that there was a beginning—a moment of creation at which the universe came into existence in a hot, violent explosion—the Big Bang. In the beginning, the universe was very hot, very dense, and very tiny. As the explosion evolved, the temperature dropped, the distribution of matter and energy thinned, and the universe expanded—all consistent with Hubble's law. From the current observed rate of expansion, we conclude that the creation event occurred between 10 and 20 billion years ago.

The steady-state theory is based on an idea called *the perfect cosmological principle*. It is *perfect* in the sense that it calls for a universe that is not only homogeneous in space—the cosmological principle—but also homogeneous in time—a concept in accord with the special theory of relativity, which places time and space on the same footing. The large-scale universe has always looked the way it looks now and always will look this same way forever in the future. This view has an appeal to a certain philosophical turn of mind to which the notion of a beginning is abhorrent. The steady-state universe had no beginning and will have no end.

In an expanding universe, the galaxies move away from each other, spreading the matter more thinly over space. On the other hand, the perfect cosmological principle requires that the density of matter in the universe remain constant over time. To make the steady-state cosmology compatible with the expanding universe, its proponents introduced the notion of *continuous creation*. As the universe expands and the galaxies move farther apart, new matter—in the form of hydrogen—is introduced into the universe. This new matter condenses to form stars and galaxies to fill in the space left by the old ones. The amount of new matter required is far too small to be measured with available instruments. Stellar and galactic evolution take place in the steady-state universe, but the general character and the overall density of the universe remain unchanged over time. In this sense, the steady-state universe itself does not evolve.

Both of these views—steady-state and Big Bang—allow for the cosmic expansion. Which gives the better description of our universe? In the 1960s, two important discoveries tipped the scales heavily in favor of the Big Bang.

21.2 The Quasars

After World War II, radio astronomy came into its own. In the 1950s, radio astronomers scanned the sky at radio wavelengths and assembled catalogues of strong radio sources. The next task was to compare these with maps of the visible sky obtained from optical telescopes. In the early 1960s, some of these radio sources that appeared as extended objects (like galaxies) were identified with point sources of visible light (like stars). They looked like stars in visible light and like galaxies in the radio spectrum. Because of this, they were called *quasi-stellar objects*—soon shortened to *quasar*.

These objects were particularly puzzling because their optical spectra showed emission lines that did not appear to be characteristic of any known chemical element. However, in 1963 an astronomer at Mt. Palomar Observatory recognized that lines in the emission spectrum from one of these objects had the same distribution as those in the visible spectrum of hydrogen, but substantially shifted to longer wavelengths. This was the key to unlocking the secret of the quasars. In Fig. 21.1, the spectrum of one of these objects is compared with a laboratory spectrum that includes hydrogen lines. This object, known as 3C273, is the 273rd item listed in the Third Cambridge Catalogue of radio sources in the sky.

The other quasars also showed emission spectra with large redshifts. If these are interpreted as cosmological redshifts (that is, arising from the expansion of the universe), then these objects are very far away—hundreds of millions to billions of light years. To appear as bright as they do at such distances, they must be extremely bright intrinsically—typically 1,000 times brighter than the average spiral galaxy.

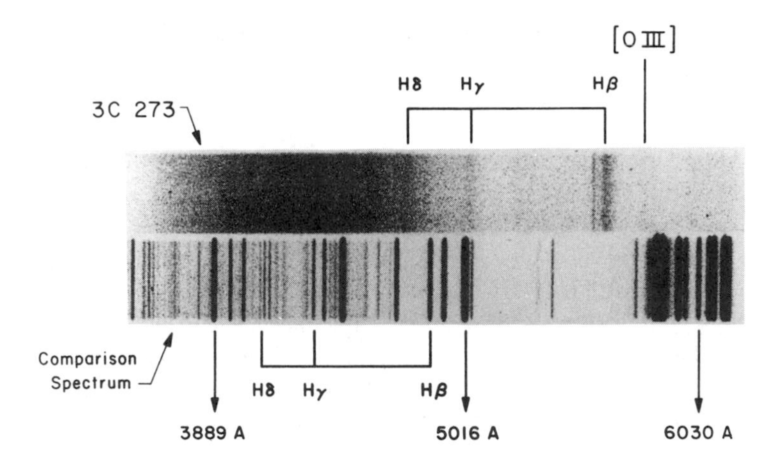

FIGURE 21.1. Spectrum of a quasar. (Courtesy of Palomar Observatory/Caltech.)

The fact that quasars vary in brightness over times of the order of weeks or months implies that they are relatively small sources of light—from a few light weeks up to a light year or so in diameter.

The distribution of quasars shows that the number increases with distance. The light now reaching us from these remote objects left them billions of years ago. The deeper we look into space, the farther back in time we go. The implication here is that earlier in the history of the universe—billions of years ago—quasars were more plentiful than they are now. Putting it another way, the universe as a whole is different now (few quasars) from what it was billions of years ago (many quasars). The universe is evolving. This is consistent with the Big Bang Cosmology, but not with the steady-state view.

21.3 The Cosmic Background Radiation

To appreciate the significance of the second important cosmological discovery of the 1960s, we consider some of the salient features of the so-called *standard model* of the Big Bang cosmology. According to this view, the universe came into existence in the gigantic explosion we call the Big Bang. From the energy of this event, particles of matter and antimatter were created according to $E = mc^2$. The mutual annihilation of the particles and antiparticles produced electromagnetic radiation. Included among these particles were electrons and quarks. About one millisecond after the creation event, the universe was cool enough for the quarks to bind together in groups of three to form neutrons and protons in a ratio of one neutron for every seven protons. Some of the protons combined with the available neutrons to produce the mass two isotope of hydrogen according to,

$$n + p \longrightarrow {}^{2}\mathrm{H} + \gamma.$$

At that time, the universe was still hot enough (about one billion degrees) and dense enough for this heavy hydrogen to form helium through nuclear fusion. A tiny amount of lithium was also produced at this time, but no other heavier elements. By the time the universe was about three minutes old, the fusion was complete. The neutrons were all bound with protons in helium nuclei leaving twelve protons for each helium nucleus produced. Since a helium nucleus has about four times the mass of a proton (hydrogen nucleus), the atomic matter in the universe at that epoch was, by mass, about 75 percent hydrogen and 25 percent helium.

The universe was still too hot for the nuclei to combine with electrons to form neutral atoms. It was filled with electrically charged particles (electrons and nuclei) and photons. The photons could not travel very far without being scattered by an electric charge. This early universe was like the cavity we described in Chapter 11—filled with radiation in thermal equilibrium with the matter embedded in it. The spectral distribution of this radiation should be that of a black body for a temperature of hundreds of millions of degrees.

The universe continued to expand and cool. After about half a million years, the temperature dropped to around 3,000 K. This was cool enough for the electrons to bind to the nuclei to form neutral atoms. The photons easily passed by the electrically neutral atoms without interacting. At an age of about 500,000 years, the universe quickly became transparent to the radiation. The distribution of the radiation had the character of cavity radiation at 3,000 K with its peak in the infrared part of the spectrum. A black body at this temperature would radiate with a dull red glow. As the universe (space) expanded, it stretched out the wavelengths of the radiation traveling through it. Now, some 15 billion years later, this stretching has moved the peak of the distribution into the microwave region of the radio part of the electromagnetic spectrum. This corresponds to a temperature of about 3 K. This radiation, often referred to as *cosmic background radiation*, has been preserved throughout the billions of years following its decoupling from the matter in the universe. It is a relic of the primordial fireball required by the Big Bang theory of the creation and evolution of the universe. Since we on earth are embedded in the fireball, this radiation must come at us uniformly from all parts of the sky. Does it exist?

In 1965, two radio astronomers working at Bell Laboratories stumbled quite by accident onto a bit of cosmic radio noise that had precisely the characteristics required. Since then, other measurements—with earth-based instruments and with instruments on board a spacecraft placed in earth orbit in 1989—have confirmed that this radiation does indeed have the spectrum of a black body at a temperature of 2.73 K. This is clearly seen in the data from these measurements shown in Fig. 21.2. The evidence is quite compelling that our universe is described by an evolving, Big Bang model and not by a steady-state theory.

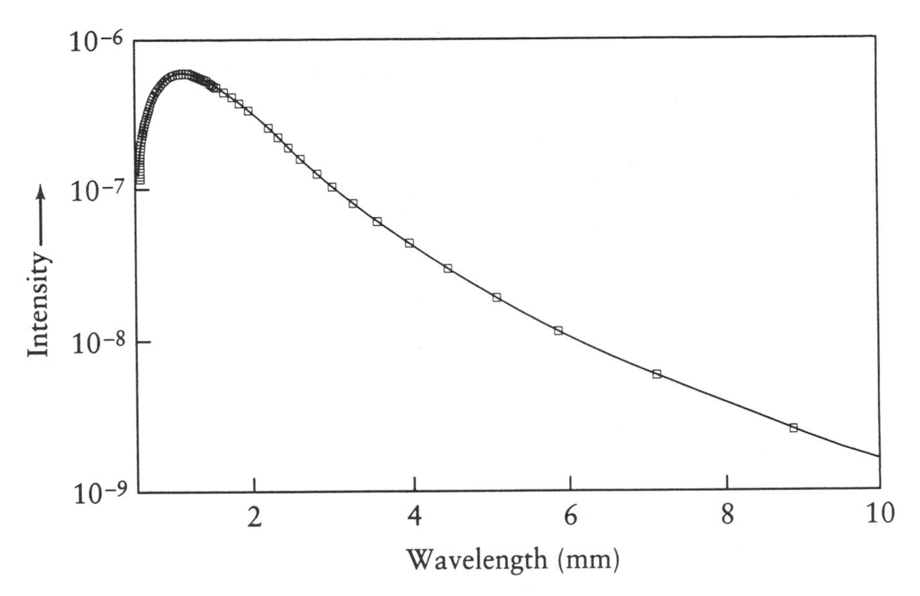

FIGURE 21.2. Data from the Cosmic Background Explorer satellite. (Reprinted with permission of W.H. Freeman and Company.)

21.4 A Final Word*

Breathtaking advances in the technology that has sharpened our image of the cosmos have led to greater understanding of the nature of the universe we live in. Yet, despite this remarkable progress, questions remain. For example, the instruments on board the spacecraft launched in 1989 to measure the cosmic background radiation established with incredible precision the fit of the radiation spectrum to that of an ideal black body at a temperature of about 2.73 K (solid line in Fig. 21.2). However, another task assigned to this orbiting laboratory was to confirm the uniformity (isotropy) of this microwave background. It succeeded exceptionally well—almost too well. Preliminary analysis of early data from this spacecraft showed a very smooth, constant background temperature in all parts of the sky—just as expected. We are inside the fireball and the background radiation should look the same in all directions.

On the other hand, we see a certain clumpiness in the universe today. The atoms of hydrogen, helium, and lithium formed in the early universe fell together under gravity to form the stars, galaxies, and clusters of galaxies we see. For that to happen, there had to be some regions where the gravity was stronger than

*Adapted from J. B. Seaborn, "Evidence for a Hot Big Bang," *The Faculty Exchange*, David Leary, editor, University of Richmond, VA, October 1992.

in other regions, that is, where matter was more concentrated. How did these concentrations come about? Ten or twenty billion years is a long time, but it is not nearly long enough for the clumpiness we see to occur if the cosmic background radiation is perfectly uniform. Density fluctuations must have existed very early in the universe. If so, some regions would be hotter than others and this would show in the microwave background as temperature fluctuations. Such warm and cool spots would provide evidence for the condensations and rarefactions of matter in the early universe that led to formation of the galaxies we see now.

In 1992, the scientists working on this project announced that further analysis of their data revealed some slight nonuniformity in the cosmic background radiation. There are, indeed, hot and cool spots in the sky—evidence of very small fluctuations in the density of matter in the early universe that could provide the seeds for galaxy formation. However, the variation in background temperature is exceedingly small—about *thirty millionths of a degree*. This departure from uniformity is so tiny that the amount of matter detectable with powerful telescopes is far too small to provide the gravity needed for the galaxies to have formed in the time since the Big Bang. This helps to drive the current intensive search for "dark matter"—some form of matter that does not reveal its presence by radiating or absorbing electromagnetic radiation.

An exciting search is underway for a new kind of matter—unseen matter to provide the gravity required to draw together the matter we *can* see to form the galaxies now populating the universe. Theorists are offering a variety of forms in which this missing matter (perhaps 80–90 percent of the matter in the universe) may be hiding, from hypothetical elementary particles (massive neutrinos, axions, WIMPs—weakly interacting massive particles) to subluminous brown dwarf stars or the dead hulks of burned-out stars. Inspired by the theorists' speculations, the observational astronomers are applying their ingenuity to build new instruments that they can turn toward the sky in search of these products of the theorists' imaginations—perhaps to find something wholly unexpected. Whatever the outcome, the search will be very exciting and the result very interesting.

Clearly, new discoveries about the universe call our attention to other secrets that, for the moment, lie tantalizingly just out of reach and push us on to find more answers, which uncover still more questions. The universe is an inexhaustable source of interest and excitement—a wonderful place to live!

Exercises

21.1. A certain quasar has an apparent magnitude of $+13.2$. Hubble's law implies that it is located 900 Mpc away from our own galaxy, which has an absolute magnitude of about -21. How many times brighter than the Milky Way is this quasar? Explain clearly.

Answer: (200 times)

21.2. The measured redshift of a certain quasar is 0.133. Assuming this is a cosmological redshift, estimate the distance to this quasar.

Answer: (2 billion ly)

21.3. Hydrogen emission lines are seen in the spectrum of a quasar. The wavelength of the green line in the Balmer series is measured at 583 nm. Estimate the distance to this quasar.

Answer: (3 billion ly)

21.4. The apparent magnitude of a quasar is observed to be +12.1. The redshift indicates that this quasar is receding from us at about 43,200 km/s. About how many times brighter than the Milky Way is this quasar? Explain clearly your reasoning.

Answer: (300 times)

21.5. The magnitude of a certain quasar is observed to be $+11.2$. The spectrum of this quasar shows hydrogen lines in the Balmer series. In the figure below, these data are compared with wavelengths of the corresponding hydrogen lines measured in the laboratory. Using **all** of the data, determine the absolute magnitude of this quasar. The absolute magnitude of our galaxy, the Milky Way, is about -21. How many times brighter than the Milky Way is this quasar? How many years does it take light emitted by this quasar to reach earth? Explain clearly your reasoning.

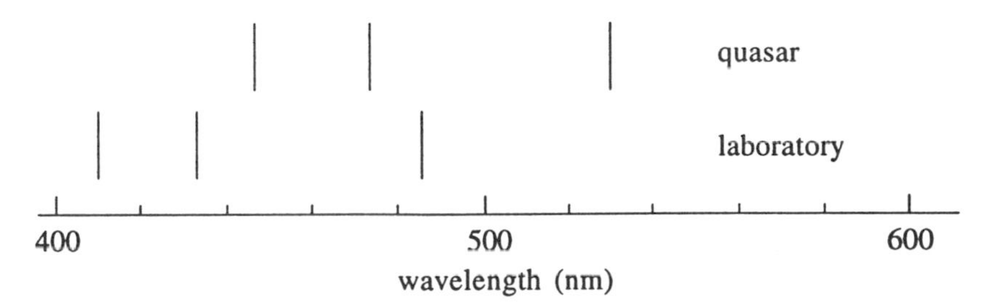

21.5. The quasar 3C273 appears in the sky as a starlike object with an apparent magnitude of $+13$. The optical spectrum shown in Fig. 21.1 reveals that its spectrum is significantly shifted toward longer wavelengths. This redshift is seen when the quasar spectrum is compared with corresponding lines in the spectrum of hydrogen (comparison spectrum) observed in the laboratory. To calibrate the photograph, use a ruler with a millimeter scale to measure the interval between the green and violet lines of hydrogen in the comparison spectrum.

scale:____ mm on the photograph = ____ nm in the laboratory

Use this scale to determine the wavelength λ of the three hydrogen lines in the quasar spectrum. Assume that the redshift $\Delta\lambda$ is due to the expansion of the universe and use Hubble's law to estimate the distance to this quasar. From the data, estimate the absolute magnitude of 3C273. How many times brighter than the sun is 3C273? How does the brightness of 3C273 compare with that of the entire Milky Way galaxy? Give a quantitative estimate. Explain clearly your reasoning.

Appendix A
Linear Graphs

Consider the equation

$$y = mx + b, \tag{A.1}$$

where m and b are constant parameters. This equation tells us that given values for m and b, we can choose any value for x and calculate the corresponding value of y. For example, let us take the values $m = 0.6$, $b = 5.2$, and calculate y for several values of x. The results are given in the table in Fig. A.1.

Now let each pair of numbers in the table correspond to a point in an xy-plot as shown in Fig. A.1. That is, we plot y versus x.* The points can be connected by passing a single straight line through all of them. Thus, the "graph" of Eq. (A.1) with $m = 0.6$ and $b = 5.2$ is seen to be a straight line. Hence, the equation is said to be a *linear* equation. This holds for any set of constant values for m and b.

It is clear that if we pick any two points in the xy-plane, there is only one straight line that can be drawn through these points. If we extend this line far enough, it will cross the y-axis at the point $(0, b)$ and the x-axis at the point $(-b/m, 0)$. Since these two points lie on the line defined by Eq. (A.1), it is clear that the two numbers, m and b, are sufficient to specify the line uniquely. An alternate way of defining the line, is to give the angle at which the line is inclined with respect to the x axis and the value of y where it crosses the vertical axis. In practice, instead of the angle itself, we use the *tangent* of the angle, which we can calculate by dividing the difference between the vertical values for any pair of points on the line, say A and B, by the difference between the corresponding horizontal values. We call this ratio the *slope* of the line and we find that it is equal to m,

$$\text{slope} = m = \frac{y_A - y_B}{x_A - x_B} = \frac{\Delta y}{\Delta x}. \tag{A.2}$$

This can easily be checked for the example above by taking the two points A and B marked on the graph at (28, 22) and (8, 10), respectively. The parameter

*Note that the expression "y versus x" means that the quantity y is plotted *vertically* and the quantity x is plotted *horizontally*.

x	y
3.0	7.0
7.0	9.4
11.0	11.8
19.0	16.6
25.0	20.2
33.0	25.0

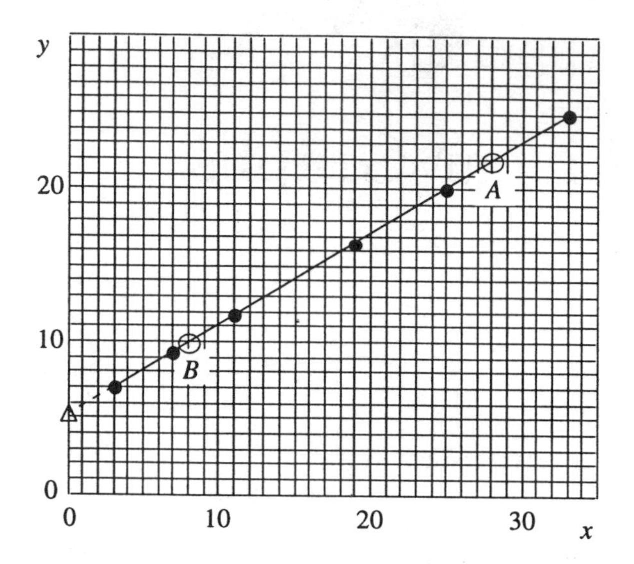

FIGURE A.1. A linear graph.

b, which is the value of y when $x = 0$, is called the *vertical intercept*, since the line intercepts the vertical axis at the point $(0, b)$. Thus, the value of b can be extracted from the graph by extending the line until it crosses the vertical axis and reading off the y-value at this point. For the example above, this value is marked with a triangle Δ in Fig. A.1. Similarly, the line intercepts the x-axis at the point $(-b/m, 0)$ and the value of x at this point is called the *horizontal intercept*. Clearly, any *two* of these three quantities, the slope (m), the vertical intercept (b), and horizontal intercept $(-b/m)$ are sufficient to determine the line uniquely. Generally, in physical applications, these quantities will have physical interpretations.

In practice, the points we plot on our graph will come from data collected in experimental observations. Because of imperfections in our instruments and our experimental techniques, these data will contain observational errors. As a result, the behavior of the system under study may be consistent with a linear graph, but some of the data points may not actually lie on the line. In fact, in some cases the experimental errors may be such that *none* of the data points touches the line.

There exist more sophisticated mathematical methods for data analysis, but for our purposes in this course a visual fit of our data to a linear graph will be adequate. For this kind of graphical analysis, our procedure will consist of five steps:

- Collect the data.
- Try to find a form for plotting the data so that the resulting graph is consistent with a straight line.
- Use a ruler to draw through the data points a straight line which gives the best fit to *all* of the data. (Remember that it isn't necessary for all of the data points to lie on the line.)

- From the graph, extract the slope and the two intercepts.
- Look for physical interpretations of the slope and the two intercepts.

Exercises

A.1. A bicycle rider pedals at constant speed along a straight, level path through a park. She observes on a stopwatch the time she passes certain markers along the path. Each marker displays its distance from the park clubhouse. Her data are given in the table. Plot the position of the marker versus the time. From your graph, determine her speed and how far she was from the clubhouse when she started her stopwatch. Write an equation expressing her position s as a function of time t. Use this equation to find how far she is from the clubhouse when her stopwatch reads exactly nine minutes.

position (m)	time (min)
300	0.98
400	1.72
500	2.48
600	3.23
700	4.00
900	5.48

A.2. A block slides down an inclined plane. A multiflash photograph of the apparatus shows the position of the block on the incline at half-second intervals. The position s is given for the corresponding time t in the table. Plot s versus t^2. Your graph should be consistent with a straight line. Calculate the slope of this line with the appropriate units. What is the physical significance of the vertical intercept? The horizontal intercept? Express each answer in a complete sentence. Write an equation expressing s as a function of t. Use your equation to find the position of the block along the incline at 3.7 seconds.

position (cm)	time (s)
30	0.5
50	1.0
80	1.5
125	2.0
180	2.5
250	3.0

A.3. A large, sealed metal tank contains a quantity of gas. The pressure of the gas is measured at various temperatures. The data are presented in the table. Plot the pressure P versus the temperature t. From your graph, obtain a linear equation for P as a function of t. Use your equation to find the temperature at which the pressure drops to zero. What is the pressure of the gas when the temperature is increased to 185°C?

P (N/m^2)	t (°C)
7.6×10^5	25
8.0×10^5	40
8.7×10^5	65
9.4×10^5	95

Appendix B

Physical and Astronomical Data

radius of the earth: 6,370 km

radius of the sun: $R_\odot = 696{,}000$ km

radius of the moon: 1,738 km

mass of the earth: 5.98×10^{24} kg

mass of the sun: $M_\odot = 1.99 \times 10^{30}$ kg

mass of the moon: 7.35×10^{22} kg

radius of earth's orbit: 1.50×10^{8} km

radius of moon's orbit: 384,400 km

speed of light: $c = 3.00 \times 10^8$ m/s

Planck's constant: $h = 4.14 \times 10^{-15}$ eV·s $= 6.63 \times 10^{-34}$ J·s

electron volt: 1 eV $= 1.60 \times 10^{-19}$ joule

Coulomb constant: $k = 9 \times 10^9$ N·m²/C²

magnetic force constant: $k' = 10^{-7}$ N·s²/C²

gravitational constant: $G = 6.67 \times 10^{-11}$ N·m²/kg²

Boltzmann's constant: $k = 1.38 \times 10^{-23}$ J/K

Avogadro's number: $N_0 = 6.02 \times 10^{23}$ molecules per mole

universal gas constant: $R = 8.314$ J/mole·K

atomic mass unit: 1 u $= 1.66056 \times 10^{-27}$ kg $= 931.50$ MeV/c^2

mass of electron: $m_e = 9.11 \times 10^{-31}$ kg $= 0.511$ MeV/c^2

mass of proton: $m_p = 1.67 \times 10^{-27}$ kg $= 938.3$ Mev/c^2

proton charge: $e = 1.60 \times 10^{-19}$ coulomb

atmospheric pressure at sea level = 1.013×10^5 N/m^2

astronomical unit: 1 AU = 1.50×10^8 km

parsec: 1 pc = 3.26 ly = 206, 265 AU = 3.086×10^{16} m

Prefixes for Powers of Ten

Factor	10^{-12}	10^{-9}	10^{-6}	10^{-3}	10^{-2}	10^3	10^6	10^9
Prefix	pico	nano	micro	milli	centi	kilo	mega	giga
Abbreviation	p	n	μ	m	c	k	M	G

Appendix C

Useful Formulas

circumference of a circle $= 2\pi r$

area of a circle $= \pi r^2$

surface area of a sphere $= 4\pi r^2$

volume of a sphere $= 4\pi r^3/3$

volume of a cylinder $= \pi r^2 h$

mathematical relations for a right triangle

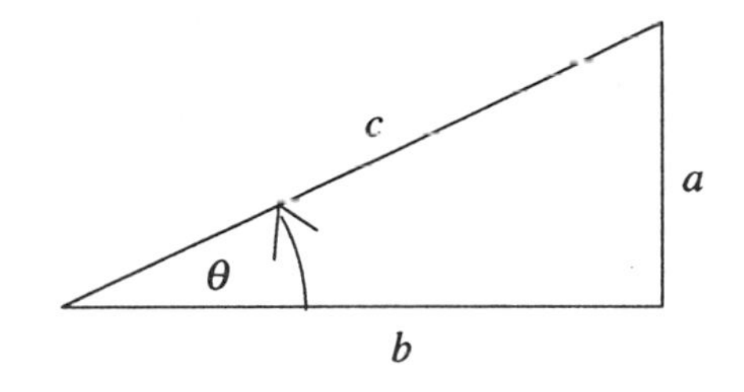

$$\sin\theta = a/c$$

$$\cos\theta = b/c$$

$$\tan\theta = a/b$$

$$c^2 = a^2 + b^2$$

Appendix D

The Chemical Elements

element	symbol	atomic number	atomic weight
actinium	Ac	89	227.03
aluminum	Al	13	26.98
americium	Am	95	(243)
antimony	Sb	51	121.75
argon	Ar	18	39.95
arsenic	As	33	74.92
astatine	At	85	(210)
barium	Ba	56	137.33
berkelium	Bk	97	(247)
beryllium	Be	4	9.01
bismuth	Bi	83	208.98
boron	B	5	10.81
bromine	Br	35	79.90
cadmium	Cd	48	112.41
calcium	Ca	20	40.08
californium	Cf	98	(251)
carbon	C	6	12.01
cerium	Ce	58	140.12
cesium	Cs	55	132.90
chlorine	Cl	17	35.45
chromium	Cr	24	52.00
cobalt	Co	27	58.93
copper	Cu	29	63.55
curium	Cm	96	(247)
dysprosium	Dy	66	162.50
einsteinium	Es	99	(252)

element	symbol	atomic number	atomic weight
erbium	Er	68	167.26
europium	Eu	63	151.96
fermium	Fm	100	(257)
fluorine	F	9	19.00
francium	Fr	87	(223)
gadolinium	Gd	64	157.25
gallium	Ga	31	69.72
germanium	Ge	32	72.59
gold	Au	79	196.97
hafnium	Hf	72	178.49
helium	He	2	4.00
holmium	Ho	67	164.93
hydrogen	H	1	1.01
indium	In	49	114.82
iodine	I	53	126.90
iridium	Ir	77	192.22
iron	Fe	26	55.85
krypton	Kr	36	83.80
lanthanum	La	57	138.90
lawrencium	Lr	103	(260)
lead	Pb	82	207.2
lithium	Li	3	6.94
lutetium	Lu	71	174.97
magnesium	Mg	12	24.30
manganese	Mn	25	54.94
mendelevium	Md	101	(258)
mercury	Hg	80	200.59
molybdenum	Mo	42	95.94
neodymium	Nd	60	144.24
neon	Ne	10	20.18
neptunium	Np	93	237.05
nickel	Ni	28	58.69
niobium	Nb	41	92.91
nitrogen	N	7	14.01
nobelium	No	102	(259)
osmium	Os	76	190.2
oxygen	O	8	16.00
palladium	Pd	46	106.42
phosphorus	P	15	30.97
platinum	Pt	78	195.08
plutonium	Pu	94	(244)

element	symbol	atomic number	atomic weight
polonium	Po	84	(209)
potassium	K	19	39.10
praseodymium	Pr	59	140.91
promethium	Pm	61	(145)
protactinium	Pa	91	231.03
radium	Ra	88	226.02
radon	Rn	86	(222)
rhenium	Re	75	186.21
rhodium	Rh	45	102.90
rubidium	Rb	37	85.47
ruthenium	Ru	44	101.07
samarium	Sm	62	150.36
scandium	Sc	21	44.96
selenium	Se	34	78.96
silicon	Si	14	28.08
silver	Ag	47	107.87
sodium	Na	11	22.99
strontium	Sr	38	87.62
sulfur	S	16	32.06
tantalum	/ta	73	180.95
technitium	Tc	43	(98)
tellurium	Te	52	127.60
terbium	Tb	65	158.92
thallium	Tl	81	204.38
thorium	Th	90	232.04
thulium	Tm	69	168.93
tin	Sn	50	118.69
titanium	Ti	22	47.88
tungsten	W	74	183.85
uranium	U	92	238.03
vanadium	V	23	50.94
xenon	Xe	54	131.29
ytterbium	Yb	70	173.04
yttrium	Y	39	88.90
zinc	Zn	30	65.38
zirconium	Zr	40	91.22

Source: *CRC Handbook of Chemistry and Physics*, 66th edition, CRC Press, Inc., Boca Raton, FL, 1985.

Appendix E

The Brightest Stars in the Sky

star	constellation	spectral type	apparent magnitude	distance (pc)
Sirius	Canis Major	A1	−1.46	2.7
Canopus	Carina	F0	−0.72	30.
Alpha Centauri	Centaurus	G2	−0.01	1.3
Arcturus	Boötes	K2	−0.06	11.
Vega	Lyra	A0	+0.04	8.0
Capella	Auriga	G8	+0.05	14.
Rigel	Orion	B8	+0.14	250.
Procyon	Canis Minor	F5	+0.37	3.5
Betelgeuse	Orion	M2	+0.41	150.
Achernar	Eridanus	B5	+0.51	20.
Beta Centauri	Centaurus	B1	+0.63	90.
Altair	Aquila	A7	+0.77	5.1
Alpha Crucis	Crux	B1	+1.39	120.
Aldebaran	Taurus	K5	+0.86	16.
Spica	Virgo	B1	+0.91	80.
Antares	Scorpius	M1	+0.92	120.
Pollux	Gemini	K3	+1.16	12.
Fomalhaut	Piscis Austrinus	A3	+1.19	7.0
Deneb	Cygnus	A2	+1.26	430.
Beta Crucis	Crux	B0	+1.28	150.

Source: Kenneth R. Lang, *Astrophysical Data: Planets and Stars*, p. 166, Springer-Verlag, New York, 1991.

Bibliography

Astronomy Texts

G. O. Abell, D. Morrison, and S. C. Wolff, *Exploration of the Universe*, 6th ed., Saunders College Publishing, Philadelphia, 1993.

W. J. Kaufmann and N. F. Comins, *Discovering the Universe*, 4th ed., W. H. Freeman, New York, 1996.

J. M. Pasachoff, *Astronomy: From the Earth to the Universe*, 4th ed., Saunders College Publishing, Philadelphia, 1993.

J. M. Pasachoff, *Contemporary Astronomy*, 3rd ed., Saunders College Publishing, Philadelphia, 1984.

W. M. Protheroe, E. R. Capriotti, and G. H. Newsom, *Exploring the Universe*, 4th ed., Merrill Publishing Company, Columbus, OH, 1989.

M. A. Seeds, *Foundations of Astronomy*, Wadsworth Publishing Company, Belmont, CA, 1994.

T. P. Snow, *The Dynamic Universe*, 4th ed., West Publishing Company, St. Paul, 1991.

History of Astronomy

A. Pannekoek, *A History of Astronomy*, Barnes and Noble, New York, 1969.

Physics Texts

B. M. Casper and R. J. Noer, *Revolutions in Physics*, W. W. Norton and Company, New York, 1972.

P. G. Hewitt, *Conceptual Physics*, 8th ed., Addison and Wesley, 1997.

D. C. Giancoli, *Physics: Principles with Applications*, 5th ed., Prentice-Hall, Upper Saddle River, NJ, 1998.

R. A. Serway and J. S. Faughn, *College Physics*, 4th ed., Saunders College Publishing, Philadelphia, 1995.

A. W. Smith and J. N. Cooper, *Elements of Physics*, 9th ed., McGraw-Hill, New York, 1979.

Index

Absolute temperature scale, 224–226
absorption, 156, 178, 179, 192–193
See also spectral analysis absorptivity, 156
acceleration, 19, 20, 57
defined, 19
instantaneous, 83
ad hoc principles, 175
Algol, 112, 212
alpha particles, 151n., 195, 242, 254
Altair, 294
ampere unit, 131
Andromeda, 273
angular momentum, 117–120
energy splitting and, 192
quantized, 190–191, 191
Antares, 214
antiparticles, 252, 279
Aquinas, St. T., 5
Arabic astronomy, 5
Arcturus, 295
Aristotle, 2–3
asteroids, 14, 100
astronomical unit(AU), 8, 198, 290
atmosphere, 97, 290
atom, 148
atomic mass, 239–240, 263
atomic number, 187, 234
atomic weight, 187n, 234
Bohr model, 176–177, 178, 183–185
classical models of, 149, 174
diameter of, 175, 233
ground state and, 183
models of, 149
Rutherford model, 174
spectra, 156, 178, 179, 193–194
structure of, 149, 172–173, 233–242
See also specific particles
AU: see astronomical unit
Avogadro's number, 289

Balmer series, 173, 177–178, 185, 283
Barnard's star, 216
Becquerel, H., 148, 150
Bessel F., 198
beta decay, 235, 237, 238
Betelgeuse, 202, 295
Big Bang cosmology, 277, 279
Big Dipper, 216
binary stars, 106, 107, 209–213
binding energy, 240, 241, 244, 246
bismuth, 238
black body radiation, 157, 158 159
black hole, 258
Bohr model
Balmer formula, 177
comets and, 185
centripetal force and, 184
hydrogen atom, 176, 176, 183
orbits in, 176, 179, 183
photons and, 177
Boltzman constant, 224, 289
Boyle's law, 219–220
Brahe, Tycho, 6–8
brown dwarfs, 282

Calendar, design of, 15
carbon monoxide, 194

carbon-nitrogen-oxygen (CNO)cycle, 252
cathode rays, 149
Cavendish experiment, 94, 99, 128
Celsius scale, 224
Centaurus, 101
center of mass, 104–105
center of gravity, 117
centripetal force, 84–85
Cepheid variable, 267–268
Ceres, 14, 100
charge carriers, 135
chemical elements, 148, 292–294
chemical energy, 131
chemical reactions, 194
Chiron, 15
chromatic aberration, 45
circular motion, 82–84
circular orbits, 5
classical electrodynamics, 170, 174–175
classical physics, 156, 158–162, 175
CNO cycle: see carbon-nitrogen-oxygen cycle
color, 157
color-magnitude diagram, 209, 216, 258–259
comets
  Bohr model and, 185
  composition of, 180
  orbit of, 14, 179, 180, 181
  origin of, 182
  photons and, 181
communications satellite, 109
conservation of energy, 72
conservation of momentum, 74, 166
coordinate systems, 1
Copernican model, 5–6, 8, 14
Copernicus, N., 5, 54
copper, 169
cosmic background radiation, 279–281, 282
cosmological principle, 277–284
Coulomb force, 244
  attractive, 190
  constant of, 128, 289
  law of, 127–128, 151, 190–191
  magnetic poles and, 132–133
  repulsive, 237, 242
  scattering, 152
coulomb unit, 128
coupling, magnetic, 137, 137
Crab Nebula, 266
Curie, M., 150, 238
curvilinear motion, 82–88

Dalton, J., 148
dark matter, 282
data fitting, 286
day, defined, 9
delta symbol, 18n.
Deneb, 215, 294
density, 56–57
diffraction, 45
dip angle, 139
dispersion, 22, 40–41
displacements, resultant of, 24
Doppler effect, 47, 210, 217, 260
dry cell, 131

Earth
  atmosphere of, 97
  core of, 139
  escape velocity of, 96–97
  kinetic energy and, 80
  magnetic field, 140
  mass of, 289
  radius of, 13, 289
eclipses, 9–12, 13–15
ecliptic plane, 2,13
Einstein, A., 166–168
  mass/energy equation, 167
  photoelectric effect, 163
  photon model, 166
  gedanken-experiment, 166
elastic collisions, 74
electric charge, 127–128
  fields from, 130
  flow of, 135
  light and, 139
  work and, 141
electric current, 131–132
  defined, 184
  induced, 137–138
  magnetic field and, 135
electric field, 128–130, 129
electromagnetism, 137–138, 161

light and, 138
matter and, 33
spectrum of, 33
waves, 162
electron-electron interaction, 192
electrons, 135, 140
bound state, 206
centripetal force on, 184
discovery of, 148
electric field and, 146
electric repulsion, 154
exclusion principle and, 191–192
ground state of, 177
intrinsic magnetism, 260
kinetic energy, 161, 164, 175
mass of, 289
orbits, discrete, 175, 176, 176
photoelectric effect, 161
potential energy, 141–142
quarks and, 279
repulsion and, 154
state of, 190
electronic potential, 131, 141–142, 176
electron volt, 164, 289
elementary particles, 131, 193, 282
See also specific particle
elements
periodic table, 194, 234
spectra of, 193
See also specific elements
ellipse, construction of, 7
emission lines, 184
emission spectrum, 179, 184
energy-level diagrams, 195
energy, 16
conservation of, 72–73, 96, 177
kinetic, 71–72, 74, 224
mass, 167
potential, 71
work and, 70
English system, 17
Eratosthenes, 2–3, 13
escape velocity, 96
exclusion principle
electrons and, 190–193
nuclear processes and, 236, 254
occupation number, 192
quantum mechanics and, 189
stars and, 257
expanding universe, 270, 277

Faraday's law, 146
Faraday, M., 137
Fahrenheit units, 224
fission, 243–244
flat earth model, 3
force, 54, 57
Franklin, B., 127
frequency, of wave, 33
fusion, 233, 244–245

Galaxies
Hubble's discovery, 266, 269–271
spectra of, 269, 270
spiral, 273–274
Galileo, G., 54, 89, 101, 109
gamma rays, 238
gas
behavior of, 218
constant, 289
interstellar dust and, 259–260
microscopic description of, 222
molecular structure of, 220
pressure of, 219
Gauss, C.F., 14
gedanken-experiment, 166
Geiger-Marsden experiments, 151, 174–175
Geiger, H., 152
geometrical constructs, 2
globular clusters, 209, 273
gold atom, 153
gravitation
acceleration due to, 60–61
atmosphere and, 97
center of mass, 104, 117
constant, 289
falling bodies and, 60, 89
field, 71
law of, 92–94
measurement of, 94–95
moon and, 61
Newton and, 89
potential energy and, 71, 73, 75
weight and, 92–94
Greek astronomy, 2–5

ground state
electron and, 177
hydrogen atom and, 183

H-R diagram: see Hertzsprung-Russell diagram
Hadar, 215
half-life, 238
Halley, E., 90, 179
halogens, 189, 194
helium, 97
ionization of, 267
scattering of, 154
Hertzsprung-Russell (H-R) diagram, 207–209, 251–252
Hooke, R., 90
Hubble constant, 271, 274
Hubble's law, 275, 277
Hubble, E., 267, 273
hydrogen, 88, 97, 146, 156
absorption lines, 204
Balmer formula, 173
Bohr model and, 176, 183
classical model, 174
ground state of, 189
ionization energy, 180, 183
radiation from, 260,
spectrum of, 172–173, 260
stars and, 262–263
hydrostatic equilibrium, 251

Ice point, 224
ideal gas, 222–223, 225
ideal radiator, 158
incidence, angle of, 35
induced currents, 137
inelastic collisions, 74
inertia, 16, 118, 127
instantaneous velocity, 23, 84
intensity, definition of, 197
International Bureau of Weights and Measures, 56
ionization energy, 180, 183
isotopes, 234, 237

Jupiter, 101, 109, 126
angular momentum, 126
mass of, 109
satellites of, 101

Kant, I., 120
Kepler, J., 257
Kelvin scale: see absolute temperature
Kepler's laws, 8, 54, 92, 106, 179
kinetic energy, 71–72, 74
temperature and, 224
Kirchhoff, G., 156–157

Laplace, P., 120
Lavoisier, A., 148
lenses
chromatic aberration, 45
focal points of, 50
pairs of, 46
telescopes and, 41–44
Lenz's law, 147
light, 33, 41
dispersion of, 173
electric charge, 139
electromagnetic waves, 138, 159
electrons and, 161
momentum of, 165–166
speed of, 289
See also photons
light year, 101, 200, 284
linear equation, 285
linear graphs, 285, 286, 287–288
linear momentum, 59–60
lithium, 193, 280
longitude, 1
lunar phases, 9–13

Macroscopic quantities, 222
magnetic fields, 133, 149
earth and, 139–141
electric current and, 135
electron in, 136
flux density, 137, 146–147
poles in, 132–133
radiation and, 150
magnetism, 132–137, 260
main-sequence stars, 207, 253, 258–259
many-body system, 74, 104
many-electron atom, 187–192, 195
Marsden, E., 152
mass, 56

center of, 90, 104
definition of, 56
density and, 56
energy and, 167
inertia and, 16
mass number, 234
matter, 16, 148–155
Maxwell theory, 138–139, 158, 161, 166, 174–175
Maxwell, J., 138–139
mean solar day, defined, 9
measure, standard, 56
measurement, units of, 4, 17
See also specific units
mechanical equilibrium, 113–115
Mercury
atmosphere of, 97
orbit, 14
meridian, 1–2
Messier, C., 266, 276
meteor, 181
metric system, 17
microwave background radiation, 282
Milky Way, 257, 266, 273, 283
mirror, 35–46
image formation, 36, 37
principal axis of, 39
small rays for, 40
spherical, 38, 39, 40, 49
virtual image, 38
mole, defined, 226
molecular weights, 226
moment of inertia, 118
moon (the earth's)
atmosphere, 97
lunar periods and, 10, 11
mass of, 289
orbit of, 13
phases of, 9
radius of, 289
synodic period, 10
sidereal period, 10
motion, measurement of, 17
Mt. Palomar Observatory, 278
Mt. Wilson Observatory, 273

Nebulae, 266
nebular hypothesis, 120
necessary conditions, 113
Neptune, 14
neutrino, 236, 236n., 261–262
neutron, 236, 257
neutron star, 257
Newton, I., 54–68, 156, 179
gravitational theory, 120
laws of motion, 54–59, 66, 92, 113, 220
Kepler and, 92
moon and, 89
newton unit, 57
nuclear force: see strong force
nuclear model, of atom, 174
nuclear reactions, 242–243, 247
nucleons, 240
See also neutrons, protons
nuclear fusion, 280
nucleus, 153, 240
atom, 174n.
binding of, 239–241
diameter of, 154
exclusion principle, 254
fusion energy, 280
stability of, 235–237
See also neutrons, protons

Occupation number, 192
"On the Revolutions of the Celestial Spheres" (Copernicus), 5
Oort, J., 182
Oort cloud, 182
orbital motion, 104
angular momentum, 126
kinetic energy and, 80
many-body system, 74, 104
motion, 104
period of, 8, 91
planetary, 5–8
two bodies, 105
orbits, planetary, 5–8
oscillators, 160

Parallax, 198, 200
parsec, 200, 290
perfect cosmological principle, 277
periodic motion
orbit and, 91
waves and, 33

Periodic Table, 193
Perseus, 112
photons
absorbed, 179
Bohr model, 177
comets and, 181
energy of, 163
existence of, 165n
photoelectric effect, 161–165, 168, 170, 177
photosphere, 203, 205–206, 254
Planck constant, 160, 169, 175, 190
Planck, M., 159
planetary model of atom, 174
planetary nebula, 255, 256
planets
atmospheres of, 97
orbits of, 5–8, 92–93
See also specific planets
plasma, the sun and, 218
Plato, 2
Pluto, mass of, 126
poles, magnetic, 132
positron, 236
potassium, 193
pound unit, 57, 61
power, 88, 91, 102
defined, 75
pressure-temperature measurement, 225
Principia (Newton), 89, 179
prism, dispersion and, 172
projectiles, 89
proton
charge on, 289
free, 237
magnetism and, 260
mass of, 167, 289
protostars, 120, 253
Proust, J., 148
Ptolemaic system, 3, 5, 200–201

Quantum mechanics
Planck and, 159–160
exclusion principle and, 190
quantum numbers, 184, 190
See also specific effects; particles
quarks, electrons and, 279
quasars, 283
emission spectra, 278–279
red shifts and, 278
spectra of, 278–279, 279, 28
quasi-stellar objects: see quasars

Radian measure, 83n
radiation
cosmic background, 279–280
electromagnetic, 137–138, 161
inverse square law and, 197
magnetic field, 150
matter and, 156
radioactivity, 148–150, 238
radio telescopes, 260
radio-waves, 174
radium, 238–239
recessional velocity, 270
red giants, 207, 254, 255
red shift, 47, 210, 217, 269–270, 274, 280
reflection, of light, 35–47
refracting telescope, 52
refraction, 35–47
Regulus, 201
Ring Nebula, 255
rockets, 61, 66
Roman Catholic Church, 5
Roman Empire, 5
rotational equilibrium, 114
rotational inertia, 119, 119
Royal Society of London, 90
RR Lyrae variables, 267
Rutherford model, 153, 174
Rutherford, E., 151

Satellite, in earth orbit, 107–108, 108
Saturn, 98
scalars, 21–24
scattering, 152–153
Scorpius, 215
screening effect, 191
seasons, 9–13
sidereal period, 10
Sirius, 112, 193, 212, 295
sodium, 193
solar day, 9
solar eclipse, 10, 13, 13, 15
solar neutrinos, 261–262
solar system

angular momentum in, 121
mass in, 120
origin of, 120
See also sun; specific planets
solstice, 3
space, 16
spectral analysis, 178, 182, 193–194
Balmer series, 172–173, 177–178, 185, 285
black body, 157, 158–159
elements and, 193
galaxies and, 269, 270
hydrogen and, 172–173, 177–178, 185, 283
quasars and, 278–279, 283
red shift, 47, 210, 217, 269, 274, 280
stars and, 203, 205, 206
spectroscopic notation, 191
spectrometer, 173, 182
spectroscopic binaries, 211
spectroscopic parallax, 208
Spica, 214, 295
spin, of electron, 190, 236n
spiral galaxies, 272, 273–274
Sputnik, 85, 126
standard model, 279
standard rays, 43–45
stars, 214, 255
absolute magnitude, 202
brightest, 196, 201, 204, 294
color of, 204
computer models of, 261
cosmic distances, 198–199
Doppler effect, 47, 210, 217, 269, 274, 280
energy produced in, 203
evolution of, 120, 196, 249, 256–258
globular clusters, 209
gravitational effects, 257
hydrogen burning, 263
magnitudes of, 201
main-sequence, 251–253
massive, 53
model of, 204
pulsating variables, 267–268
spectra of, 203, 205, 206
See also specific types
statistical mechanics, 159
steady-state theory, 277–278
stellar nebula, 250–251
strong force, 235
sufficient conditions, 113
sun, the, 1–2, 218
angular momentum of, 121
ecliptic and, 2
evolution of, 249
magnetic field of, 121
mass of, 106, 289
nuclear fusion, 233–248
radius of, 289
solstice, 3
temperature of, 232
white dwarf, 63
supergiants, 207, 257,
supernova, 257, 258, 259
synodic period, 10

Tangential acceleration, 83
telescopes
astronomical, 45–47
construction of, 51
converging lenses, 42
focal lengths, 42
lenses and, 41–44
reflecting, 44, 45–47
refracting, 44, 45–47
temperature, 223, 224
See also specific scales
thin lenses, 43–45
Thomson model, 152, 154
Thomson, J.J., 148
time, 16
Titan, 98
torque, 115–117, 119
translational equilibrium, 113
Trifid nebula, 250
triple-alpha process, 254

Unified electromagnetic theory, 138, 156
uniform circular motion, 82–84
uniform motion, 18
universal gas constant, 227
universal gravitation, 94
uranium, 150, 239, 243
Uranus, orbit of, 8

Vectors, 21–24, 128
Vega, 216
velocity, 18, 82, 84
Venus, 27
virtual image, 37
volt unit, 142
Vulcan, 14

Water waves, 34
watt unit, 75
waves, 33–35
  diffraction and, 45
  electromagnetic radiation and, 33
  equation for, 138–139
  frequency of, 33
  perpendicular components, 33
  propagation velocity, 34
  stationary source of, 47
weak force, 236
weakly interacting massive particles, 282
weight, 60–62
white dwarfs, 63, 208, 254–255
WIMPs: see weakly interacting massive particles
work
  function, 164–165, 168
  measure of, 70
  mechanical, 69–81
  power and, 75
Wren, C., 90

Year, defined, 9

Zodiac, 1